Synthesis and Application of Biomass-Derived Carbon-Based Nanomaterials

Synthesis and Application of Biomass-Derived Carbon-Based Nanomaterials

Editor

Dapeng Wu

Basel • Beijing • Wuhan • Barcelona • Belgrade • Novi Sad • Cluj • Manchester

Editor
Dapeng Wu
School of Environment
Henan Normal University
Xinxiang
China

Editorial Office
MDPI
St. Alban-Anlage 66
4052 Basel, Switzerland

This is a reprint of articles from the Special Issue published online in the open access journal *Nanomaterials* (ISSN 2079-4991) (available at: www.mdpi.com/journal/nanomaterials/special_issues/biomass_nanocarbon).

For citation purposes, cite each article independently as indicated on the article page online and as indicated below:

Lastname, A.A.; Lastname, B.B. Article Title. *Journal Name* **Year**, *Volume Number*, Page Range.

ISBN 978-3-0365-8757-8 (Hbk)
ISBN 978-3-0365-8756-1 (PDF)
doi.org/10.3390/books978-3-0365-8756-1

© 2023 by the authors. Articles in this book are Open Access and distributed under the Creative Commons Attribution (CC BY) license. The book as a whole is distributed by MDPI under the terms and conditions of the Creative Commons Attribution-NonCommercial-NoDerivs (CC BY-NC-ND) license.

Contents

Mengmeng Zhang, Pengfei Li and Dapeng Wu
Editorial for Special Issue: "Synthesis and Application of Biomass-Derived Carbon-Based Nanomaterial"
Reprinted from: *Nanomaterials* 2023, 13, 2020, doi:10.3390/nano13132020 1

Thanakorn Yeamsuksawat, Luting Zhu, Takaaki Kasuga, Masaya Nogi and Hirotaka Koga
Chitin-Derived Nitrogen-Doped Carbon Nanopaper with Subwavelength Nanoporous Structures for Solar Thermal Heating
Reprinted from: *Nanomaterials* 2023, 13, 1480, doi:10.3390/nano13091480 4

Anying Long, Hailin Liu, Shengrui Xu, Suling Feng, Qin Shuai and Shenghong Hu
Polyacrylic Acid Functionalized Biomass-Derived Carbon Skeleton with Highly Porous Hierarchical Structures for Efficient Solid-Phase Microextraction of Volatile Halogenated Hydrocarbons
Reprinted from: *Nanomaterials* 2022, 12, 4376, doi:10.3390/nano12244376 15

Mengmeng Zhang, Kexin Huang, Yi Ding, Xinyu Wang, Yingli Gao and Pengfei Li et al.
N, S Co-Doped Carbons Derived from *Enteromorpha prolifera* by a Molten Salt Approach: Antibiotics Removal Performance and Techno-Economic Analysis
Reprinted from: *Nanomaterials* 2022, 12, 4289, doi:10.3390/nano12234289 26

Xingang Bai, Luyang Xing, Ning Liu, Nana Ma, Kexin Huang and Dapeng Wu et al.
Humulus scandens-Derived Biochars for the Effective Removal of Heavy Metal Ions: Isotherm/Kinetic Study, Column Adsorption and Mechanism Investigation
Reprinted from: *Nanomaterials* 2021, 11, 3255, doi:10.3390/nano11123255 40

Le Wang, Bingyu Jiao, Yan Shen, Rong Du, Qipeng Yuan and Jinshui Wang
Co-Immobilization of Lactase and Glucose Isomerase on the Novel g-C_3N_4/CF Composite Carrier for Lactulose Production
Reprinted from: *Nanomaterials* 2022, 12, 4290, doi:10.3390/nano12234290 62

Ye Chen, Miao Tian and Xupo Liu
Supramolecular Self-Assembly Strategy towards Fabricating Mesoporous Nitrogen-Rich Carbon for Efficient Electro-Fenton Degradation of Persistent Organic Pollutants
Reprinted from: *Nanomaterials* 2022, 12, 2821, doi:10.3390/nano12162821 80

Yanbo Wang, Yiqing Chen, Hongwei Zhao, Lixiang Li, Dongying Ju and Cunjing Wang et al.
Biomass-Derived Porous Carbon with a Good Balance between High Specific Surface Area and Mesopore Volume for Supercapacitors
Reprinted from: *Nanomaterials* 2022, 12, 3804, doi:10.3390/nano12213804 94

Bony Thomas, Mohini Sain and Kristiina Oksman
Sustainable Carbon Derived from Sulfur-Free Lignins for Functional Electrical and Electrochemical Devices
Reprinted from: *Nanomaterials* 2022, 12, 3630, doi:10.3390/nano12203630 108

Chao Ma, Mengmeng Zhang, Yi Ding, Yan Xue, Hongju Wang and Pengfei Li et al.
Green Production of Biomass-Derived Carbon Materials for High-Performance Lithium–Sulfur Batteries
Reprinted from: *Nanomaterials* 2023, 13, 1768, doi:10.3390/nano13111768 124

Mengdan Yan, Yuchen Qin, Lixia Wang, Meirong Song, Dandan Han and Qiu Jin et al.
Recent Advances in Biomass-Derived Carbon Materials for Sodium-Ion Energy Storage Devices
Reprinted from: *Nanomaterials* **2022**, *12*, 930, doi:10.3390/nano12060930 . . . . **148**

Editorial

Editorial for Special Issue: "Synthesis and Application of Biomass-Derived Carbon-Based Nanomaterial"

Mengmeng Zhang [1], Pengfei Li [1] and Dapeng Wu [1,2,*]

[1] Key Laboratory of Green Chemistry Medias and Reactions, Ministry of Education, School of Business, Henan Normal University, Xinxiang 453007, China; zhangmengmeng@htu.edu.cn (M.Z.); lipengfei@htu.edu.cn (P.L.)
[2] School of Environment, Henan Normal University, Xinxiang 453007, China
* Correspondence: dapengwu@htu.edu.cn

Biomass-derived carbon-based nanomaterials represent a group of green and high-quality materials which can be potentially employed in the fields of environmental protection, energy conversion and clean energy storage. The unique composition and tissue structures of the biomass grant these carbon-based nanomaterials outstanding features, such as a high surface area, well-developed porous texture and active heteroatom doping sites. Meanwhile, the abundant sources of biomass further endows these materials with feasibility for potential large-scale applications, which has arouse great research interest from both the science and industrial communities. Our Special Issue consists of eight articles and two reviews, which are contributions to the topics of photothermal conversion, environmental protection, clean energy storage, and catalytic application. For example, Yeamsuksawat et al. prepared chitin-derived nitrogen-doped carbon nanopaper with sub-wavelength nano-porous structures for photothermal energy conversion. They found that its solar energy absorption could be improved by nitrogen doping, which greatly exceeds the performances of the commercial product [1].

In the field of environmental protection, Long et al. proposed a novel polyacrylic-acid-modified carbon skeleton derived from grapefruit peel, which could serve as solid-phase extraction coating to capture and determine volatile halogenated hydrocarbons in water. Due to their large specific surface area and abundant surface functional groups, the detection exhibits a wide linear range and good reproducibility [2]. In addition, Zhang et al. adopted Entermorpha prolifera (EP) as a precursor to prepare N–S co-doped bio-carbon with a layered porous structure and high specific surface area using a molten salt method to degrade sulfamethoxazole via persulfate activation. It was found that the performance of potassium-chloride-derived biochar was better than that of sodium-chloride-derived biochar, and the removal rate reached 99.6% in actual water treatment. Moreover, economic evaluation confirmed that this EP-derived biochar is more competitive than commercial activated carbon in cost, which demonstrates that the EP could be employed as an abundant biowaste that produces bio-carbons with high performances, and the molten salt strategy could be further optimized by adopting low-cost salt to reduce the cost of manufacture [3]. Based on this concept, Humulus scandens was firstly used as a biomass precursor to prepare biochar through a molten salt method by Bai et al. The as-derived biochar exhibited a high specific surface area, abundant oxygen-containing functional groups, and a good adsorption performance for heavy metal ions such as Cu^{2+} and Pb^{2+}. In addition, the optimized biochar demonstrated good anti-interference ability and outstanding removal efficiency in simulated wastewater. The mechanism study and DFT calculation show that the oxygen functional group enhanced the binding energy of metal ions and plays a dominant role in the adsorption process [4].

Due to the well-developed porous structure and high conductivity as well as rich heteroatom doping sites, the biomass-derived carbons usually exhibit high performances

Citation: Zhang, M.; Li, P.; Wu, D. Editorial for Special Issue: "Synthesis and Application of Biomass-Derived Carbon-Based Nanomaterial". *Nanomaterials* **2023**, *13*, 2020. https://doi.org/10.3390/nano13132020

Received: 30 June 2023
Accepted: 4 July 2023
Published: 7 July 2023

Copyright: © 2023 by the authors. Licensee MDPI, Basel, Switzerland. This article is an open access article distributed under the terms and conditions of the Creative Commons Attribution (CC BY) license (https://creativecommons.org/licenses/by/4.0/).

in energy-storage devices. In this section, Wang et al. proposed a salt sealing technology combined with potassium hydroxide activation to convert pre-carbonized wheat shells into high-performance carbon materials. This novel method could achieve a good balance between a high surface area and mesoporous volume of biomass-derived porous carbon due to the molten salt and the activation of potassium hydroxide, which provide both channels for fast ion transfer and abundant active sites for charge storage [5]. Thomas et al. produced carbon particles from process lignin, sulfate lignin, soda lignin, lignin boost lignin and hydrolyzed lignin at different carbonization temperatures of 1000 °C and 1400 °C. It was found that the lignin source and carbonation temperature significantly affected the carbon quality and microstructure of carbon particles, which could in turn determine their performances in energy-storage devices [6]. To better illustrate the application of biomass-derived carbon materials in energy-storage devices, Ma et al. [7] and Yan et al. [8] comprehensively reviewed the recent progresses in the design, preparation and application of biomass carbon in the fields of lithium–sulfur batteries and sodium-ion energy-storage devices, respectively.

The carbon materials with a highly porous structure as well as rich heteroatom doping sites usually possess high catalytic performances. For example, Wang et al. adopted a g-C_3N_4/carbon fiber composite as a carrier to immobilize lactase and glucose isomerase, which improved the efficiency of immobilized lactulose production. The g-C_3N_4/carbon fiber composite showed positive effects on the stability of the enzyme and endowed it with a high producing stability at a wide pH range [9]. In addition, Chen et al. adopted a supramolecular self-assemble strategy to synthesize supramylamine–cyanuric acid supramolecular aggregates. The as-fabricated carbon materials have an ordered layered microstructure, highly specific surface area, rich mesoporous distribution, and high N doping. In the Fenton system, the hydrogen peroxide production could be promoted, leading to the selective degradation of organic pollutants [10].

These works presented in this Special Issue demonstrated that the carbon-based materials prepared from biomass could not only pave a new way to yield high-performance materials for various applicational fields, but also contribute to the effective utilization of the abundant biomass resources. In addition, it was also proposed that bio-carbon manufacture techniques are of great significance to determine the performances of the final carbon materials. The further development of green, economic, sustainable and standardized techniques for the mass production of biomass carbon could provide more opportunities for the final commercialization of the biomass-derived carbon materials.

Conflicts of Interest: The authors declare no conflict of interest.

References

1. Yeamsuksawat, T.; Zhu, L.T.; Kasuga, T.; Nogi, M.; Koga, H. Chitin-Derived Nitrogen-Doped Carbon Nanopaper with Subwavelength Nanoporous Structures for Solar Thermal Heating. *Nanomaterials* **2023**, *13*, 1480. [CrossRef] [PubMed]
2. Long, A.Y.; Liu, H.L.; Xu, S.R.; Feng, S.L.; Shuai, Q.; Hu, S.H. Polyacrylic Acid Functionalized Biomass-Derived Carbon Skeleton with Highly Porous Hierarchical Structures for Efficient Solid-Phase Microextraction of Volatile Halogenated Hydrocarbons. *Nanomaterials* **2022**, *12*, 4376. [CrossRef] [PubMed]
3. Zhang, M.; Huang, K.; Ding, Y.; Wang, X.; Gao, Y.; Li, P.; Zhou, Y.; Guo, Z.; Zhang, Y.; Wu, D. N, S Co-Doped Carbons Derived from Enteromorpha prolifera by a Molten Salt Approach: Antibiotics Removal Performance and Techno-Economic Analysis. *Nanomaterials* **2022**, *12*, 4289. [CrossRef] [PubMed]
4. Bai, X.; Xing, L.; Liu, N.; Ma, N.; Huang, K.; Wu, D.; Yin, M.; Jiang, K. Humulus scandens-Derived Biochars for the Effective Removal of Heavy Metal Ions: Isotherm/Kinetic Study, Column Adsorption and Mechanism Investigation. *Nanomaterials* **2021**, *11*, 3255. [CrossRef] [PubMed]
5. Wang, Y.; Chen, Y.; Zhao, H.; Li, L.; Ju, D.; Wang, C.; An, B. Biomass-Derived Porous Carbon with a Good Balance between High Specific Surface Area and Mesopore Volume for Supercapacitors. *Nanomaterials* **2022**, *12*, 3804. [CrossRef] [PubMed]
6. Thomas, B.; Sain, M.; Oksman, K. Sustainable Carbon Derived from Sulfur-Free Lignins for Functional Electrical and Electrochemical Devices. *Nanomaterials* **2022**, *12*, 3630. [CrossRef] [PubMed]
7. Ma, C.; Zhang, M.; Ding, Y.; Xue, Y.; Wang, H.; Li, P.; Wu, D. Green Production of Biomass-Derived Carbon Materials for High-Performance Lithium–Sulfur Batteries. *Nanomaterials* **2023**, *13*, 1768. [CrossRef] [PubMed]

8. Yan, M.; Qin, Y.; Wang, L.; Song, M.; Han, D.; Jin, Q.; Zhao, S.; Zhao, M.; Li, Z.; Wang, X.; et al. Recent Advances in Biomass-Derived Carbon Materials for Sodium-Ion Energy Storage Devices. *Nanomaterials* **2022**, *12*, 930. [CrossRef] [PubMed]
9. Wang, L.; Jiao, B.; Shen, Y.; Du, R.; Yuan, Q.; Wang, J. Co-Immobilization of Lactase and Glucose Isomerase on the Novel g-C_3N_4/CF Composite Carrier for Lactulose Production. *Nanomaterials* **2022**, *12*, 4290. [CrossRef] [PubMed]
10. Chen, Y.; Tian, M.; Liu, X. Supramolecular Self-Assembly Strategy towards Fabricating Mesoporous Nitrogen-Rich Carbon for Efficient Electro-Fenton Degradation of Persistent Organic Pollutants. *Nanomaterials* **2022**, *12*, 2821. [CrossRef] [PubMed]

Disclaimer/Publisher's Note: The statements, opinions and data contained in all publications are solely those of the individual author(s) and contributor(s) and not of MDPI and/or the editor(s). MDPI and/or the editor(s) disclaim responsibility for any injury to people or property resulting from any ideas, methods, instructions or products referred to in the content.

Article

Chitin-Derived Nitrogen-Doped Carbon Nanopaper with Subwavelength Nanoporous Structures for Solar Thermal Heating

Thanakorn Yeamsuksawat, Luting Zhu, Takaaki Kasuga, Masaya Nogi and Hirotaka Koga *

SANKEN (The Institute of Scientific and Industrial Research), Osaka University, 8-1 Mihogaoka, Ibaraki 567-0047, Osaka, Japan; y.thanakorn@eco.sanken.osaka-u.ac.jp (T.Y.); sharollzhu@eco.sanken.osaka-u.ac.jp (L.Z.); tkasuga@eco.sanken.osaka-u.ac.jp (T.K.); nogi@eco.sanken.osaka-u.ac.jp (M.N.)
* Correspondence: hkoga@eco.sanken.osaka-u.ac.jp; Tel.: +81-6-6879-8442; Fax: +81-6-6879-8444

Abstract: Sustainable biomass-derived carbons have attracted research interest because of their ability to effectively absorb and convert solar light to thermal energy, a phenomenon known as solar thermal heating. Although their carbon-based molecular and nanoporous structures should be customized to achieve enhanced solar thermal heating performance, such customization has insufficiently progressed. In this study, we transformed a chitin nanofiber/water dispersion into paper, referred to as chitin nanopaper, with subwavelength nanoporous structures by spatially controlled drying, followed by temperature-controlled carbonization without any pretreatment to customize the carbon-based molecular structures. The optimal carbonization temperature for enhancing the solar absorption and solar thermal heating performance of the chitin nanopaper was determined to be 400 °C. Furthermore, we observed that the nitrogen component, which afforded nitrogen-doped carbon structures, and the high morphological stability of chitin nanofibers against carbonization, which maintained subwavelength nanoporous structures even after carbonization, contributed to the improved solar absorption of the carbonized chitin nanopaper. The carbonized chitin nanopaper exhibited a higher solar thermal heating performance than the carbonized cellulose nanopaper and commercial nanocarbon materials, thus demonstrating significant potential as an excellent solar thermal material.

Keywords: chitin nanofiber; biomass-derived carbon; subwavelength nanoporous structures; nitrogen-doped carbon; solar thermal heating

1. Introduction

Solar thermal heating has been receiving increasing attention from researchers as a promising process for using solar light as thermal energy. Solar thermal heating requires photothermal materials, which absorb light and convert it into heat [1,2]. Examples of photothermal materials include plasmonic metal nanoparticles, metal oxide semiconductors, carbon materials [2], and plasmonic metamaterials [3,4]. Among these, carbon materials exhibit a broad light absorption range [5], which covers the wavelength range of solar light (300–2500 nm, ASTM G173-03, Air Mass 1.5 Global spectrum (AM1.5G) [6]). Biomass-derived carbons have been used as sustainable photothermal materials for a variety of solar thermal heating applications, such as solar steam generation [7], desalination, wastewater purification [8,9], and photothermal catalysis [10].

The rational structural design of biomass-derived carbons is desirable for further enhancing their solar thermal heating performance [2]. For instance, carbon-based molecular structures, including sp^2-hybridized carbon and heteroatom-doped carbon structures with a customized distribution of the highest occupied molecular orbital and lowest unoccupied molecular orbital, should be designed. The design of such carbon-based molecular structures can influence not only light absorption but also the conversion of the absorbed light to heat by vibration relaxation [11,12]. Furthermore, subwavelength nanoporous structures should be designed to suppress light reflection and facilitate light absorption via light confinement, as reported for plasmonic metal nanoparticles and metal oxide semiconductors [13].

We previously reported that the carbon material derived from wood cellulose nanofiber paper, referred to as cellulose nanopaper, whose molecular and subwavelength nanoporous structures can be customized, acts as a photothermal material for solar thermal heating [14]. The carbon-based molecular structures of cellulose nanopaper were customized by controlling the carbonization temperatures; the sp^2-hybridized carbon structures formed by semicarbonization at 500 °C afforded high solar light absorption by adequately balancing solar light absorption and reflection. Subwavelength nanopore structures have also been constructed within cellulose nanopaper-derived carbon by expanding the pore spaces between cellulose nanofibers via treatment with low-surface-tension *tert*-butyl alcohol (*t*-BuOH) [15], thereby suppressing solar light reflection. The resulting cellulose nanopaper-derived carbon with customized molecular and subwavelength nanoporous structures exhibited effective solar thermal heating performance, which was higher than those of previously reported biomass-derived carbons and conventional nanocarbon materials, including graphene and carbon nanotube films [14]. However, cellulose nanopaper requires iodine gas pretreatment (100 °C) for a long duration (24 h) to retain its customized subwavelength nanoporous structures after carbonization because the morphology of cellulose nanofibers collapses after high-temperature treatment [15]. Hence, discovering alternative biomass nanofibers that can retain their original morphology without pretreatment, even after carbonization, with adequate carbon-based molecular structures for excellent solar thermal heating is desirable.

Chitin (β-(1→4)-linked *N*-acetyl anhydroglucosamine) is among the most abundant biomass materials on earth. Its molecular structure is similar to that of cellulose, except for the presence of an acetyl-amino group instead of a hydroxyl group on C-2 in cellulose. Chitin can be extracted as nanofibers from the exoskeletons of crustacean wastes, such as crabs, squid pens, and prawns [16–19]. Chitin nanofibers intrinsically contain nitrogen derived from their acetyl-amino group, providing an opportunity to prepare N-doped carbon structures by carbonization [20,21]. N-doped carbon structures have demonstrated enhanced functionality in various applications such as energy storage [20,22,23], adsorption and catalysis [21], photosensing [23], and microwave absorption [24]. Moreover, chitin nanofibers exhibit higher thermal stability than cellulose nanofibers against the collapse of their morphology during carbonization [25]. Thus, carbonized chitin nanofibers could be a superior alternative to carbonized cellulose nanofibers as high-performance photothermal materials. However, to the best of our knowledge, the photothermal heating properties of carbonized chitin nanofiber materials have not yet been explored.

In this study, chitin-nanofiber-derived N-doped carbon was fabricated and evaluated as a photothermal material for solar thermal heating. Chitin nanofibers were transformed into a nanopaper with subwavelength nanoporous structures by expanding the pore spaces between the nanofibers via *t*-BuOH treatment. Subsequently, carbonization was performed without pretreatment at controlled temperatures to customize the carbon-based molecular structures. Moreover, to elucidate the significance of the carbonized chitin nanopaper, its solar thermal heating performance was compared with that of a carbonized cellulose nanopaper.

2. Materials and Methods

2.1. Materials

Aqueous dispersions of chitin nanofibers (2 wt%, SFo-20002) and cellulose nanofibers (2 wt%, WFo-10002) were obtained from Sugino Machine Ltd., Namerikawa, Japan. *t*-BuOH (>99% purity) was supplied by Nacalai Tesque Inc., Kyoto, Japan.

2.2. Preparation and Carbonization of Cellulose and Chitin Nanopapers

A water dispersion of chitin or cellulose nanofibers (0.2 wt%, 200 mL) was vacuum filtered on a hydrophilic polytetrafluorethylene membrane (pore diameter: 0.2 μm, H020A090C, Advantec Toyo Kaisha, Ltd., Tokyo, Japan). The nanofibers on the membrane were then treated by gently pouring t-BuOH onto it while vacuum filtering. The resulting wet nanopaper was peeled from the membrane, stored in a refrigerator (SJ-23T-S, Sharp, Corp., Osaka, Japan) at -18 °C for 0.5 h, and then freeze-dried overnight (FDU-2200, Tokyo Rikakikai Co., Ltd., Tokyo, Japan). Subsequently, the as-prepared nanopaper with a thickness of ~300 μm was cut into a square sheet with an area of 1.5×1.5 cm^2. A molybdenum block (area: 1.5×1.5 cm^2, thickness: ~1.0 cm, weight: ~20 g, MO-293771, The Nilaco Corp., Tokyo, Japan) was placed on it. Then, carbonization was performed in a furnace (KDF-75, DENKEN-HIGHDENTAL Co., Ltd., Kyoto, Japan) under a N$_2$ gas flow at a flow rate of ~500 mL min^{-1} in three stages [14]: (1) the temperature was increased from room temperature to 240 °C at 2 °C min^{-1} and maintained for 17 h; (2) the temperature was increased from 240 °C to the target temperature (300–1100 °C) at 2 °C min^{-1} and maintained for 1 h; and (3) the temperature was decreased to room temperature at 2 °C min^{-1}.

2.3. Solar Thermal Heating Performances

Following our previous study [14], we evaluated the solar thermal heating performances of the original and carbonized nanopapers by measuring the changes in their surface temperatures during solar light irradiation. Prior to the evaluation, the emissivity of each nanopaper was evaluated using a commercial black tape (emissivity: 0.95, HB-250, OPTEX Co., Ltd., Otsu, Japan) as a reference. Briefly, black tape and a carbonized nanopaper were heated to 75 °C on a thermo-controller (SBX-303, Sakaguchi E.H VOC Corp., Tokyo, Japan). The emissivity of each nanopaper was estimated using a thermal imaging camera (FLIR ETS320, FLIR Systems. Inc., Wilsonville, OR, USA) by adjusting the temperature according to the reference temperature of black tape. Subsequently, surface temperature measurements were performed using a solar simulator (HAL-320W, Asahi Spectra Co., Ltd., Tokyo, Japan). The nanopaper with an area of less than 1.5×1.5 cm^2 was placed on an acrylic plate (3×3 cm^2) with a rectangular hole (0.7×0.7 cm^2). Thereafter, it was irradiated by simulated solar light (AM1.5G, light intensity: 1.0 kW m^{-2} (1 sun)) such that the area of light illumination was larger than that of the nanopaper. The surface temperature of the nanopaper was recorded using a thermal imaging camera, and its equilibrium surface temperature was evaluated from the average temperature during a solar illumination time of 500–600 s (~850 plots). More than five samples were prepared and evaluated under each condition. The surface temperature measurements were conducted at 25 °C and 65% relative humidity.

2.4. Optical Properties

The light absorption, transmittance, and reflection of the carbonized chitin and cellulose nanopapers were evaluated using an ultraviolet−visible−near-infrared (UV–vis–NIR) spectrometer (UV-3600i Plus, Shimadzu Corp., Kyoto, Japan) equipped with an ISR-603 integrating sphere (Shimadzu Corp., Kyoto, Japan). More than five samples were prepared and evaluated under each condition. Light absorption was calculated from the total light transmittance and reflection spectra. Solar absorption was calculated using Equation (1) [26] as follows:

$$\overline{\alpha}\,(\%) = \frac{\int_{\lambda_{min}}^{\lambda_{max}} I_{solar}(\lambda)\cdot\alpha_{solar}(\lambda)d\lambda}{\int_{\lambda_{min}}^{\lambda_{max}} I_{solar}(\lambda)d\lambda} \times 100, \qquad (1)$$

where $\overline{\alpha}$ is the solar absorption (%); λ is the wavelength (nm); λ_{min} and λ_{max} are 300 and 2500 nm, respectively; $I_{solar}(\lambda)$ is the solar spectral irradiance (AM1.5G) at λ; and $\alpha_{solar}(\lambda)$ is the light absorption (%) at λ. The optical bandgap values were also calculated from the

UV–vis–NIR absorption spectra according to a previously reported method [15] and Tauc's equation [27] (Equation (2)):

$$(\alpha h\nu)^{1/n} = A(h\nu - E_g), \qquad (2)$$

where α, $h\nu$, A, and E_g are the absorbance, photon energy, constant, and optical band gap, respectively. The optical bandgap was estimated by plotting $(\alpha h\nu)^{1/n}$ vs. photon energy ($h\nu$) and extrapolating the linear region of the curve to the X-axis (Figure S1). The parameter n was set to 2 for the indirect transition of the carbonized nanopapers because of their amorphous carbon structures.

2.5. Molecular Structures

Laser Raman spectroscopic analyses were performed using a RAMAN-touch VISNIR-OUN spectrometer (Nanophoton Corp., Osaka, Japan) with an incident laser wavelength of 532 nm. Elemental analyses were performed using a 2400II instrument (PerkinElmer Japan Co., Ltd., Kanagawa, Japan). X-ray photoelectron spectroscopy (XPS) profiles were recorded using a JPS-9010 photoelectron spectrometer with a monochromatic Al Kα X-ray source (1486.6 eV) (JEOL, Ltd., Tokyo, Japan) at 15 kV voltage and 20 mA current.

2.6. Nanoporous Structures

Surface structures of the original and carbonized chitin and cellulose nanopapers were observed using field-emission scanning electron microscopy (FE-SEM) (SU-8020, Hitachi High-Tech Science Corp., Tokyo, Japan) at an accelerating voltage of 2 kV. Prior to FE-SEM, platinum sputtering of the samples was conducted using an E-1045 Ion Sputter (Hitachi High-Tech Science Corp., Tokyo, Japan) at a current of 20 mA for 10 s. Pore size distribution curves were obtained using nitrogen adsorption analysis at $-196\ °C$ based on the Brunauer–Emmett–Teller and density functional theory models (NOVA 4200e, Quantachrome Instruments, Kanagawa, Japan).

3. Results and Discussion

The fabrication of the carbonized chitin nanopaper is schematically illustrated in Figure 1a. A water dispersion of crab-shell-derived chitin nanofibers (0.2 wt%, 200 mL) was suction filtered. Then, it was treated by gently pouring t-BuOH (200 mL) onto the resulting wet sheet and vacuum filtered. Next, it was freeze-dried overnight and carbonized at 400 °C under a N_2 atmosphere without any pretreatment. The color of the chitin nanopaper changed from white to black after carbonization. Similarly, a carbonized cellulose nanopaper was fabricated using a water dispersion of wood-derived cellulose nanofibers (0.2 wt%, 200 mL). Although the chitin and cellulose nanopapers became somewhat brittle after carbonization, they were freestanding and allowed easy handling for characterization and evaluation.

The solar thermal heating properties of the carbonized chitin and cellulose nanopapers were evaluated and compared. As shown in Figure 1b, the change in the surface temperature of the carbonized chitin or cellulose nanopaper under simulated solar illumination (AM1.5G, light intensity: 1 sun) was monitored using a thermal imaging camera. The surface temperatures of the carbonized chitin and cellulose nanopapers rapidly increase upon 1-sun illumination and are saturated within 600 s (Figure 1c). The equilibrium surface temperatures of the original chitin and cellulose nanopapers (before carbonization) are 37.7 ± 0.40 and $37.1 \pm 1.40\ °C$, respectively, while those of the carbonized chitin and cellulose nanopapers are 75.9 ± 1.27 and $66.9 \pm 1.40\ °C$, respectively (Figure 1d), suggesting that the carbonized chitin nanopaper exhibits a higher solar thermal heating performance than the carbonized cellulose nanopaper.

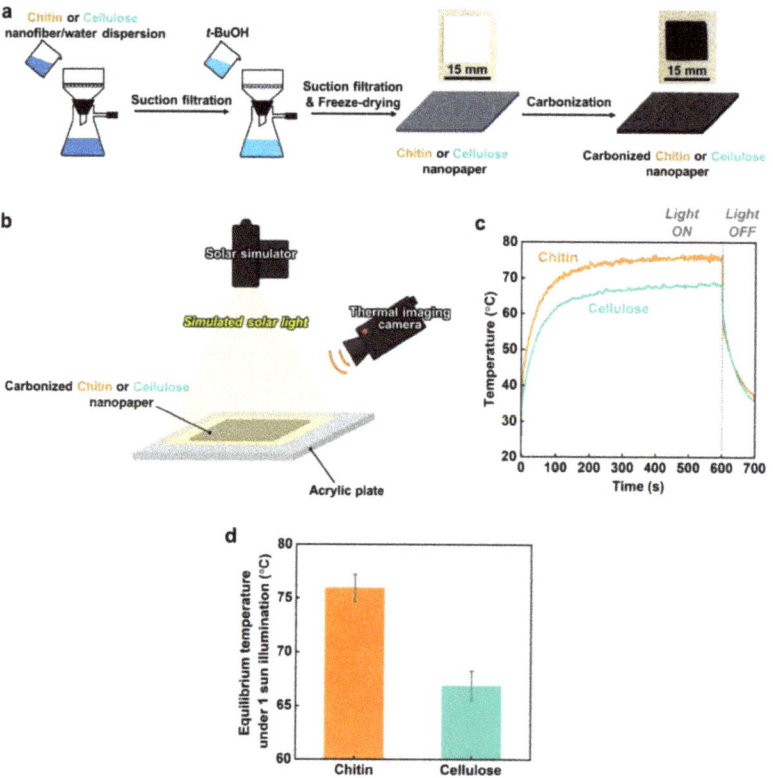

Figure 1. Preparation and solar thermal heating performance of carbonized chitin and cellulose nanopapers. (**a**) Schematic of the preparation of the original and carbonized chitin or cellulose nanopaper and optical images of the original and carbonized chitin nanopaper; (**b**) schematic of the experimental setup for the measurement of the surface temperature during simulated solar light illumination; (**c**) surface temperature evolution and (**d**) equilibrium surface temperature of carbonized chitin and cellulose nanopapers under 1-sun illumination. Carbonization temperature: 400 °C.

Solar thermal heating by photothermal materials depends on their ability to absorb solar light and convert it into heat [1,2]. Therefore, to observe the difference in the solar thermal heating properties of the carbonized chitin and cellulose nanopapers, their light absorption properties were compared (Figure 2). As shown in Figure 2a, the carbonized chitin nanopaper exhibits a higher light absorption than the carbonized cellulose nanopaper in the wavelength range of solar light (AM1.5G, 300–2500 nm) [6]. The higher light absorption of the carbonized chitin nanopaper is attributed to (1) the light transmission (transmittance: ~0%) wavelength being extended to a longer wavelength region (Figure 2b) and (2) light reflection being suppressed in the entire solar wavelength region (300−2500 nm) (Figure 2c). Furthermore, for a clearer comparison of the carbonized chitin and cellulose nanopapers, their solar absorptions, transmittances, and reflections were calculated from UV−vis−NIR absorption, transmittance, and reflection spectra, respectively, and the AM1.5G solar spectral irradiance, according to a previously reported method [26]. As shown in Figure 2d–f, the carbonized chitin nanopaper provides a higher solar absorption (97.0% ± 0.19%) than the carbonized cellulose nanopaper (90.9% ± 0.43%) owing to the slight suppression of solar transmittance and large suppression of reflection. Thus, the carbonized chitin nanopaper exhibits a higher solar thermal heating performance than the carbonized cellulose nanopaper owing to higher solar absorption.

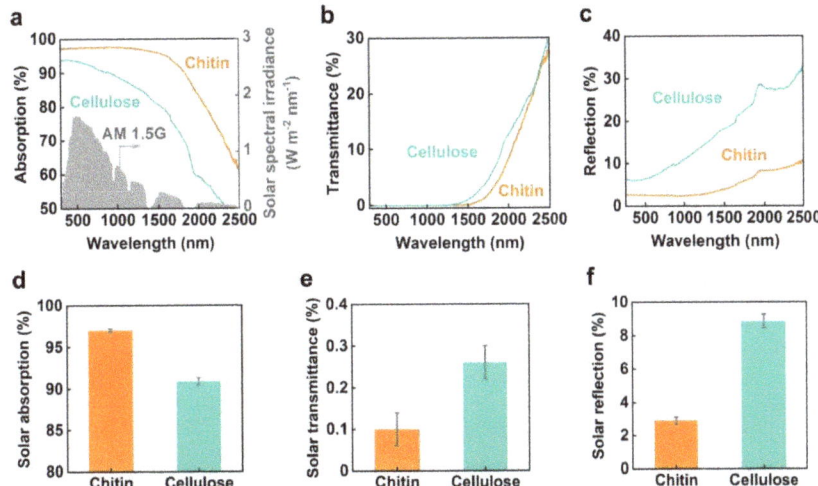

Figure 2. Light absorption properties of the carbonized chitin and cellulose nanopapers. (**a**) AM1.5G solar spectral irradiance and UV–vis–NIR absorption, (**b**) transmittance, and (**c**) reflection spectra; solar light (**d**) absorption, (**e**) transmittance, and (**f**) reflection. Carbonization temperature: 400 °C.

The UV–vis–NIR absorption and transmission spectra of the carbonized chitin nanopaper exhibit light absorption over longer wavelength regions compared to those of the carbonized cellulose nanopaper (Figure 2a,b). This characteristic of the carbonized chitin nanopaper is beneficial for the suppression of light transmission and reflection in the longer wavelength region. We attributed this phenomenon to the lower optical bandgap of the carbonized chitin nanopaper (0.74 eV) compared with that of the carbonized cellulose nanopaper (1.01 eV) (see Supplementary Information, Figure S1). A lower bandgap can facilitate light absorption at lower energies (longer wavelengths). To confirm that the carbonized chitin nanopaper exhibited a lower optical bandgap, its molecular structures were analyzed and compared with those of the carbonized cellulose nanopaper (Figure 3). The original chitin and cellulose nanopapers before carbonization have the wide σ−σ* bandgap due to their sp^3-hybridized carbon structures, resulting in low light absorption (i.e., high light transmission and reflection) (Figure S2). The Raman spectra of the carbonized chitin and cellulose nanopapers display G and D bands (Figure 3a), which are associated with graphitic sp^2-hybridized carbon domains and defective carbon structures [28], respectively, indicating that both chitin and cellulose nanopapers formed graphitic and defective carbon structures after carbonization at 400 °C. Owing to the graphitic carbon structures (i.e., π-orbital), the carbonized chitin and cellulose nanopapers had the π−π* bandgap, which lies within the σ−σ* bandgap. Hence, the carbonized nanopapers could promote light absorption as compared with the original nanopapers, while suppressing light transmission and reflection (Figures 2a–c and S2). Elemental analyses show that the carbonized chitin nanopaper contains C (69.9 wt%), O (17.8 wt%), H (3.90 wt%), and N (8.40 wt%) and that the carbonized cellulose nanopaper contains C (75.5 wt%), O (20.3 wt%), and H (4.20 wt%) (Figure 3b). The presence of N in the carbonized chitin nanopaper was verified by XPS (Figure 3c). The C 1s spectrum of the carbonized chitin nanopaper can be deconvoluted into five peaks: C–C or C=C (284.6 eV), C=N (285.8 eV), C–O (286.1 eV), C–N (287.4 eV), and C=O (287.8 eV) [29,30], whereas that of the carbonized cellulose nanopaper shows three peaks: C–C or C=C, C–O, and C=O (Figure 3d). The N 1s spectrum of the carbonized chitin nanopaper suggests the formation of pyridinic N (398.4 eV), pyrrolic N (399.9 eV), and graphitic N (401.0 eV) [31] (Figure 3e). These results indicate that the carbonized cellulose nanopaper possessed O-doped defective carbon structures, whereas the carbonized chitin nanopaper possessed N- and O-doped defective carbon structures. The carbonized cellulose nanopaper had graphitic sp² -hybridized carbon domains (π-orbital) and defective regions such as O-containing functional groups (n-orbital), in which the n energy level lies within the π–π* energy gaps and reduces the optical bandgap [32]. The carbonized chitin nanopaper had additional N-containing functional groups (n-orbital) in the defective regions, which could further reduce the optical bandgap. Thus, the N- and O-doped defective carbon structures of the carbonized chitin nanopaper result in a decreased optical bandgap, facilitating light absorption at lower energies (longer wavelengths) and consequently enhancing its solar absorption performance.

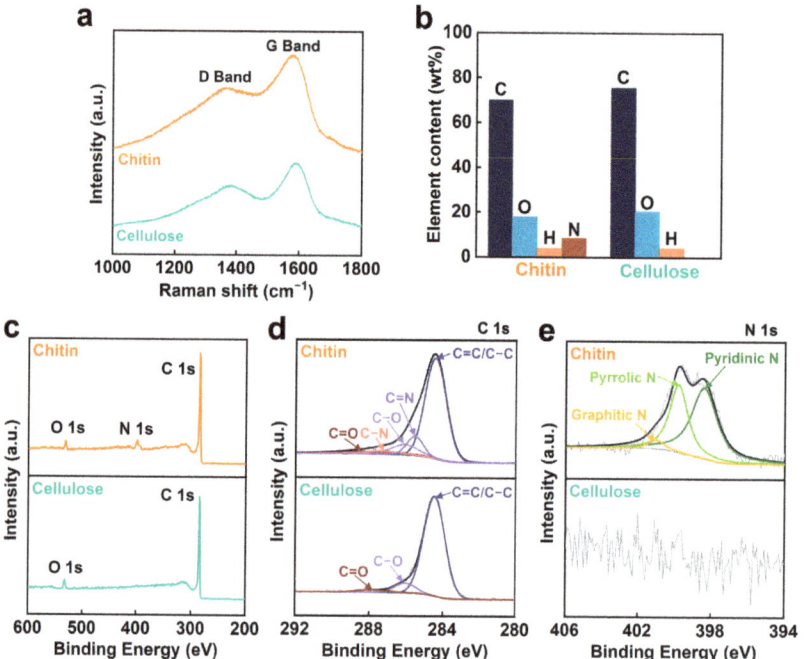

Figure 3. Molecular structures of the carbonized chitin and cellulose nanopapers. (**a**) Raman spectra, (**b**) element contents, (**c**) wide XPS profiles, (**d**) C 1s XPS, and (**e**) N 1s XPS. Carbonization temperature: 400 °C.

Unlike the carbonized cellulose nanopaper, the carbonized chitin nanopaper suppresses light reflection over the entire solar wavelength region (Figure 2c), demonstrating excellent solar absorption. To determine the reason for this suppressed light reflection, the morphologies of the carbonized chitin and cellulose nanopapers were analyzed and compared (Figure 4). The original chitin and cellulose nanopapers prepared in this study constitute subwavelength nanoporous structures that can suppress light reflection via the light confinement effect [13]. The subwavelength nanoporous structures of both nanopapers have similar pore size distributions (Figure 4a,b,e). However, the cellulose nanopaper shrinks considerably after carbonization, closing its nanopores and forming microscale wrinkles on its surface (Figure 4d). The resulting dense and microscale structures cause light reflection [14]. By contrast, the shrinkage in chitin nanopaper after carbonization is lower than that in the cellulose nanopaper, which helps maintain the nanoporous structure and prevents the formation of microscale wrinkles in chitin nanopaper (Figure 4c,f). The morphological stability of the chitin nanopaper against carbonization could be attributed to the acetyl-amino group of chitin [33]. Hence, the carbonized chitin nanopaper demonstrated the light confinement effect owing to its subwavelength nanoporous structures, which suppress light reflection, thereby facilitating solar absorption.

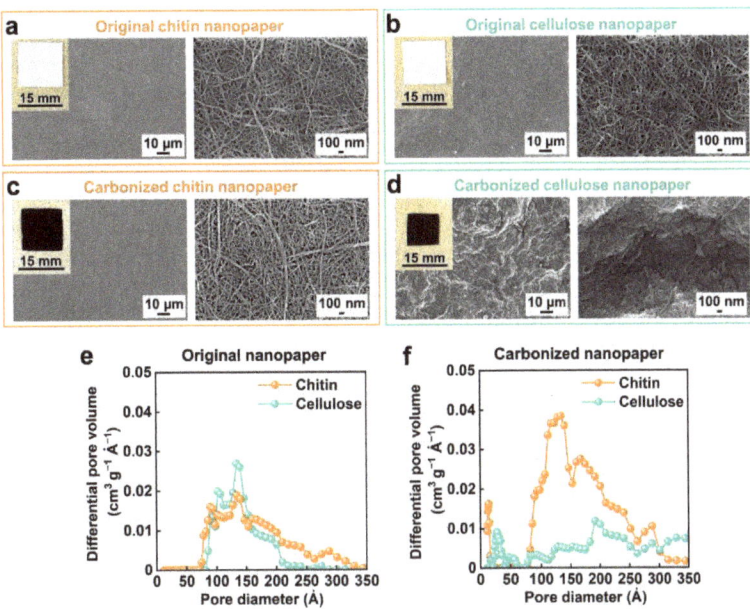

Figure 4. Morphologies of the carbonized chitin and cellulose nanopapers. Optical and field-emission scanning electron microscopy images of the original (**a**) chitin and (**b**) cellulose nanopapers and the carbonized (**c**) chitin and (**d**) cellulose nanopapers; pore size distribution curves of the (**e**) original and (**f**) carbonized chitin and cellulose nanopapers. Carbonization temperature: 400 °C.

Finally, the morphologies and solar thermal heating properties of the carbonized chitin and cellulose nanopapers were compared at different carbonization temperatures. At the carbonization temperatures of 300–1100 °C, the chitin nanopaper retains more area and volume than the cellulose nanopaper (Figure 5a,b), thus retaining the subwavelength nanoporous structures even at 1100 °C (Figure S3). Furthermore, the carbonized chitin nanopaper suppresses solar reflection and demonstrates higher solar absorption than does the carbonized cellulose nanopaper at all carbonization temperatures (300–1100 °C) (Figure 5c–e). Although the thicknesses of the carbonized chitin and cellulose nanopapers were gradually decreased with increasing carbonization temperatures (Figure S4), the carbonized nanopapers exhibited very low solar transmittance, regardless of their carbonization temperatures (Figure 5d). These results suggested that the thickness of the carbonized nanopapers investigated in this study is not a dominant factor for their solar absorption properties. Notably, the chitin nanopaper carbonized at 400 °C exhibits the highest solar absorption and the highest equilibrium surface temperature under 1-sun illumination (Figure 5c,f). The solar absorption and surface temperature under 1-sun illumination of the carbonized chitin nanopaper gradually decrease with increasing carbonization temperature above 400 °C. The lower solar absorption at higher carbonization temperatures (Figure 5c) is attributed to higher light reflection (Figure 5e); the light reflection increases with increasing carbonization temperatures, possibly due to the gradual growth of graphitic sp^2-hybridized carbon domains by removing N and O [23,24]. The increased light reflection could be derived from a graphitic carbon domain-induced metallic luster, as reported for carbonized cellulose nanopaper [14], graphite films [34], and graphene papers [35]. The gradual formation of microscale wrinkles on the surfaces of the carbonized chitin nanopaper with increasing carbonization temperatures (Supplementary Information, Figure S3a) could also increase its light reflection. The lower surface temperature under 1-sun illumination at higher carbonization temperatures could be attributed to the increased through-plane thermal conductivity resulting from the growth of the graphitic sp^2-hybridized carbon domain [14] in addition to the decreased solar absorption. The optimal carbonization temperature for solar absorption and surface temperature under 1-sun illumination are slightly different for the chitin (400 °C) and cellulose nanopapers (500 °C) (Figure 5c,f), which could be ascribed to the balance of their carbon-based molecular structures and morphologies. Moreover, the chitin nanopaper exhibits the best solar thermal heating performance at a lower carbonization temperature of 400 °C than that of the cellulose nanopaper (500 °C). Furthermore, the carbonized chitin nanopaper exhibits a higher solar thermal heating performance than the carbonized cellulose nanopaper regardless of the carbonization temperature. The solar thermal heating performance (surface temperature under 1-sun illumination: 75.9 ± 1.27 °C) of the chitin nanopaper carbonized at 400 °C is superior to those of commercial nanocarbon materials, such as carbon nanotube black body (55.0 °C), graphite sheet (64.5 °C), graphene paper (65.2 °C), and graphene oxide film (69.4 °C) [14].

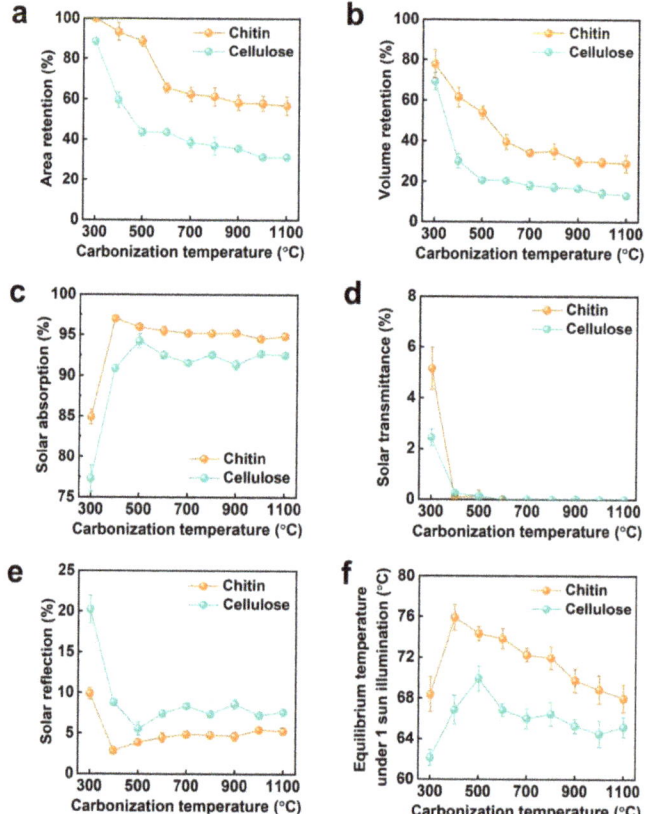

Figure 5. Morphology retention, solar absorption properties, and solar thermal heating performances of the chitin and cellulose nanopapers carbonized at different temperatures. (**a**) Area and (**b**) volume retention; solar (**c**) absorption, (**d**) transmittance, and (**e**) reflection; and (**f**) equilibrium surface temperature under 1-sun illumination. Carbonization temperature: 300–1100 °C.

4. Conclusions

In this study, chitin and cellulose nanopapers were prepared and carbonized. The carbonized chitin nanopaper demonstrated excellent solar thermal heating properties owing to its N-doped carbon structures and subwavelength nanoporous structures, which facilitated effective solar thermal heating. The optimal carbonization temperature for the chitin nanopaper was 400 °C, which balanced its carbon-based molecular structure and nano/microscale morphology. The solar thermal heating performance of the optimized carbonized chitin nanopaper (solar absorption: 97.0 ± 0.19%, surface temperature under 1 sun illumination: 75.9 ± 1.27 °C) was higher than that of the carbonized cellulose nanopaper (solar absorption: 94.3 ± 0.85%, surface temperature under 1-sun illumination: 69.9 ± 1.20 °C). This was attributed to the lower optical bandgap in the carbonized chitin nanopaper derived from its N-doped carbon structures and higher morphological stability against carbonization that maintained its subwavelength nanoporous structures without any pretreatment. Thus, carbonized chitin nanopaper is expected as a promising photothermal material towards the effective use of solar energy. Further design of nanoporous structures within carbonized chitin nanopaper can enhance its solar thermal heating performance. Moreover, the strategies presented herein can promote the functionalization of carbonized bionanofiber materials with heteroatom-doped and nanostructured carbons for various applications.

Supplementary Materials: The following supporting information can be downloaded at: https://www.mdpi.com/article/10.3390/nano13091480/s1, Figure S1: Tauc plots and estimated optical bandgap values of chitin and cellulose nanopapers carbonized at 400 °C; Figure S2: Light absorption properties of the original chitin and cellulose nanopapers before carbonization; Figure S3: Morphologies of chitin and cellulose nanopapers carbonized at 500 and 1100 °C; Figure S4: Thickness of the chitin and cellulose nanopapers carbonized at different temperatures.

Author Contributions: Conceptualization, H.K.; methodology, T.Y., L.Z. and H.K.; investigation, formal analysis, and validation, T.Y.; data curation and visualization, T.Y. and L.Z.; resources, T.K., M.N. and H.K.; supervision, M.N. and H.K.; project administration and funding acquisition, H.K.; writing—original draft preparation, T.Y. and H.K.; writing—review and editing, L.Z., T.K., M.N. and H.K. All authors have read and agreed to the published version of the manuscript.

Funding: This research was partially supported by the JST FOREST Program (Grant No. JPMJFR2003 to H.K.), "Nanotechnology Platform Project (Nanotechnology Open Facilities in Osaka University)" of the Ministry of Education, Culture, Sports, Science and Technology, Japan (MEXT) (Grant No. JPMXP1222OS1040 to H. K.) and JICA Innovative Asia Program 4th Batch (Grant No. 201905897J023 to T.Y.).

Data Availability Statement: Data presented in this study are available in this article.

Acknowledgments: The authors thank the Comprehensive Analysis Center and the Flexible 3D System Integration Laboratory, SANKEN, Osaka University, for elemental and XPS analyses and FE-SEM observation.

Conflicts of Interest: The authors declare no conflict of interest.

References

1. Guney, M.S. Solar power and application methods. *Renew. Sustain. Energy Rev.* **2016**, *57*, 776–785. [CrossRef]
2. Gao, M.; Zhu, L.; Peh, C.K.; Ho, G.W. Solar absorber material and system designs for photothermal water vaporization towards clean water and energy production. *Energy Environ. Sci.* **2019**, *12*, 841–864. [CrossRef]
3. Liang, Y.; Koshelev, K.; Zhang, F.; Lin, H.; Lin, S.; Wu, J.; Jia, B.; Kivshar, Y. Bound States in the Continuum in Anisotropic Plasmonic Metasurfaces. *Nano Lett.* **2020**, *20*, 6351–6356. [CrossRef] [PubMed]
4. Loh, J.Y.Y.; Safari, M.; Mao, C.; Viasus, C.J.; Eleftheriades, G.V.; Ozin, G.A.; Kherani, N.P. Near-Perfect Absorbing Copper Metamaterial for Solar Fuel Generation. *Nano Lett.* **2021**, *21*, 9124–9130. [CrossRef] [PubMed]
5. Han, B.; Zhang, Y.-L.; Chen, Q.-D.; Sun, H.-B. Carbon-based photothermal actuators. *Adv. Funct. Mater.* **2018**, *28*, 1802235. [CrossRef]
6. ASTM G173-03; Standard Tables for Reference Solar Spectral Irradiances: Direct Normal and Hemispherical on 37° Tilted Surface. ASTM International: West Conshohocken, PA, USA, 2012.
7. Fillet, R.; Nicolas, V.; Fierro, V.; Celzard, A. A review of natural materials for solar evaporation. *Sol. Energy Mater. Sol. Cells* **2021**, *219*, 110814. [CrossRef]
8. Guan, W.; Guo, Y.; Yu, G. Carbon materials for solar water evaporation and desalination. *Small* **2021**, *17*, e2007176. [CrossRef]
9. Saleque, A.M.; Nowshin, N.; Ivan, M.N.A.S.; Ahmed, S.; Tsang, Y.H. Natural porous materials for interfacial solar steam generation toward clean water production. *Sol. RRL* **2022**, *6*, 2100986. [CrossRef]

10. Zhou, S.; Zhou, L.; Zhang, Y.; Sun, J.; Wen, J.; Yuan, Y. Upgrading earth-abundant biomass into three-dimensional carbon materials for energy and environmental applications. *J. Mater. Chem. A* **2019**, *7*, 4217–4229. [CrossRef]
11. Feng, G.; Zhang, G.Q.; Ding, D. Design of superior phototheranostic agents guided by Jablonski diagrams. *Chem. Soc. Rev.* **2020**, *49*, 8179–8234. [CrossRef]
12. Tao, P.; Ni, G.; Song, C.; Shang, W.; Wu, J.; Zhu, J.; Chen, G.; Deng, T. Solar-driven interfacial evaporation. *Nat. Energy* **2018**, *3*, 1031–1041. [CrossRef]
13. Zheng, X.; Zhang, L. Photonic nanostructures for solar energy conversion. *Energy Environ. Sci.* **2016**, *9*, 2511–2532. [CrossRef]
14. Yeamsuksawat, T.; Morishita, Y.; Shirahama, J.; Huang, Y.; Kasuga, T.; Nogi, M.; Koga, H. Semicarbonized subwavelength-nanopore-structured nanocellulose paper for applications in solar thermal heating. *Chem. Mater.* **2022**, *34*, 7379–7388. [CrossRef]
15. Koga, H.; Nagashima, K.; Suematsu, K.; Takahashi, T.; Zhu, L.; Fukushima, D.; Huang, Y.; Nakagawa, R.; Liu, J.; Uetani, K.; et al. Nanocellulose paper semiconductor with a 3D network structure and its nano–micro–macro trans-scale design. *ACS Nano* **2022**, *16*, 8630–8640. [CrossRef]
16. Shamshina, J.L.; Berton, P.; Rogers, R.D. Advances in functional chitin materials: A review. *ACS Sustain. Chem. Eng.* **2019**, *7*, 6444–6457. [CrossRef]
17. Ifuku, S.; Saimoto, H. Chitin nanofibers: Preparations, modifications, and applications. *Nanoscale* **2012**, *4*, 3308–3318. [CrossRef]
18. Fan, Y.; Saito, T.; Isogai, A. Preparation of chitin nanofibers from squid pen β-chitin by simple mechanical treatment under acid conditions. *Biomacromolecules* **2008**, *9*, 1919–1923. [CrossRef]
19. Ifuku, S.; Nogi, M.; Abe, K.; Yoshioka, M.; Morimoto, M.; Saimoto, H.; Yano, H. Preparation of chitin nanofibers with a uniform width as α-chitin from crab shells. *Biomacromolecules* **2009**, *10*, 1584–1588. [CrossRef]
20. Nguyen, T.-D.; Shopsowitz, K.E.; MacLachlan, M.J. Mesoporous nitrogen-doped carbon from nanocrystalline chitin assemblies. *J. Mater. Chem. A* **2014**, *2*, 5915. [CrossRef]
21. Gao, Y.; Chen, X.; Zhang, J.; Yan, N. Chitin-derived mesoporous, nitrogen-containing carbon for heavy-metal removal and styrene epoxidation. *ChemPlusChem* **2015**, *80*, 1556–1564. [CrossRef]
22. Ding, B.; Huang, S.; Pang, K.; Duan, Y.; Zhang, J. Nitrogen-enriched carbon nanofiber aerogels derived from marine chitin for energy storage and environmental remediation. *ACS Sustain. Chem. Eng.* **2018**, *6*, 177–185. [CrossRef]
23. Zhu, L.; Huang, Y.; Morishita, Y.; Uetani, K.; Nogi, M.; Koga, H. Pyrolyzed chitin nanofiber paper as a three-dimensional porous and defective nanocarbon for photosensing and energy storage. *J. Mater. Chem. C* **2021**, *9*, 4444–4452. [CrossRef]
24. Li, X.; Zhu, L.; Kasuga, T.; Nogi, M.; Koga, H. Chitin-derived-carbon nanofibrous aerogel with anisotropic porous channels and defective carbon structures for strong microwave absorption. *Chem. Eng. J.* **2022**, *450*, 137943. [CrossRef]
25. Nogi, M.; Kurosaki, F.; Yano, H.; Takano, M. Preparation of nanofibrillar carbon from chitin nanofibers. *Carbohydr. Polym.* **2010**, *81*, 919–924. [CrossRef]
26. Mandal, J.; Wang, D.; Overvig, A.C.; Shi, N.N.; Paley, D.; Zangiabadi, A.; Cheng, Q.; Barmak, K.; Yu, N.; Yang, Y. Scalable, "dip-and-dry" fabrication of a wide-angle plasmonic selective absorber for high-efficiency solar-thermal energy conversion. *Adv. Mater.* **2017**, *29*, 1702156. [CrossRef]
27. Tauc, J.; Grigorovici, R.; Vancu, A. Optical properties and electronic structure of amorphous germanium. *Phys. Status Solidi (B)* **1966**, *15*, 627–637. [CrossRef]
28. Dresselhaus, M.S.; Jorio, A.; Souza Filho, A.G.; Saito, R. Defect characterization in graphene and carbon nanotubes using Raman spectroscopy. *Philos. Trans. A Math. Phys. Eng. Sci.* **2010**, *368*, 5355–5377. [CrossRef]
29. Yu, H.; Shang, L.; Bian, T.; Shi, R.; Waterhouse, G.I.N.; Zhao, Y.; Zhou, C.; Wu, L.Z.; Tung, C.H.; Zhang, T. Nitrogen-doped porous carbon nanosheets templated from g-C_3N_4 as metal-free electrocatalysts for efficient oxygen reduction reaction. *Adv. Mater.* **2016**, *28*, 5080–5086. [CrossRef]
30. Sheng, Z.H.; Shao, L.; Chen, J.J.; Bao, W.J.; Wang, F.B.; Xia, X.H. Catalyst-free synthesis of nitrogen-doped graphene via thermal annealing graphite oxide with melamine and its excellent electrocatalysis. *ACS Nano* **2011**, *5*, 4350–4358. [CrossRef]
31. Yang, H.B.; Miao, J.; Hung, S.F.; Chen, J.; Tao, H.B.; Wang, X.; Zhang, L.; Chen, R.; Gao, J.; Chen, H.M.; et al. Identification of catalytic sites for oxygen reduction and oxygen evolution in N-doped graphene materials: Development of highly efficient metal-free bifunctional electrocatalyst. *Sci. Adv.* **2016**, *2*, e1501122. [CrossRef]
32. Li, M.; Cushing, S.K.; Zhou, X.; Guo, S.; Wu, N. Fingerprinting photoluminescence of functional groups in graphene oxide. *J. Mater. Chem.* **2012**, *22*, 23374. [CrossRef]
33. Simsir, H.; Eltugral, N.; Karagoz, S. Hydrothermal carbonization for the preparation of hydrochars from glucose, cellulose, chitin, chitosan and wood chips via low-temperature and their characterization. *Bioresour. Technol.* **2017**, *246*, 82–87. [CrossRef] [PubMed]
34. Isayama, M.; Nomiyama, K.; Kunitake, T. Template synthesis of a large, self-supporting graphite film in montmorillonite. *Adv. Mater.* **1996**, *8*, 641–644. [CrossRef]
35. Chen, H.; Müller, M.B.; Gilmore, K.J.; Wallace, G.G.; Li, D. Mechanically strong, electrically conductive, and biocompatible graphene paper. *Adv. Mater.* **2008**, *20*, 3557–3561. [CrossRef]

Disclaimer/Publisher's Note: The statements, opinions and data contained in all publications are solely those of the individual author(s) and contributor(s) and not of MDPI and/or the editor(s). MDPI and/or the editor(s) disclaim responsibility for any injury to people or property resulting from any ideas, methods, instructions or products referred to in the content.

Article

Polyacrylic Acid Functionalized Biomass-Derived Carbon Skeleton with Highly Porous Hierarchical Structures for Efficient Solid-Phase Microextraction of Volatile Halogenated Hydrocarbons

Anying Long [1,2], Hailin Liu [3], Shengrui Xu [3,*], Suling Feng [3], Qin Shuai [4] and Shenghong Hu [1,*]

1 State Key Laboratory of Biogeology and Environmental Geology, School of Earth Sciences, China University of Geosciences, Wuhan 430074, China
2 113 Geological Brigade, Bureau of Geology and Mineral Exploration and Development Guizhou Province, Liupanshui 553000, China
3 Key Laboratory of Green Chemical Media and Reactions, Ministry of Education, Collaborative Innovation Center of Henan Province for Green Manufacturing of Fine Chemicals, School of Chemistry and Chemical Engineering, Henan Normal University, Xinxiang 453007, China
4 Faculty of Materials Science and Chemistry, China University of Geosciences, Wuhan 430074, China
* Correspondence: xushengrui@126.com (S.X.); shhu@cug.edu.cn (S.H.)

Abstract: In this study, polyacrylic acid functionalized N-doped porous carbon derived from shaddock peels (PAA/N-SPCs) was fabricated and used as a solid-phase microextraction (SPME) coating for capturing and determining volatile halogenated hydrocarbons (VHCs) from water. Characterizations results demonstrated that the PAA/N-SPCs presented a highly meso/macro-porous hierarchical structure consisting of a carbon skeleton. The introduction of PAA promoted the formation of polar chemical groups on the carbon skeleton. Consequently, large specific surface area, highly hierarchical structures, and abundant chemical groups endowed the PAA/N-SPCs, which exhibited superior SPME capacities for VHCs in comparison to pristine N-SPCs and commercial SPME coatings. Under the optimum extraction conditions, the proposed analytical method presented wide linearity in the concentration range of 0.5–50 ng mL^{-1}, excellent reproducibility with relative standard deviations of 5.8%–7.2%, and low limits of detection varying from 0.0005 to 0.0086 ng mL^{-1}. Finally, the proposed method was applied to analyze VHCs from real water samples and observed satisfactory recoveries ranging from 75% to 116%. This study proposed a novel functionalized porous carbon skeleton as SPME coating for analyzing pollutants from environmental samples.

Keywords: biomass-derived carbon skeleton; hierarchical structure; polyacrylic acid; solid-phase microextraction; volatile halogenated hydrocarbons

1. Introduction

Volatile halogenated hydrocarbons (VHCs) such as dichloromethane, trichloromethane, tetrachloromethane, trichloroethylene, tribromethane, etc., have been widely used in industrial applications such as disinfectants, chemical intermediates, and organic solvents [1,2]. However, the waste VHCs would migrate to water, soil, and the atmosphere, thereby resulting in serious harmful issues to human when ingested due to its high toxicity and difficulty of degradation [3,4]. VHCs in water have been listed as priority pollutants by both Chinese and American governments. Therefore, the development of a sensitive and effective analytical method for the monitoring of concentration levels of VHCs in water is essential.

In general, the determination of volatile organic compounds from aqueous samples is implemented by gas chromatography (GC) or gas chromatography–mass spectrometry (GC–MS) [5]. However, sample pretreatments are of the essence prior to detection by instruments [6–8].

Among various sample pretreatment techniques, solid-phase microextraction (SPME) exhibited remarkable advantages in the determination of organic compounds from aqueous samples owing to its solvent-free sampling, straightforward operation, and the integration of sampling, extraction, and injection into a single step [9–12]. In view of these superiorities, SPME has been widely used in environmental analysis [13–15], food analysis [16–18], pharmaceutical analysis [19–21], and biological analysis [22–25]. The performance of SPME was related to the coating materials as the extraction was based on the equilibrium partitioning of analytes between the sample matrix and fiber coating [26–28]. To date, various coating materials have been extensively investigated to improve the extraction efficiencies of SPME, such as activated carbon, metal oxides, graphene, metal–organic frameworks, covalent organic frameworks, ionic liquids, polymers, and other composites [26,29–32].

Among the developed materials, activated carbon presented unique advantages as a SPME coating for efficiently capturing organic compounds from environmental samples, owing to its high surface area, hydrophobic interaction, low cost, and ease to obtain from agricultural waste [33–36]. For instance, Ji et al. [37] fabricated a type of nitrogen-doped porous carbon derived from marine algae with a large specific surface area and wide pore size distribution, and used it as an SPME coating for the determination of chlorobenzenes from water. Results showed that the as-prepared SPME fiber exhibited outstanding extraction capacities and the proposed method presented ultra-sensitivity for analyzing chlorobenzenes. Yin et al. [38] proposed a facile method for fabricating SPME coating based on peanut shell-derived biochar, and confirmed its excellent performances for extracting polycyclic aromatic hydrocarbons from water samples. The abovementioned studies demonstrated the effectiveness of biomass-derived carbon as an SPME coating, owing to its high surface area. However, the biomass-derived carbon had undergone a treatment of high temperature, which led to the diminution of chemical groups during the pyrolysis process, thereby resulting in insufficient extraction of slightly polar compounds [39]. Therefore, the decoration of the activated carbon surface with polar groups is an effective means to promote the extraction capacities of VHCs.

Polyacrylic acid (PAA) is a type of eco-friendly polymer with carboxyl (C=O) groups on every two carbon atoms of its main chain. Abundant polar chemical groups endow PAA to be of great potential for adsorbing compounds with polar groups by hydrogen-bond interaction and polar binding [40–43]. Therefore, we proposed PAA as the donor of polar groups to modify the biomass-derived porous carbon for SPME of VHCs. In addition, discarded shaddock peels are rich in lignin and cellulose, and possess abundant intrinsic pores, thereby becoming an inspiring raw material for the fabrication of biomass-derived porous carbon.

In this study, the nitrogen-doped porous carbons (N-SPCs) were prepared by pyrolysis of discarded shaddock peels at 800 °C under nitrogen protection. Before the pyrolysis process, the shaddock peels were treated by sodium hydroxide and urea solution, where sodium hydroxide acted as activation reagent during the pyrolysis process for the formation of a highly porous structure, and urea was used as the donor of nitrogen. Subsequently, the N-SPCs was modified by PAA—which was labeled as PAA/N-SPCs—and used as SPME coating for determination of VHCs from water coupled with GC–MS (Scheme 1). Notably, the as-prepared PAA/N-SPCs exhibited superior extraction capacities for VHCs compared with pristine N-SPCs and commercial SPME coatings owing to its high surface area and abundant chemical groups. Finally, the proposed analytical method was successfully applied for the determination of VHCs from real water samples. To our best knowledge, the as-prepared PAA/N-SPCs was first used as an SPME coating material and applied for the capturing and analyzing of VHCs from water.

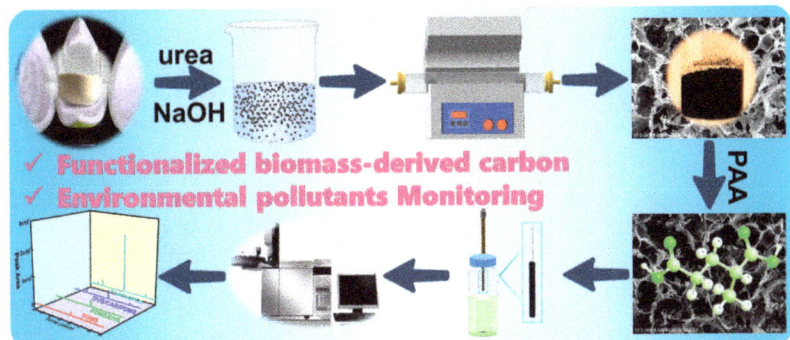

Scheme 1. Synthesis of PAA/N-SPCs and SPME procedures for VHCs determination.

2. Experimental Section

2.1. Reagents, Materials, and Instruments

The VHCs standards, including trichloromethane ($CHCl_3$), tetrachloromethane (CCl_4), trichlorethylene (C_2HCl_3), tetrachloroethylene (C_2Cl_4), and tribromethane ($CHBr_3$), were provided by the Beijing North Weiye Institute of Measuring and Testing Technology (Beijing, China). Urea, sodium hydroxide, and sodium chloride were purchased from Damao Chemical Reagent Factory (Tianjin, China). PAA was obtained from Shanghai Aladdin Biochemical Technology Co., Ltd. (Shanghai, China). Stainless steel wire with a diameter of 100 μm was obtained from Shenzhen Hubei Baofeng Industrial Co., Ltd. (Shenzhen, China). Sylard184 silicone elastomer was purchased from Dow Silicones Corporation (Seneffe, Belgium). Commercial SPME coatings, including PDMS, PDMS/DVB, and DVB/CAR/PDMS, were provided by ANPEL Laboratory Technologies Inc. (Shanghai, China). Shaddock peels were collected from a local market located in Xinxiang, China.

The details of instruments used for characterizations of as-prepared materials including scanning electron microscope (SEM), X-ray diffraction spectrometer (XRD), Fourier-transform infrared spectrometer (FTIR), X-ray photoelectron spectroscopy (XPS), and nitrogen adsorption/desorption apparatus are described in Section S1. The detection of VHCs was performed by GC–MS (Agilent 7890B–7000D, Santa Clara, CA, USA).

2.2. Preparation of PAA/N-SPCs

As displayed in Scheme 1, the N-SPCs were prepared by pyrolysis of modified shaddock peels under nitrogen atmosphere. First, the outer yellow layers of shaddock peels were removed to obtain the homogeneous precursors. After that, the shaddock peels were cut into small fragments and dried in a freeze dryer. Then, the shaddock peel fragments were ground into powder. Afterwards, 10 g of shaddock peels powder, 3 g of urea, and 10 g of sodium hydroxide were dispersed in 50-mL water and thoroughly mixed by stirring. After drying in an oven, the modified shaddock peel powder was placed in a horizontal quartz tube furnace (BTF-1200CC-S, Anhui BEQ Equipment Technology Co., Ltd., Hefei, China) for thermal treatment at 800 °C for 2 h with a heating rate of 5 °C/min under nitrogen protection. The obtained powder was washed by hydrochloric acid solution (0.1 M) and deionized water until neutrality. The N-SPCs was then observed after drying in an oven at 105 °C. Then, 1 g of N-SPCs powder was added into PAA aqueous solution for stirring 24 h at room temperature. Finally, the PAA/N-SPCs-x was obtained after being washed by deionized water and ethanol three times, respectively, and dried in an oven, where x represented the mass percentage content of PAA in aqueous solution.

2.3. Fabrication of PAA/N-SPCs Coated SPME Fiber

The fabrication procedures were carried out according to our previous studies [13,23,27,44]. Typically, the cleaned stainless-steel wire (3–4 cm in length) was immersed into silicone sealant solution with PDMS polymer and curing agent of 10:1 in mass ratio. Then, the stainless-steel wire was immersed into PAA/N-SPCs powder to form a uniform coating after curing at 120 °C in an oven. Finally, the coated stainless-steel wire was assembled onto an empty SPME needle. Prior to being used for extraction, the as-prepared SPME fiber was aged in the GC–MS injector for 20 min at 250 °C in order to remove the potential interfering compounds.

2.4. SPME Procedures and GC–MS Analysis

The SPME process was operated with headspace (HS) mode in a 20-mL commercial vial. First, 10 mL of aqueous solution was added into sample vial, and then the as-prepared SPME fiber was injected into the vial headspace for extraction. During the extraction process, the temperatures were set at 30–60 °C; the extraction time was kept within 20–60 min; the solution acidity was adjusted by hydrochloric acid (0.1 M) and sodium hydroxide solution (0.1 M) with pH values ranging from 3–9; the ionic strength of the solution was controlled by sodium chloride with contents varying from 0 to 20%. After extraction, the SPME fiber was immediately inserted into the GC–MS injector for desorption and analysis. The operating parameters of GC–MS and characteristic ions for analyzing VHCs were detailed in Tables S1 and S2.

2.5. Collection and Analysis of Real Water Samples

The real water samples were collected from campus tap water (1#) located in Xinxiang and unknown lake water (2#, 3#) located in Anshun, China. The collected water samples were sealed by Parafilm to avoid compounds loss and stored in a refrigerator at 4 °C before analysis by the proposed method.

3. Results and Discussion

3.1. Characterizations of PAA/N-SPCs

The micro morphologies of both N-SPCs and PAA/N-SPCs were investigated by SEM. As shown in Figure 1a, N-SPCs exhibit a highly porous hierarchical structure assembled by carbon skeletons under the activation of sodium hydroxide. After the modification of PAA, the morphology of PAA/N-SPCs was not changed significantly (Figure 1b). The generated highly porous structure endowed both N-SPCs and PAA/N-SPCs with sufficient specific surface area and accessible contact sites for analytes. It can be seen from optical microscope image (Figure 1c) that a uniform coating of PAA/N-SPCs was formed on the surface of the stainless-steel wire. The thickness of the coating was detected to be 60 μm according to the dimensions before and after coating.

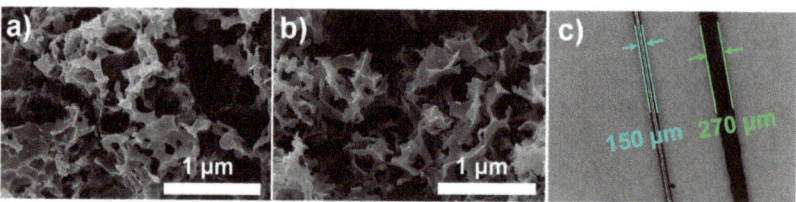

Figure 1. SEM images of N-SPCs (**a**) and PAA/N-SPCs (**b**), and optical microscope image of as-prepared SPME fiber before and after coating (**c**).

To demonstrate the structural compositions of as-prepared PAA/N-SPCs, XRD analysis was performed (Figure 2a). It can be seen that the XRD patterns of both pristine N-SPCs and PAA/N-SPCs display two peaks at 24.7° and 29.4°, which correspond to the diffraction peaks of amorphous carbon and NaO_2, respectively [45]. The XRD results suggest that the as-prepared porous carbon skeletons have not undergone significant structural change before and after the modification. FTIR analysis was carried out to investigate the surface functional groups and the impact of PAA addition on the chemical structure of the N-SPCs (Figure 2b). As shown in the spectrum of N-SPCs, the broad band at 3437 cm^{-1} is due to the O–H stretching, and the peaks at 2927 cm^{-1} and 2852 cm^{-1} correspond to the presence of –CH$_2$– bond. The absorption peak at 2355 cm^{-1} belongs to the stretching of C≡N [46]. It can be found that, compared to the characteristic stretching vibrations of C=O located at 1620 cm^{-1} for N-SPCs, the C=O band shifted to a much lower wavelength (1554 cm^{-1}) for the PAA/N-SPCs, demonstrating the strong hydrogen bonding between PAA and N-SPCs [47]. Moreover, XPS analysis was employed to further explore the chemical groups of PAA/N-SPCs. The XPS survey spectrum (Figure 2c) confirms the presence of C, O, and N at 283.8, 532.3, and 400.1 eV with atom contents of 78.0%, 19.7%, and 2.3%, respectively. High-resolution spectrum of C 1s (Figure 2d) were deconvoluted into four peaks at binding energies of 284.0, 284.7, 285.5, and 288.7 eV, which were ascribed to C=C, C–C, C–N, and COOH, respectively [48]. The XPS spectrum of O 1s (Figure 2e) was divided into two peaks at 531.7 and 532.4 eV, referring to C–O and O–C=O groups, respectively [49]. N 1s spectrum (Figure 2f) was deconvoluted into two peaks for C–N and N–H at 398.9 and 400.3 eV, respectively [50]. The generated oxygen/nitrogen-containing groups promoted the binding with compounds containing polar groups [39,44].

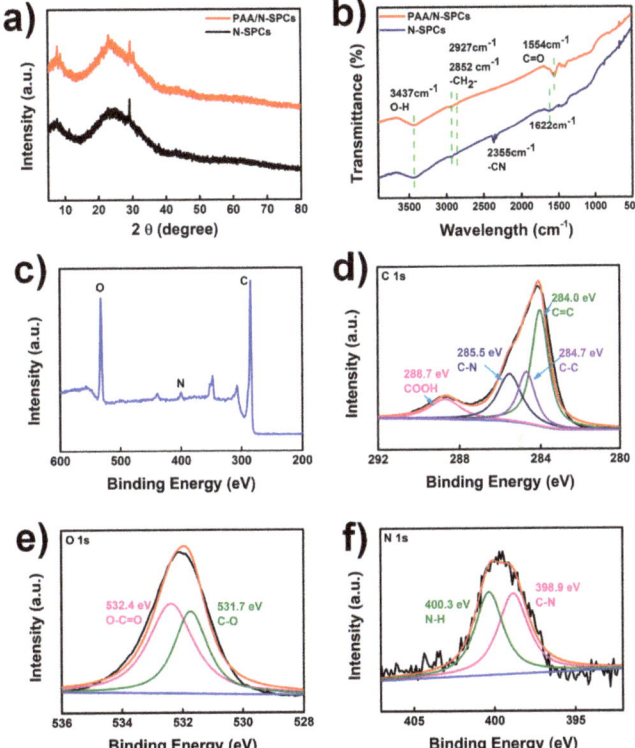

Figure 2. XRD patterns (a), FTIR spectra (b) of N-SPCs (a) and PAA/N-SPCs, XPS survey spectrum (c), and high resolution XPS spectra of C 1s (d), O 1s (e), and N 1s (f) for PAA/N-SPCs.

In general, high surface area and porosity are conducive to provide more active adsorption sites for analytes. Therefore, the specific surface areas and pore size distributions of both N-SPCs and PAA/N-SPCs were evaluated using N$_2$ adsorption–desorption and BJH isotherms. As displayed in Figure 3a, the specific surface areas of N-SPCs and PAA/N-SPCs were calculated to be 517.7 and 485.7 m^2/g, respectively. The average pore diameters of both N-SPCs and PAA/N-SPCs (Figure 3b) were observed with values of 3.6 and 405.2 nm, respectively, which confirmed the formation of both meso- and macro-porous structures.

The mesopore and macropore volumes were obtained with values of 0.15 and 0.63 cm^3/g for N-SPCs, and 0.13 and 0.58 cm^3/g for PAA/N-SPCs, respectively. Plentiful porosity provided numerous channels for analytes transferring from the outer surface to inside. Although the specific surface area of PAA/N-SPCs was slightly lower than that of N-SPCs, the formed abundant chemical groups on PAA/N-SPCs, which was verified in FTIR and XPS analysis, improved the adsorption capacities for VHCs by polar interaction. Moreover, the thermal stability of as-prepared materials is essential as the desorption of SPME is performed in a GC–MS injector at 250 °C. Herein, the desorption of the as-prepared new SPME fiber in the GC–MS injector at 250 °C was carried out with scan mode (Figure S1). Results showed that no distinct interfering compounds were found, indicating the excellent thermal stability of the as-prepared SPME coating.

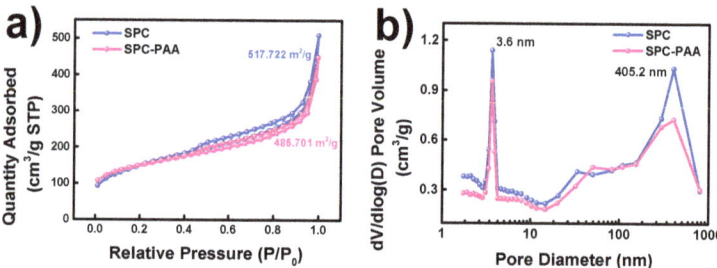

Figure 3. N$_2$ adsorption–desorption isotherms (**a**) and pore diameter distributions (**b**) of N-SPCs (**a**) and PAA/N-SPCs.

3.2. SPME Capacities of PAA/N-SPCs Coating for VHCs

The extraction efficiencies of as-prepared SPCs, N-SPCs, and PAA/N-SPCs coatings for five VHCs from water were investigated. As shown in Figure 4a, the extraction efficiencies of PAA/N-SPCs-5% exhibit an evident enhancement for VHCs compared with pristine N-SPCs and SPCs—especially for C$_2$Cl$_2$, owing to the functionalization of polar chemical groups by PAA. Typically, excess polar chemical groups on the materials would decrease the hydrophobic interaction with analytes. Figure 4b displays the GC–MS chromatograms of VHCs observed by PAA/N-SPCs-5% and commercial SPME coatings. Results demonstrate that the extraction efficiencies of PAA/N-SPCs are much higher than that of commercial SPME coatings including PDMS, PDMS/DVB, and DVB/CAR/PDMS. The outstanding extraction performance of PAA/N-SPCs toward VHCs can be ascribed to two factors: (i) high specific surface area and meso/maro-porous hierarchical structures promoted the adsorption active sites; (ii) polar chemical groups (C–O, O–C=O, –CN) enhanced the interaction with VHCs by hydrogen bond and polar binding; where, the extraction capacity of commercial coatings was mainly attributed to the hydrophobic crosslinking. It is worth noting that although polar binding was formed between the analytes and the coating material, the adsorbed analytes can be easily desorbed in the GC–MS injector at 250 °C.

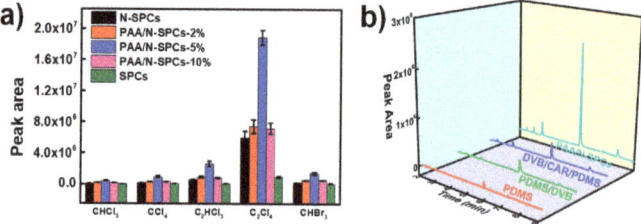

Figure 4. Extraction efficiencies of N-SPCs and PAA/N-SPCs coatings (**a**); GC-MS chromatograms of VHCs by PAA/N-SPCs-5% and commercial SPME coatings (**b**).

3.3. Optimization of SPME Conditions

To observe the finest extraction efficiencies, the effects of SPME conditions including extraction temperature, extraction time, ionic strength, and pH of solution were studied using a spiked aqueous solution with a VHCs concentration of 10 ng mL^{-1}. Figure 5a presents the effect of temperature on the extraction efficiencies. It can be seen that the extraction efficiencies exhibit a decreased trend with the temperature increase from 30 to 60 °C. Although higher temperature can promote the release of analytes to the headspace, the distribution coefficients of analytes on the fiber coating diminish. Further, as a type of highly volatile compound, VHCs can be released to the headspace from water at a low temperature. Therefore, the extraction efficiencies of PAA/N-SPCs decrease gradually in the temperature range of 30–60 °C. SPME with headspace mode is an equilibrium-based process of analytes among the sample matrix, headspace, and fiber coating. The effect of the extraction time ranging from 20 to 60 min on extraction efficiencies was evaluated (Figure 5b). Results imply that the extraction process reaches equilibrium within 40 min. High ionic strength of solution can enhance the analytes' release from the matrix, resulting in an increase in extraction efficiency. Herein, sodium chloride was used to control the ionic strength of water. As displayed in Figure 5c, sodium chloride in water with a content of 5% improves the extraction efficiency, whereas excess contents of sodium chloride weaken the extraction capacity due to the adhesion of sodium chloride on the coating surface. The effect of the solution pH ranging from 3 to 9 (Figure 5d) presents an unapparent change in extraction efficiencies of VHCs, except for tetrachloroethene, which demonstrates that the PAA/N-SPCs coating can be applied in a wide acidity range. To sum up, the optimum extraction conditions of PAA/N-SPCs coating with a temperature of 30 °C, extraction time of 40 min, sodium chloride content of 5%, and pH of 7 were performed for method validation.

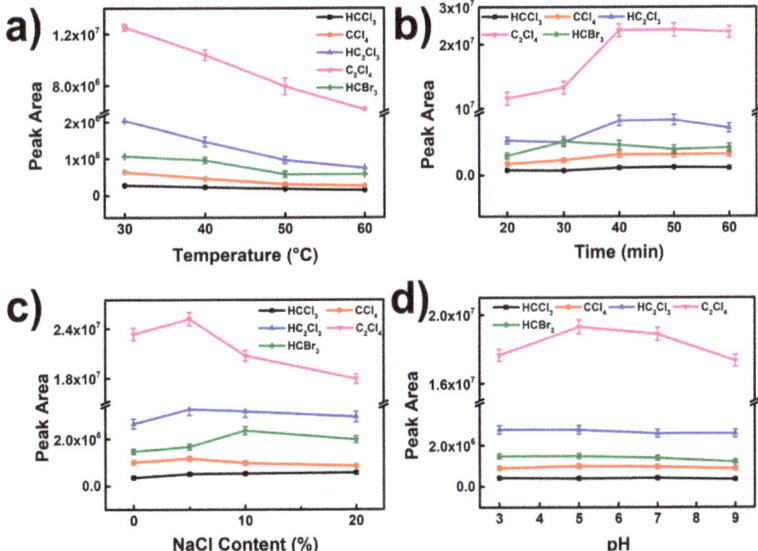

Figure 5. Effect of extraction temperature (**a**), extraction time (**b**), salt concentration (**c**), and pH (**d**) on extraction efficiencies of PAA/N-SPCs coating for SPME of VHCs.

3.4. Analytical Method Performance

Under the optimal extraction conditions, the method performances for analyzing VHCs from water were evaluated by means of linearity, limits of detection (LODs), limits of quantitation (LOQs), and relative standard deviations (RSDs). As listed in Table 1, the proposed method presents wide linearity in the VHCs concentration of 0.5–50 ng mL^{-1}, with linear coefficient (R^2) of 0.9879–0.9973. The LODs and LOQs were calculated to be 0.0005–0.0086 ng mL^{-1} and 0.0015–0.029 ng mL^{-1}, according to three and 10 times of signal-to-noise, respectively. The reproducibility of method was evaluated by RSDs, with results that the RSDs with five replicates range from 5.8% to 7.2% for analyze VHCs from water samples. Excellent linearity, high sensitivity, and good reproducibility suggest the proposed method has great potential in real water analysis.

Table 1. Analytical performance of PAA/N-SPCs SPME coating for VHCs.

Analytes	Linear Ranges (ng mL^{-1})	R^2	LODs (ng mL^{-1})	LOQs (ng mL^{-1})	RSDs (%) (n = 5)
CHCl$_3$	0.5–50	0.9973	0.0054	0.0181	7.1
CCl$_4$	0.5–50	0.9879	0.0086	0.0285	5.8
C$_2$HCl$_3$	0.5–50	0.9918	0.0031	0.0102	6.8
C$_2$Cl$_4$	0.5–50	0.9922	0.0005	0.0015	7.2
CHBr$_3$	0.5–250	0.9938	0.0060	0.0199	6.1

3.5. Real Water Samples Analysis

Finally, the proposed method based on PAA/N-SPCs SPME fiber coupled with GC–MS was applied to measure VHCs from real water samples. As listed in Table 2, no VHCs were found in campus tap water (1#). Trichloromethane and tribromomethane were detected with concentrations of 8.0 and 8.5 ng mL^{-1} in Sample 2#, respectively; trichloromethane and tetrachloroethene were found with concentration of 7.5 and 12.0 ng mL^{-1} in Sample 3#, respectively, which were collected from an unknown lake. The recoveries were obtained in a range of 75%–116% with spiked concentration of 0.5 ng mL^{-1} for sample 1#, and 5 ng mL^{-1} for the other samples. Satisfactory recoveries confirmed the effectiveness of the proposed method in analyzing VHCs from real water samples.

Table 2. Analytical results and recoveries for determination of VHCs from real water samples.

Analytes	Campus Tap Water (1#)			Lake Water (2#)			Lake Water (3#)		
	Found (ng mL^{-1})	RSDs (%, n = 3)	Recoveries (%, Spiked with 0.5 ng mL^{-1})	Found (ng mL^{-1})	RSDs (%, n = 3)	Recoveries (%, Spiked with 5 ng mL^{-1})	Found (ng mL^{-1})	RSDs (%, n = 3)	Recoveries (%, Spiked with 5 ng mL^{-1})
CHCl$_3$	ND	5.9	96	2.23	8.0	75	1.31	7.5	76
CCl$_4$	ND	8.2	102	ND	–	103	ND	–	106
C$_2$HCl$_3$	ND	7.5	110	ND	–	103	ND	–	109
C$_2$Cl$_4$	ND	8.1	103	ND	–	101	0.027	12.0	111
CHBr$_3$	ND	15.0	106	0.43	8.5	116	ND	–	112

4. Conclusions

In summary, PAA-functionalized N-SPCs with highly meso/macro-porous hierarchical structure, large surface area, and abundant chemical groups, was fabricated and innovatively used as an SPME coating for extracting VHCs from water. The decoration of PAA enhanced the extraction capacities of VHCs compared with N-SPCs pristine, owing to its abundant polar chemical groups that promoted the interaction with polar group-containing compounds by hydrogen bonding and polar binding. In view of these distinct advantages, the proposed analytical method based on PAA/N-SPCs SPME fiber presented good linearity, excellent reproducibility, and high sensitivity for analyzing VHCs from water. This study confirmed the effectiveness of PAA/N-SPCs as an SPME coating for the capture and analysis of trace pollutants from aqueous samples.

Supplementary Materials: The following supporting information can be downloaded at: https://www.mdpi.com/article/10.3390/nano12244376/s1, Section S1: Apparatus used for characterizations of materials [51]; Figure S1: GC-MS chromatogram detected by a new PAA/N-SPCs fiber in the GC–MS injector at 250 °C; Table S1: Operating parameters of of GC–MS; Table S2: Retention time and characteristic ions of five VHCs.

Author Contributions: A.L.: investigation, and writing—original draft preparation; H.L.: validation and formal analysis; S.X.: conceptualization, methodology, and writing—review and editing; S.F.: project administration; Q.S.: resouces; S.H.: supervision. All authors have read and agreed to the published version of the manuscript.

Funding: This work was financially supported by funding from the National Natural Science Foundation of China (21976052), Natural Science Foundation of Guizhou Province of China ([2019]1423), and Foundation of Bureau of Geology and Mineral Exploration and Development Guizhou Province ([2019]16).

Data Availability Statement: This study presents novel concepts and did not report any data.

Acknowledgments: The authors would like to thank Yuxiang Wu from Shiyanjia Lab (www.shiyanjia.com) for the BET analysis accessed on 29 September 2022.

Conflicts of Interest: The authors declare no conflict of interest.

References

1. Liu, R.; Wu, X.; Zhang, W.; Chen, Y.; Fu, J.; Ou, H. Volatile organic compounds generation pathways and mechanisms from microplastics in water: Ultraviolet, chlorine and ultraviolet/chlorine disinfection. *J. Hazard. Mater.* **2023**, *441*, 129813. [CrossRef]
2. Wang, J.; Wu, S.; Yang, Q.; Gu, Y.; Wang, P.; Li, Z.; Li, L. Performance and mechanism of the in situ restoration effect on VHCs in the polluted river water based on the orthogonal experiment: Photosynthetic fluorescence characteristics and microbial community analysis. *Environ. Sci. Pollut. R.* **2022**, *29*, 43004–43018. [CrossRef] [PubMed]
3. Lin, X.; Xu, C.; Zhou, Y.; Liu, S.; Liu, W. A new perspective on volatile halogenated hydrocarbons in Chinese agricultural soils. *Sci. Total Environ.* **2020**, *703*, 134646. [CrossRef]
4. Dowty, B.; Carlisle, D.; Laseter, J.L.; Storer, J. Halogenated hydrocarbons in New Orleans drinking water and blood plasma. *Science* **1975**, *187*, 75–77. [CrossRef] [PubMed]
5. Ueta, I. Gas chromatographic determination of volatile compounds. *Anal. Sci.* **2022**, *38*, 737–738. [CrossRef] [PubMed]
6. López-Lorente, Á.I.; Pena-Pereira, F.; Pedersen-Bjergaard, S.; Zuin, V.G.; Ozkan, S.A.; Psillakis, E. The ten principles of green sample preparation. *TrAC Trend. Anal. Chem.* **2022**, *148*, 116530. [CrossRef]
7. Xu, J.; Li, F.; Xia, F.; Zhu, T.; Wu, D.; Chingin, K.; Chen, H. High throughput online sequential extraction of natural rare earth elements and determination by mass spectrometry. *Sci. China Chem.* **2021**, *64*, 642–649. [CrossRef]
8. Guzowska, M.; Wasiak, W.; Wawrzyniak, R. Comparison of extraction techniques for the determination of volatile organic compounds in liverwort samples. *Molecules* **2022**, *27*, 2911. [CrossRef] [PubMed]
9. Li, H.; Zhou, Y.; Xiao, Y.; Fan, J.; Feng, S.; Xu, S. Application of cooling-assisted solid phase microextraction in analysis of complex matrix sample. *Chin. J. Anal. Chem.* **2022**, *50*, 1289–1298.
10. Piri-Moghadam, H.; Ahmadi, F.; Pawliszyn, J. A critical review of solid phase microextraction for analysis of water samples. *Trac-Trend. Anal. Chem.* **2016**, *85*, 133–143. [CrossRef]
11. Reyes-Garces, N.; Gionfriddo, E.; Gmez-Rios, G.A.; Alam, M.N.; Boyaci, E.; Bojko, B.; Singh, V.; Grandy, J.; Pawliszyn, J. Advances in solid phase microextraction and perspective on future directions. *Anal. Chem.* **2018**, *90*, 302–360. [CrossRef] [PubMed]
12. Xu, S.; Li, H.; Wu, H.; Xiao, L.; Dong, P.; Feng, S.; Fan, J. A facile cooling-assisted solid-phase microextraction device for solvent-free sampling of polycyclic aromatic hydrocarbons from soil based on matrix solid-phase dispersion technique. *Anal. Chim. Acta* **2020**, *1115*, 7–15. [CrossRef]
13. Xu, S.; Liu, H.; Chen, C.; Feng, S.; Fan, J. Ultrasound-assisted one-step reduction and self-assembly of carbon dots-reduced graphene oxide: Mechanism investigation and solid phase microextraction of ultra-trace organochlorine pesticides. *Chem. Eng. J.* **2023**, *451*, 138569. [CrossRef]
14. Xu, S.; Shuai, Q.; Pawliszyn, J. Determination of polycyclic aromatic hydrocarbons in sediment by pressure-balanced cold fiber solid phase microextraction. *Anal. Chem.* **2016**, *88*, 8936–8941. [CrossRef] [PubMed]
15. Peng, S.; Shen, M.; Li, X.; Tong, Y.; Guo, J.; Lin, W.; Ye, Y.; Xu, J.; Zhou, N.; Zhu, F.; et al. Rational design of ordered porous nanoparticles for selective extraction of nitrobenzene compounds. *J. Hazard. Mater.* **2023**, *441*, 129971. [CrossRef]
16. Zang, X.; Chang, Q.; Li, H.; Zhao, X.; Zhang, S.; Wang, C.; Wang, Z. Construction of a ringent multi-shelled hollow MIL-88B as the solid-phase microextraction fiber coating for the extraction of organochlorine pesticides. *Sep. Purif. Technol.* **2023**, *304*, 122350. [CrossRef]
17. Xu, S.; Li, H.; Dong, P.; Wang, M.; Chen, C.-P.; Feng, S.; Fan, J. High-throughput profiling volatiles in edible oils by cooling assisted solid-phase microextraction technique for sensitive discrimination of edible oils adulteration. *Anal. Chim. Acta* **2022**, *1221*, 340159. [CrossRef]
18. Tuzen, M.; Hazer, B.; Elik, A.; Altunay, N. Synthesized of poly(vinyl benzyl dithiocarbonate-dimethyl amino ethyl methacrylate) block copolymer as adsorbent for the vortex-assisted dispersive solid phase microextraction of patulin from apple products and dried fruits. *Food Chem.* **2022**, *395*, 133607. [CrossRef]
19. Gao, Y.; Sheng, K.; Bao, T.; Wang, S. Recent applications of organic molecule-based framework porous materials in solid-phase microextraction for pharmaceutical analysis. *J. Pharmaceut. Biomed.* **2022**, *221*, 115040. [CrossRef]

20. Filipiak, W.; Bojko, B. SPME in clinical, pharmaceutical, and biotechnological research—How far are we from daily practice? *TrAC Trend. Anal. Chem.* **2019**, *115*, 203–213. [CrossRef]
21. Wang, Y.; Jie, Y.; Hu, Q.; Yang, Y.; Ye, Y.; Zou, S.; Xu, J.; Ouyang, G. A polymeric solid-phase microextraction fiber for the detection of pharmaceuticals in water samples. *J. Chromatogr. A* **2020**, *1623*, 461171. [CrossRef]
22. Song, L.; Chingin, K.; Wang, M.; Zhong, D.; Chen, H.; Xu, J. Polarity-specific profiling of metabolites in single cells by probe electrophoresis mass spectrometry. *Anal. Chem.* **2022**, *94*, 4175–4182. [CrossRef] [PubMed]
23. Xu, S.; Liu, Q.; Wang, C.; Xiao, L.; Feng, S.; Li, N.; Chen, C.-P. Three-dimensional pompon-like Au/ZnO porous microspheres as solid phase microextraction coating for determination of volatile fatty acids from foot odor. *Talanta* **2020**, *209*, 120519. [CrossRef] [PubMed]
24. Zhu, W.; Zhang, J.; Zhang, X.; Han, L.; Qin, P.; Tian, S.; Zhou, Q.; Zhang, X.; Lu, M. Preparation of Al-doped mesoporous crystalline material-41 as fiber coating material for headspace solid-phase microextraction of polycyclic aromatic hydrocarbons from human urine. *J. Chromatogr. A* **2020**, *1626*, 461354. [CrossRef]
25. Zhang, J.; Li, W.; Zhu, W.; Qin, P.; Lu, M.; Zhang, X.; Miao, Y.; Cai, Z. Mesoporous graphitic carbon nitride@NiCo$_2$O$_4$ nanocomposite as a solid phase microextraction coating for sensitive determination of environmental pollutants in human serum samples. *Chem. Commun.* **2019**, *55*, 10019–10022. [CrossRef] [PubMed]
26. Peng, S.; Huang, X.; Huang, Y.; Huang, Y.; Zheng, J.; Zhu, F.; Xu, J.; Ouyang, G. Novel solid-phase microextraction fiber coatings: A review. *J. Sep. Sci.* **2022**, *45*, 282–304. [CrossRef]
27. Xu, S.; Liu, H.; Long, A.; Li, H.; Chen, C.; Feng, S.; Fan, J. Carbon dot-decorated graphite carbon nitride composites for enhanced solid-phase microextraction of chlorobenzenes from water. *Nanomaterials* **2022**, *12*, 335. [CrossRef]
28. Xu, S.; Dong, P.; Liu, H.; Li, H.; Chen, C.; Feng, S.; Fan, J. Lotus-like Ni@NiO nanoparticles embedded porous carbon derived from MOF-74/cellulose nanocrystal hybrids as solid phase microextraction coating for ultrasensitive determination of chlorobenzenes from water. *J. Hazard. Mater.* **2022**, *429*, 128384. [CrossRef]
29. Vazquez-Garrido, I.; Flores-Aguilar, J.F.; Miranda, J.M.; Santos, E.M.; Jardinez, C.; Ibarra, I.S. Carbonaceous materials in sample treatment techniques in the determination of pesticides in food and environmental analysis. A review. *Int. J. Environ. Anal. Chem.* **2022**, 1–35. [CrossRef]
30. Kang, J.Y.; Shi, Y.P. Recent advances and application of carbon nitride framework materials in sample preparation. *TrAC Trend. Anal. Chem.* **2022**, *153*, 116661. [CrossRef]
31. Shahhoseini, F.; Azizi, A.; Bottaro, C.S. A critical evaluation of molecularly imprinted polymer (MIP) coatings in solid phase microextraction devices. *TrAC Trend. Anal. Chem.* **2022**, *156*, 116695. [CrossRef]
32. Torabi, E.; Mirzaei, M.; Bazargan, M.; Amiri, A. A critical review of covalent organic frameworks-based sorbents in extraction methods. *Anal. Chim. Acta* **2022**, *1224*, 340207. [CrossRef] [PubMed]
33. Wang, Y.; Chen, J.; Ihara, H.; Guan, M.; Qiu, H.D. Preparation of porous carbon nanomaterials and their application in sample preparation: A review. *TrAC Trend. Anal. Chem.* **2021**, *143*, 116421. [CrossRef]
34. Kusmierek, K.; Borucka, M.; Swiatkowski, A.; Dabek, L. Evaluation of different carbon materials in adsorption and solid-phase microextraction of 2,4,6-trichlorophenol from water. *Desalin. Water Treat.* **2019**, *157*, 129–137. [CrossRef]
35. Xie, X.; Yang, H.; Han, J.; Tong, Y.; Hu, Y.; Ouyang, S.; Cui, S.; Zheng, J.; Ouyang, G. Nitrogen, oxygen-codoped hierarchically porous biochar for simultaneous enrichment and ultrasensitive determination of o-xylene and its hydroxyl metabolites in human urine by solid phase microextraction-gas chromatography-mass spectrometry. *Microchem. J.* **2022**, *178*, 107384. [CrossRef]
36. Behbahan, A.K.; Mahdavi, V.; Roustaei, Z.; Bagheri, H. Preparation and evaluation of various banana-based biochars together with ultra-high performance liquid chromatography-tandem mass spectrometry for determination of diverse pesticides in fruiting vegetables. *Food Chem.* **2021**, *360*, 130085. [CrossRef] [PubMed]
37. Ji, R.T.; Wu, Y.R.; Bian, Y.R.; Song, Y.; Sun, Q.; Jiang, X.; Zhang, L.J.; Han, J.G.; Cheng, H. Nitrogen-doped porous biochar derived from marine algae for efficient solid-phase microextraction of chlorobenzenes from aqueous solution. *J. Hazard. Mater.* **2021**, *407*, 124785. [CrossRef]
38. Yin, L.; Hu, Q.; Mondal, S.; Xu, J.; Ouyang, G. Peanut shell-derived biochar materials for effective solid-phase microextraction of polycyclic aromatic hydrocarbons in environmental. *Talanta* **2019**, *202*, 90–95. [CrossRef]
39. Fang, Q.; Chen, B.; Lin, Y.; Guan, Y. Aromatic and hydrophobic surfaces of wood-derived biochar enhance perchlorate adsorption via hydrogen bonding to oxygen-containing organic groups. *Environ. Sci. Technol.* **2014**, *48*, 279–288. [CrossRef]
40. Gokmen, F.O.; Yaman, E.; Temel, S. Eco-friendly polyacrylic acid based porous hydrogel for heavy metal ions adsorption: Characterization, adsorption behavior, thermodynamic and reusability studies. *Microchem. J.* **2021**, *168*, 106357. [CrossRef]
41. Bibi, A.; Bibi, S.; Abu-Dieyeh, M.; Al-Ghouti, M.A. New material of polyacrylic acid-modified graphene oxide composite for phenol remediation from synthetic and real wastewater. *Environ. Technol. Inno.* **2022**, *27*, 102795. [CrossRef]
42. Li, R.; Wen, Y.; Liu, M.; Su, L.; Wang, J.; Li, S.; Zhong, M.-e.; Zhou, Z.; Zhou, N. Simultaneous removal of organic inorganic composite contaminants by in situ double modified biochar: Performance and mechanisms. *J. Taiwan Inst. Chem. E.* **2022**, *139*, 104523. [CrossRef]
43. Salami, M.; Talebpour, Z.; Alizadeh, R. Fabrication of a new SPME fiber based on Polyacrylic acid/ MIL-88(Fe)-NH$_2$ composite as a self-healing coating for the analysis of breast cancer biomarkers in the urine sample. *J. Pharmaceut. Biomed.* **2022**, *219*, 114902. [CrossRef]

44. Xu, S.; Dong, P.; Qin, M.; Liu, H.; Long, A.; Chen, C.; Feng, S.; Wu, H. Core-shell structured Fe_2O_3/CeO_2@MnO_2 microspheres with abundant surface oxygen for sensitive solid-phase microextraction of polycyclic aromatic hydrocarbons from water. *Microchim. Acta* **2021**, *188*, 337. [CrossRef]
45. Wu, S.; Qiao, Y.; Jiang, K.; He, Y.; Guo, S.; Zhou, H. Tailoring sodium anodes for stable sodium–oxygen batteries. *Adv. Funct. Mater.* **2018**, *28*, 1706374. [CrossRef]
46. Jing, X.; Mi, H.-Y.; Peng, X.-F.; Turng, L.-S. Biocompatible, self-healing, highly stretchable polyacrylic acid/reduced graphene oxide nanocomposite hydrogel sensors via mussel-inspired chemistry. *Carbon* **2018**, *136*, 63–72. [CrossRef]
47. Wang, L.; Wang, W.; Fu, Y.; Wang, J.; Lvov, Y.; Liu, J.; Lu, Y.; Zhang, L. Enhanced electrical and mechanical properties of rubber/graphene film through layer-by-layer electrostatic assembly. *Compos. Part B: Eng.* **2016**, *90*, 457–464. [CrossRef]
48. Zhang, Y.; Liu, X.; Wang, Y.; Lou, Z.; Shan, W.; Xiong, Y. Polyacrylic acid-functionalized graphene oxide for high-performance adsorption of gallium from aqueous solution. *J. Colloid. Interf. Sci.* **2019**, *556*, 102–110. [CrossRef]
49. Xu, G.; Yan, Q.; Kushima, A.; Zhang, X.; Pan, J.; Li, J. Conductive graphene oxide-polyacrylic acid (GOPAA) binder for lithium-sulfur battery. *Nano Energy* **2017**, *31*, 568–574. [CrossRef]
50. Jian, X.; Liu, X.; Yang, H.; Li, J.; Song, X.; Dai, H.; Liang, Z. Construction of carbon quantum dots/proton-functionalized graphitic carbon nitride nanocomposite via electrostatic self-assembly strategy and its application. *Appl. Surf. Sci.* **2016**, *370*, 514–521. [CrossRef]
51. Singh, G.; Aher, S.C.; Varma, P.V.S.; Sathe, D.B.; Bhatt, R.B. Analysis of BET specific surface area in several recycled oxide powders. *J. Radioanal. Nucl. Chem.* **2021**, *327*, 555–564. [CrossRef]

Article

N, S Co-Doped Carbons Derived from *Enteromorpha prolifera* by a Molten Salt Approach: Antibiotics Removal Performance and Techno-Economic Analysis

Mengmeng Zhang [1], Kexin Huang [2], Yi Ding [1], Xinyu Wang [2], Yingli Gao [1], Pengfei Li [1,*], Yi Zhou [2], Zheng Guo [3], Yi Zhang [3] and Dapeng Wu [2,*]

1 School of Business, Henan Normal University, Xinxiang 453007, China
2 Key Laboratory of Green Chemistry Medias and Reactions, Ministry of Education, School of Environment, Henan Normal University, Xinxiang 453007, China
3 College of Textiles, Zhongyuan University of Technology, Zhengzhou 451191, China
* Correspondence: lipengfei@htu.edu.cn (P.L.); dapengwu@htu.edu.cn (D.W.)

Abstract: N, S co-doped bio-carbons with a hierarchical porous structure and high surface area were prepared using a molten salt method and by adopting *Entermorpha prolifera* (EP) as a precursor. The structure and composition of the bio-carbons could be manipulated by the salt types adopted in the molten salt assisted pyrolysis. When the carbons were used as an activating agent for peroxydisulfate (PDS) in SMX degradation in the advanced oxidation process (AOP), the removal performance in the case of KCl derived bio-carbon (EPB-K) was significantly enhanced compared with that derived from NaCl (EPB-Na). In addition, the optimized EPB-K also demonstrated a high removal rate of 99.6% in the system that used local running water in the background, which proved its excellent application potential in real water treatment. The degradation mechanism study indicated that the N, S doping sites could enhance the surface affinity with the PDS, which could then facilitate $^1O_2$ generation and the oxidation of the SMX. Moreover, a detailed techno-economic assessment suggested that the price of the salt reaction medium was of great significance as it influenced the cost of the bio-carbons. In addition, although the cost of EPB-K was higher (USD 2.34 kg^{-1}) compared with that of EPB-Na (USD 1.72 kg^{-1}), it was still economically competitive with the commercial active carbons for AOP water treatment.

Keywords: bio-carbon; advanced oxidation; persulfate; sulfamethoxazole; molten salt method

1. Introduction

Sulfamethoxazole (SMX) has played a great role in the treatment and prevention of bacterial and fungal infections in clinical health care and in the treatment of animal infections; however, due to its high thermal stability, it was difficult to degrade SMX naturally. This is because it was discharged into the water environment, which could breed drug-resistant bacteria, thus reducing its capability in disease treatment and resulting in potential harm to the ecosystem. At present, traditional water treatment based on biological processes is not effective for removing persistent organic pollutants such as SMX from water; therefore, advanced oxidation processes (AOPs) were developed for the treatment of persistent antibiotics [1].

The AOPs that were based on persulfate ($SO_4^{\bullet-}$) and hydroxyl (HO·) were effective technologies that removed persistent antibiotic pollutants [2]. Compared with ·OH, $SO_4^{\bullet-}$ has greater oxidization potential, it is environmentally resistant, and it has a prolonged half-life time, which thus endows persulfate-based AOPs with a promising water treatment strategy. Among the persulfates, PDS (peroxydisulfate) is used more widely than PMS (peroxymonosulfate) due to its higher stability and lower price [3]; however, considering that PDS hardly reacts with persistent pollutants, light or heat irradiation, catalytic materials, or activating ions are commonly used to activate the PDS molecules for the production of active oxygen content species (ROSs). These include sulfate radicals, hydroxyl radicals, singlet oxygen, and so on. [4]. Among these PDS activating strategies, carbon materials, with abundant surface defects and edges, were considered as low cost, effective, and environmentally benign catalytic mediums to initiate PDS activation [5].

Biomasses, especially biowastes that are automatically generated from the forestry, agricultural, and food industries, were regarded as abundant and easily available bio-resources that yield low-cost carbon materials for water treatment [6–12]. As for AOPs, wood chips and ammonium ferric citrate were adopted as precursors to prepare bio-carbons through a multi-step pyrolysis, which had the potential to effectively remove ~99% of SMX in the water system [13]. Peanut shells were modified with iron salt to prepare magnetic bio-carbons with a rich pore structure and high specific surface area; this gave an excellent performance in terms of sulfapyridine and ciprofloxacin degradation [14]. In addition, sugarcane waste [15] and poplar wood powder [16] were also employed as precursors to yield bio-carbons; therefore, they could serve as high performance catalysts to activate PDS to effectively remove bisphenol A.

Enteromorpha prolifera (EP) is a marine alga which can survive a wide range of temperature and salinity fluctuations [17]. The massive reproduction of EP has caused a great deal of damage in the ecosystem and to the economies of the fishing and tourism industries. EP biomass is rich in carbon, nitrogen, and iron, which have been employed as precursors to prepare carbon materials for applications in energy storage, heavy metal adsorption, and organic pollutant degradation [18–21]. For example, EP was adopted to prepare Fe/N co-doped carbon using in situ pyrolysis in a N_2 atmosphere for the degradation of acetaminophen [20]. EP biomass was also used as precursor to obtain a high performance from Fe_3C/C composites in a Fenton reaction to degrade methylene blue [21]. Although EP could be employed as an abundant and low-cost biowaste for the preparation of high-performance carbon materials, few studies have prepared carbonous catalysts based on EP for AOPs [22–24].

The molten salt method is an effective strategy that yields carbon materials by adopting a low melting point inorganic salt as the protecting medium instead of commonly used inert atmospheres; this means that it is a low cost and energy saving method that prepares carbon materials for different applications [25–30]. Based on previous studies, the etching effect of molten salt and the oxygen enriched reaction medium could effectively introduce a hierarchical porous structure as well as rich surface defects, which are favorable for persulfate activation [31–34]; however, to the best of our knowledge, few studies have devoted themselves to the preparation of bio-carbons, based on the molten salt method, in order to obtain AOPs and study its economic feasibility by systematic tech-economic analysis.

In this study, EP derived bio-carbons were prepared using a facile molten salt method and the effects of a molten salt medium (NaCl and KCl) on the structure and composition of bio-carbons were investigated. When the prepared N, S co-doped porous bio-carbon is adopted in PDS activation for SMX degradation, the bio-carbon that was derived from KCl (EPB-K) shows a higher catalytic removal rate than the bio-carbon derived from NaCl (EPB-Na). The EPB-K also demonstrates good resistance in a varied pH range and it demonstrates great potential in actual water treatment due to the $^1O_2$ mediated mechanism. Techno-economic analysis indicated that the cost of the molten salt system comprises the greatest portion of carbon price. Although the cost of EPB-K is higher compared with EPB-Na, it is economically competitive with commercial active carbon for AOP water treatment.

2. Experiment Section

2.1. Preparation of Bio-Carbons

The EP biomass was obtained from Tsingtao, Shandong Province of China, and it was washed with double distilled water and dried at 60 °C, in an oven, overnight. Then, the completely dried EP biomass was ground and sieved with a 100 mesh for later use. The chemical reagents of KCl (AR > 99.5%), NaCl (AR > 99.5%), Pb(NO$_3$)$_2$ (AR > 99.0%) and Cu(NO$_3$)$_2$·3H$_2$O (AR > 99.0%) were purchased from Sinopharm Chemical Reagent Co., Ltd. (Shanghai, China).

The bio-carbon was prepared using a typical molten salt method wherein 7.0 g of EP powder was mixed with 15.0 g of salt (NaCl or KCl). The mixture was transferred into a porcelain crucible and covered with a layer of salt (1.0 g) to seal the reaction system. Afterwards, the crucible was covered and transferred to a muffle furnace. The temperature was heated to 700 °C at a rate of 10 °C min^{-1}, and then it was maintained for 1 h. After it had cooled to room temperature, the yielded bio-carbon was rinsed repeatedly with double distilled water and filtered to remove the salt. The filtered product was also soaked in 1 mol L^{-1} diluted hydrochloric acid for 12 h and filtered and rinsed repeatedly until neutral. The final product was dried to obtain the bio-carbons, which were labeled as EBC-Na and EBC-K, respectively.

The EBC-K was also prepared under different temperatures of 500, 600, 700, and 800 °C with similar pyrolysis profiles in the KCl molten salt medium. The final products were denoted as EBC-500, EBC-600, EBC-700, and EBC-800.

2.2. Material Characterization

The morphology and structure were characterized by field emission scanning electron microscopy (SEM, SU8010, Hitachi, Tokyo, Japan) and transmission electron microscopy equipped with energy dispersive spectroscopy (TEM, TF20, JEOL 2100F, Akishima, Japan). The crystalline structure of the sample was determined using X-ray powder diffraction (XRD, Rigaku Dmax-2000, Rigaku, Tokyo, Japan). The graphitization of the bio-carbons was ascertained using Raman spectroscopy (LABRAMHR EVO, HORIBA France SAS, Longjumeau, France), and the surface functional groups and element compositions of EBCs were analyzed using infrared spectrometry (FTIR, Nexus 470, GMI Inc., Lebanon, OH, USA) and X-ray diffraction (XPS, ESCALAB 250Xi, Thermo Fisher Scientific, Waltham, MA, USA). The surface area and porous structure were measured by N$_2$ adsorption–desorption analyzer (BET, ASAP2020, Micromeritics Inc., Norcross, GA, USA).

2.3. Advanced Oxidation Performances

Moreover, 0.05 g of the prepared bio-carbon was added into 100 mL and 50 mg L^{-1} SMX solutions in 250 mL conical flasks, respectively, and they were placed on a shaking table with a rotation speed of 180 r min^{-1}. In addition, 2 mL of the reaction solution was sampled at 5, 10, 30 and 60 min, respectively, and the solution was filtered with a 0.22 μm membrane to remove the catalysts. The solution was then measured using an UV spectrophotometer (260 nm) in order to monitor the concentration variation; this is related to the adsorption performances of the bio-carbons. After the adsorption equilibrium, 10 mL of the 50 mmol L^{-1} PDS solution was added into the reaction system to start the AOP degradation. Then, 2 mL of the solution was withdrawn at 65, 70, 90, 120, and 180 min. Before the UV spectrophotometer measurement, the sample was filtered with a 0.22 μm membrane and 100 μL methanol was added to quench the reaction. All the tests were averaged using three control trials.

In addition, the performances of the optimized EBC-K were also tested using different initial pH levels, catalyst dosages, and SMX concentrations. In order to further explore potential applications in real water treatment scenarios, SMX was also added to the running water in the background to test the degradation rate of EBC-K. Please find the experimental details in the Supplementary Materials.

3. Result and Discussion

3.1. Structure and Composition Characterization

Figure 1a,b shows how the Enteromorpha prolifera (EP) invaded the seashore as well as the collected EP biomass in the salvage yard at Tsingtao City, Shandong Province of

China. As depicted in Figure 1c, the fresh EP biomass exhibits a smooth surface without porous tissue structures; however, after the molten salt treatment (KCl), the prepared bio-carbon (EBC-K) exhibited sheet-like patterns with well-developed porous structures (Figure 1d). The SEM image with high magnification, shown in Figure 1e, shows the rich porous structures with diameters of several tens of nanometers; this was possibly caused by the etching effects of both the KCl and the oxygen which is penetrated through the molten salt [32]. Figure 1f depicts how the EBC-Na that was derived from NaCl molten salt demonstrates sheet-like patterns with porous structures that have diameters of ~100 nm. In addition, based on the TEM images displayed in Figure 1g,h, the EBC-K possesses regularly distributed mesoporous structures with diameters of 2–5 nm. The corresponding selected area, as shown in the electron diffraction (SAED) image, discloses two sets of indistinct diffraction rings that are attributed to the carbon lattice; this indicates that the EBC-K has a medium graphite structure. As indicated by the HAADF-STEM and EDS mappings of EBC-K (Figure 1i), C, O, N and S are found homogeneously distributed on the surface of the bio-char. As no N or S contented regents were adopted in the synthesis, the rich N and S originated from the intrinsic composition of the EP biomass. Moreover, slight amounts of Si and K could be also found embedded in the bio-carbon; this may also be ascertained from the intrinsic composition of the biomass (Figure S1).

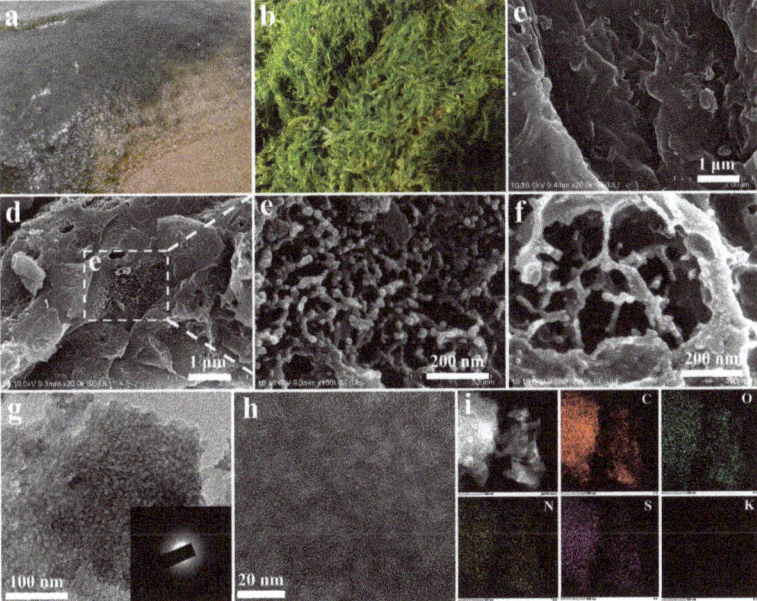

Figure 1. The photographs of (**a**) an EP invaded seashore at Tsingtao City, Shandong Province of China, and (**b**) the fresh EP biomass. The SEM images of (**c**) fresh EP, (**d**,**e**) EBC-K, and (**f**) EBC-Na. The TEM images with (**g**) low and (**h**) high magnification of EBC-K; the inset is the SAED image. (**i**) HAADF-STEM image and EDS element distribution of EBC-K (red: carbon, green: oxygen, yellow: nitrogen, magenta: sulfur and blue: potassium).

Figure S2a depicts the XRD characterization of the bio-carbons, which possess typical diffraction peaks that can be ascribed to the (002) and (100) planes of carbon, respectively [31,32]. In addition, several sharp peaks belonging to the insolvable mineral products could be also detected; this indicates that the mineral contents in EP could be converted into insoluble ceramics during the high temperature treatment. As shown in Figure 2a, the FTIR peak centered at ~1049 cm^{-1} can be ascribed to the COOH or CHO bending vibration, whereas the characteristic peak centered at ~1622 cm^{-1} occurs due to the C=O stretching vibration [33,34]. The peak intensities of EBC-K correspond with these oxygen content function groups, and they are slightly increased compared with EBC-Na, thus indicating that the population in the surface function group is greatly enhanced. As shown in Figure 2b, the Raman spectra demonstrate two characteristic peaks that correspond with the D band (~1300 cm^{-1}); these originated from defects and from the G band (~1600 cm^{-1}) in the in-plane sp^2 vibrational mode of the carbon. The normalized Raman curves indicate that the I_D/I_G ratio could be measured as 1.01 and 0.97 for EBC-K and EBC-Na; this indicates that fewer surface defects are formed in the EBC-Na [35–38]. This phenomenon may possibly be ascribed to the fact that the KCl with a lower melting point could react better with the carbon skeleton to create abundant surface defect sites on the carbon skeleton. The N_2 adsorption–desorption isotherms and the pore size distribution of EBC-K and EBC-Na are displayed in Figure 2c,d. The curves both demonstrated typical IV curves, which indicate the co-existence of micro- and meso-porous structures [39]. In addition, the hysteresis loops could be observed in the isotherms of both of the bio-carbons, thus suggesting the existence of meso-porosity which possibly results from the tissue structure of the EP; however, the N_2 adsorption of the EBC-K was much enhanced compared with that of the EBC-Na, which gives rise to the greater BET surface area of 356.8 m^2 g^{-1}, compared with that of EBC-Na (257.4 m^2 g^{-1}). Furthermore, the N_2 adsorption that occurred with relative pressure from 0–0.2 and 0.4–0.8 were related to the microporous and mesoporous structures, respectively. It is obvious that the two bio-carbons exhibited a similar microporous structure, but the EBC-K was dominated by mesopores [40]. As shown in Figure 2d, EBC-K possesses a higher pore size distribution in both the micro- and meso-porous range. In addition, the average pore size of the EBC-K was greatly increased to 2.9 nm compared with that of the EBC-Na (2.3 nm); this results in an increased number of increased active sites for both PDS activation and SMX adsorption. In addition, EBC-K also demonstrates a higher pore volume of 0.258 m^3 g^{-1} compared with that of EBC-Na (0.151 m^3 g^{-1}); this indicates that EBC-K exhibits higher mass diffusion dynamics during SMX degradation.

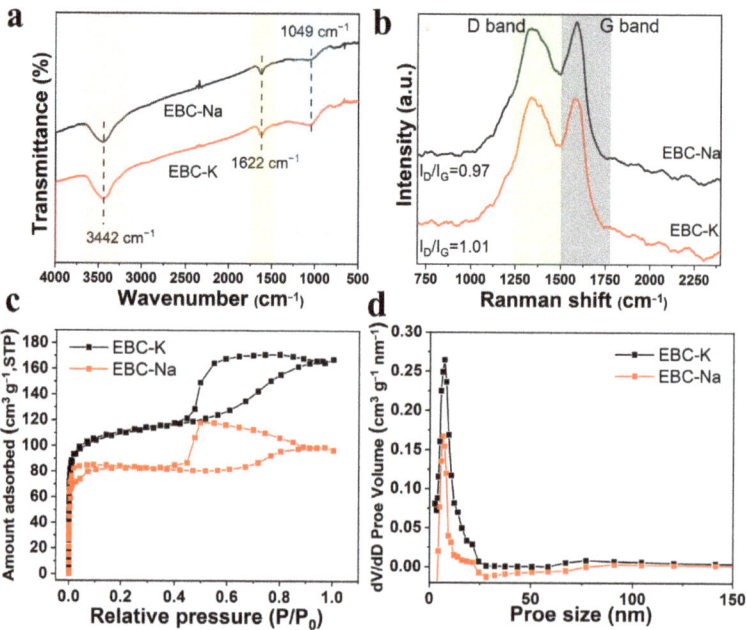

Figure 2. (**a**) FTIR and the (**b**) Raman curves of the EBC-Na and EBC-K. (**c**) N_2 adsorption/desorption isotherms and (**d**) pore size distribution of EBC-Na and EBC-K.

The compositions of the bio-carbons were also characterized by X-ray photoelectron spectroscopy (XPS). As shown in Figure S3 and Table S1, the survey spectra of the two samples exhibit prominent C, O, N and S signals, thus indicating that the N, S elements originated from the EP biomass. Moreover, EBC-K possesses relatively high N (4.85%), S (2.35%), and O (24.07%) contents compared with that of EBC-Na (N (4.77%), S (1.96%) and O (19.09%)), thus suggesting that the KCl molten salt system could better preserve the heteroatoms in the bio-carbon; this is in accordance with the Raman spectra. Further analysis of the high-resolution peaks is depicted in Figure 3a–d. The C1s signals could be deconvolved into four peaks, which correspond with C-C/C=C, C-O, C=O and COOH, respectively (Figure 3a) [41,42]. As shown in Figure 3b and Table 1, the greater O content of EBC-K is largely derived from the enhanced COOH on the surface; this is considered to be a reliable adoption site for SMX. In addition, the high-resolution N spectra in Figure 3c could be separated into four peaks belonging to pyridinic N, pyrrolic N, graphitic N, and oxidized N, respectively [43,44]. EBC-K shows a much higher pyridinic N content (74.59%) than that of the EBC-Na (68.64%), which means that it could serve as an active sit for PDS activation. On the other hand, it was reported that the S doping sites could give rise to the asymmetric electron spin and charge polarization on the carbon surface, which could accelerate the adsorption of PDS to facilitate the AOP process. As depicted in Figure 3d and Table 1, the high-resolution S 2p spectra of EBC-K indicates that the two peaks centered at ~163 eV and ~164 eV are assigned to S $2p_{3/2}$ and S $2p_{1/2}$ (thiophene-S), respectively, and the other two that are centered at ~167 eV and 168 eV correspond with the oxidized S (-SO_x) [45,46]. Compared with EBC-Na, the EBC-K with an enhanced portion of lower valence thiophene-S (42.21%) and oxidized S (24.07%) are expected to better activate the PDS for the AOPs.

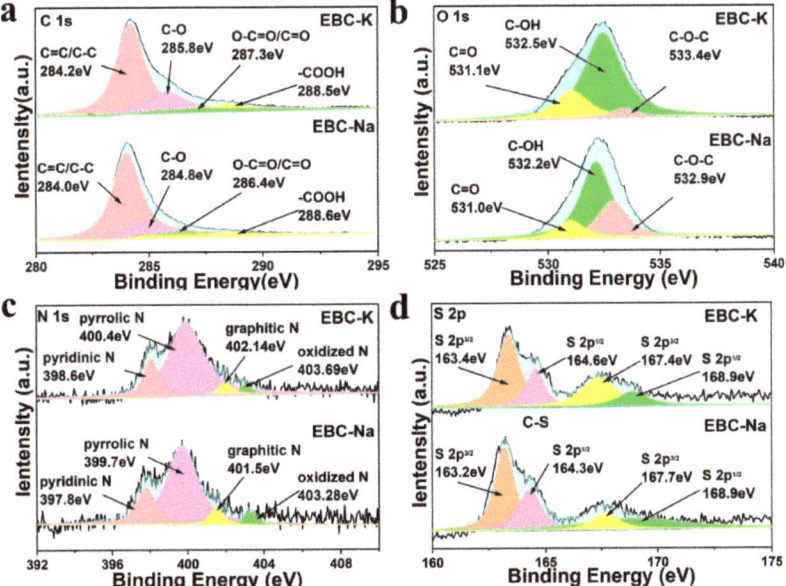

Figure 3. High resolution XPS spectrum of EBC-Na and EBC-K, (a) C 1s, (b) O 1s, (c) N 1s, and (d) S 2p.

Table 1. The fitting results of the high resolution XPS spectra of C, O, N, and S for EBC-K and EBC-Na.

Bio-Carbons	C Distribution (%, atm)				O Distribution (%, atm)			
	C=C/C-C	C-O	O-C=O/C=O	COOH	C=O	C-OH		C-O-C
EBC-K	75.29	19.75	0.79	4.17	22.37	72.39		5.24
EBC-Na	76.95	18.34	3.42	1.29	13.13	59.39		27.48

	N Distribution (%, atm)				S Distribution (%, atm)			
	Pyridinic	Pyrrolic	Graphitic	Oxidized	C-S		Oxidized Sulfur	
					S 2p$_{3/2}$	S 2p$_{1/2}$	S 2p$_{3/2}$	S 2p$_{1/2}$
EBC-K	17.35	74.59	4.78	3.28	42.21	19.27	27.03	11.49
EBC-Na	20.76	68.64	6.36	4.24	39.32	23.28	11.32	26.08

3.2. AOP Performances

The PDS activation performances of the bio-carbons were explored by using SMX as the simulating pollutant. In order to better illustrate the adsorption and PDS degradation, the measurements were divided into two phases. After the adsorption reaches equilibrium after 60 min, the PDS was subsequently added into the system to start the AOP. As shown in Figure 4a, EBC-K shows a much higher adsorption capability as well as degradation process compared with that of EBC-Na. The overall removal rate could achieve up to 92.7% a significant improvement compared with the removal rate of EBC-Na (65.9%); therefore, the detailed optimization of the preparation parameters was carried out in the KCl system. Based on Figure 4b, it was found that without a catalyst, the SMX experiences no obvious degradation with the addition of PDS, thus indicating that the self-activation of PDS is rather low; however, for the degradation system without PDS, the removal of SMX is solely caused by the adsorption process, and the high adsorption capability of EBC-K (~50%) could facilitate the later degradation process. As demonstrated in Figure 4c, the EBC-700 that was prepared under 700 °C possessed the best performance. The higher pyrolysis temperature inevitably leads to the evaporation of the slat medium, which leads to lower adsorption and catalytic performances.

In addition, the pH condition is of great significance to the SMX degradation rate as it affects the surface adsorption and reactive species formation process. As shown in Figure 4d, although the high pH leads to much lower adsorption performances, an overall degradation rate of up to ~75% can nevertheless be achieved. Moreover, Figure 4e indicates that SMX degradation could be further increased to ~100% by increasing the catalytic dosages to 1.00 g. Figure 4f shows that even with lower SMX concentrations, the adsorption and degradation rates experience obvious increases as expected, and the maximum removal rate could reach up to 99.7% when the concentration is reduced to 20 mg L^{-1}.

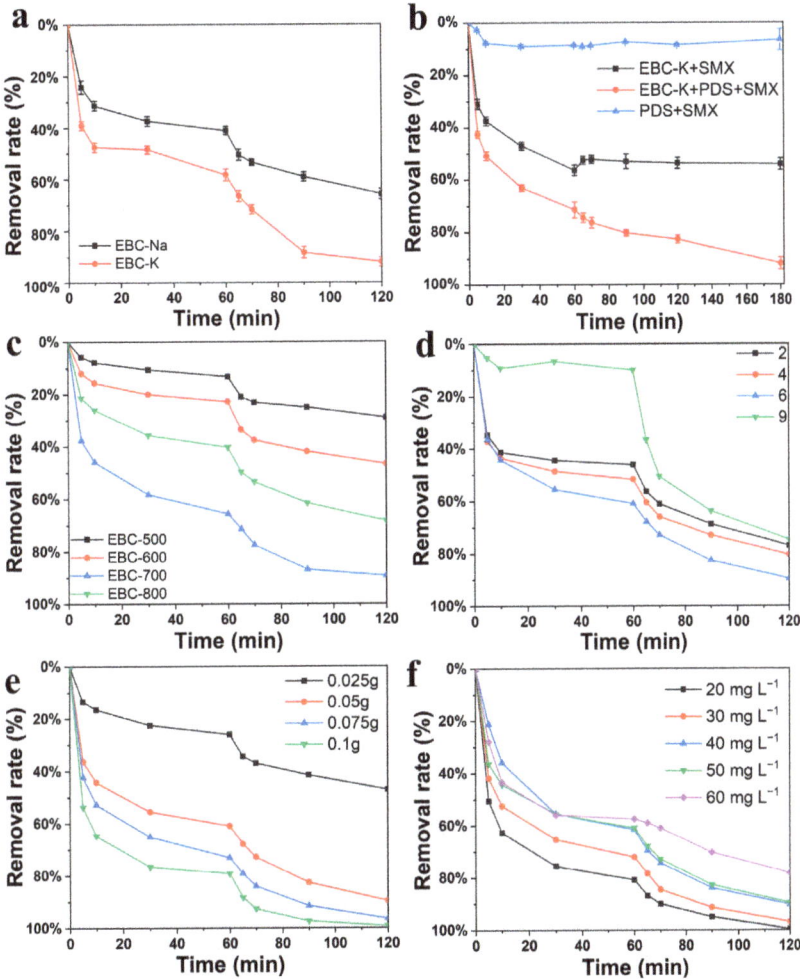

Figure 4. (a) The SMX degradation performances of EBC-K and EBC-Na, (b) the degradation under different degradation conditions, the detail degradation performances of (c) EBC-Ks prepared under varied pyrolysis temperatures, and the optimized EBC-K measured at varied (d) initial pH levels, (e) bio-carbon dosages, and (f) SMX concentrations.

3.3. Degradation Mechanism

In order to determine the reactive species generated by the PDS activation, systematical radical quenching experiments were performed. As shown in Figure 5a, MeOH was adopted to salvage the $SO_4^{\bullet-}$ and $^{\bullet}OH$ active species. The degradation performance of EBC-K was barely influenced. To further confirm the role of $^{\bullet}OH$, TBA was introduced to quench the $^{\bullet}OH$ without affecting the $SO_4^{\bullet-}$ (Figure 5b). Although the overall level of degradation was slightly reduced, this reduction possibly resulted from the surface adsorption sites which were partially occupied by the TBA; this finally gave rise to the lower adsorption accommodation for SMX [47–49]. This evidence indicates that $SO_4^{\bullet-}$ and $^{\bullet}OH$ were not the dominating reactive species in SMX degradation. In addition, L-histidine was adopted to quench $^1O_2$ during the degradation process and the results were shown in Figure 5c. After increasing the L-histidine amounts, SMX degradation drops to 48.3%. In order to further confirm the generation of $^1O_2$, FFA was adopted to trap $^1O_2$

(Figure 5d). After increasing the FFA concentration, the SMX degradation rate decreased to 50.4%, thus confirming the fact that $^1O_2$ serves as the dominating reactive species in SMX degradation [50]. To better understand the role of EBC-K in real AOP scenarios, running water from the local water supply company was used in the background in order to simulate the SMX pollution. As shown in Figure S4, although the intrinsic organic compounds and intervening ions inhibit the degradation process, it was found that over 85% of the degradation rate could be achieved (50 mg L^{-1}), which is relatively lower than the rate that can be achieved with pure water. If the SMX concentration is further reduced to 20 mg L^{-1}, the removal rate could reach to 99.6%, which is similar to the rate achieved with the pure water system.

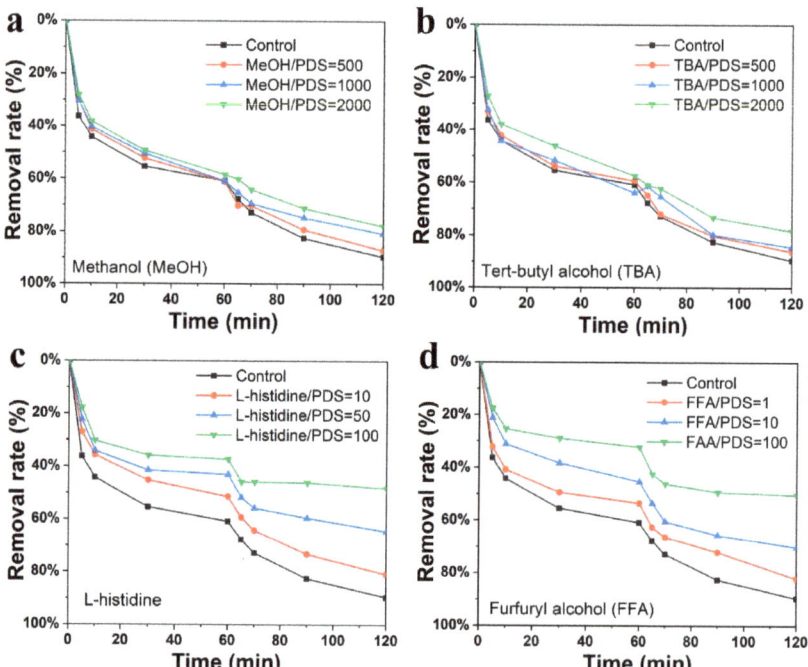

Figure 5. SMX degradation performances of EBC-K measured with different scavengers: (**a**) methanol (MeOH), (**b**) tert-butyl alcohol (TBA), (**c**) L-histidine, and (**d**) furfuryl alcohol (FFA). (Control, in the inset legends, represents the test without a scavenger, and the values represent the molar ratio between the scavenger and the PDS.).

3.4. Tech-Economic Assessment

The annual explosive growth of EP between spring and summer brings forth a tremendous hazard to the local ecosystem as well as the fishing and tourism industries in Tsingtao City, Shandong Province of China; therefore, the local government offers millions of dollars in subsidies to clean them from the sea, and millions of tons of EP biomass are collected and stored. Only a small portion can be utilized as fertilizer or biomass for bio-oil production; therefore, it is imperative to develop an economic treatment for EP biomass in order to minimize the potential hazards. Based on these considerations, the bio-carbon factory will be established 100 km away from the EP storage yard in Tsingtao City. The cost of manufacturing (COM) is expected to anticipate the economic cost of bio-carbon produced by EP in the process of industrial manufacturing. Based on previous reports, COM is composed of five main parts: the fixed capital investment (FCI), costs of labors (C_{OL}), raw materials (C_{RM}), waste treatment (C_{WT}) and utilities (C_{UT}). Moreover, the COM of bio-carbon can therefore be calculated with the following equation, and the detailed calculation is presented with supporting information [51].

$$COM = 0.230 FCI + 2.73 C_{OL} + 1.23(C_{RM} + C_{WT} + C_{UT})$$

As shown in Figure 6a, the production procedure of bio-carbon includes three steps: EP transportation, mixing process of biomass and salt, and the high temperature pyrolysis process. The industrial furnace, which has the capacity of ~3000 L, a temperature of 800 °C, and a maximum power of 150 kW, is necessary. Moreover, a mixing machine with a capacity of ~3000 L and a maximum power of 30 kW is equipped with the furnace. Four sets of furnaces and mixing machines will be installed in the factory. The price of a furnace is USD 3500 and a mixing machine is USD 3000. The FCI could be estimated as costing USD 26,000. The average wage when working in the manufacturing industry in Shandong Province is USD 3.57 h^{-1} (National Bureau of Statistics of China, 2021). Factories tend to operate 330 days a year with 24 h shifts which require four employees per shift. Based on this, the total annual C_{OL} could be calculated as USD 226,195.2. The C_{RM} includes the EP biomass and the molten salt, which are directly used for bio-carbon production. As the local government will subsidize the cost of EP collection and storage, the cost of the EP could be estimated by using the transportation fee from the storage yard to the factory. The local transportation price is USD 0.0917 per ton-kilometer, and the price of the EP is calculated as USD 9.17 ton^{-1}. The factory processes 5280 tons of EP every year, thus, the annual cost of biomass is USD 48,417.6. During the molten salt process, the weight ratio of salt and biomass is 1:1, and 50% of the salt can be recycled. In addition, the price of KCl and NaCl is USD 260 and USD 80 ton^{-1}, respectively; therefore, the costs of KCl and NaCl are estimated as USD 686,400 and USD 211,200 every year, respectively; therefore, the total annual C_{RM} is USD 734,817.6 and USD 259,617.6. As the factory recycles approximately half of the water and salt in the production process, the C_{WT} can be ignored. The C_{UT} refers to the electricity cost consumed by the furnaces and the mixing machines. The industrial and commercial electricity price is USD 0.099 h^{-1} in Shandong Province; therefore, the total annual C_{UT} is USD 564,357.6. The carbon yield in the production process is estimated as ~18%, and the COMs of the bio-carbon using KCl and NaCl in the production process are USD 2.34 kg^{-1} and USD 1.72 kg^{-1}, respectively; these prices can compete with the commercially active carbons. As shown in Figure 6b,c, the sensitivity analysis of COM remains stable as the market price fluctuates. In addition, as depicted in Figure 6d, a 10% fluctuation of C_{RM} leads to about a 4.27% COM variation, indicating that, in the case of KCl, the C_{RM} is the primary contributor to the COM. On the other hand, Figure 6e further suggests that when the NaCl is used as the reaction medium, a 10% fluctuation in C_{UT} leads to 4.07% COM variation, thus indicating that the C_{UT} is the main contributor of the COM in the case of NaCl. The tech-economic analysis concludes that the COM is significantly impacted by the cost of salt and the electricity; therefore, the molten salt method with low-price salt, as well as with a low temperature operation process, should be developed to further reduce the COM and the emissions from bio-carbon production.

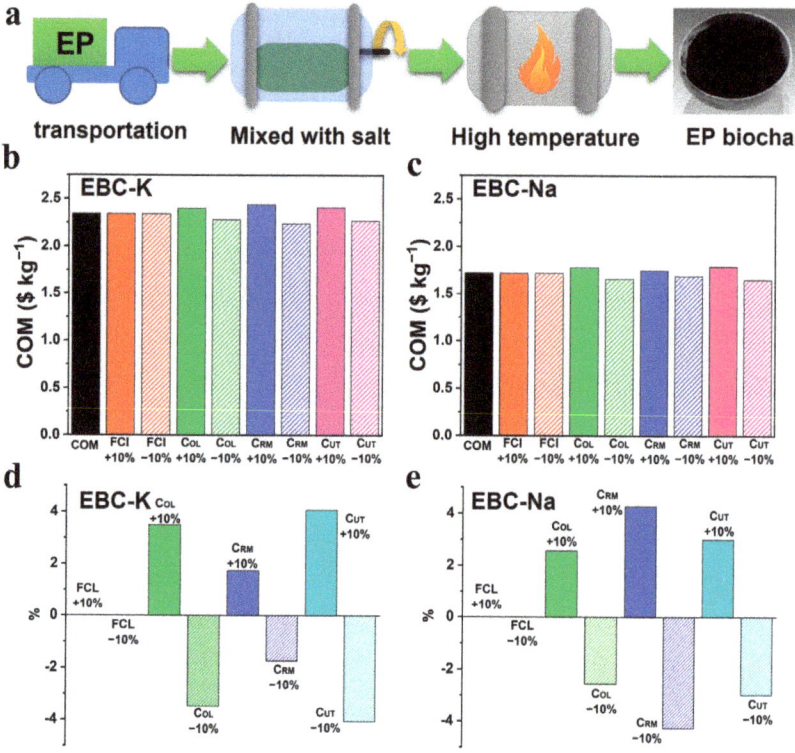

Figure 6. (a) The manufacturing processes of the bio-carbons derived from EP using the molten salt method; the COM of (b) EBC-K and (c) EBC-Na; the sensitivity analysis of the COM variation of (d) EBC-K and (e) EBC-Na upon the cost fluctuation with 10% increase or decrease.

4. Conclusions

Entermorpha prolifera, a widely distributed harmful algae, was employed as precursor for the first time in order to prepare N, S co-doped porous bio-carbon with a hierarchical porous structure and high surface area (356.8 m^2 g^{-1}), and KCl was adopted as the molten salt medium. It was found that the KCl with a low melting point was more reactive, thus achieving a high surface area and introducing rich surface function groups; this gives rise to high PDS adsorption and rapid $^1O_2$ generation levels. Therefore, the optimized EBC-K demonstrated a high SMX removal rate in both the experimental and applicational scenarios. Moreover, the techno-economic assessment indicated although the cost of EPB-K is higher (USD 2.34 kg^{-1}) compared with the EPB-Na (USD 1.72 kg^{-1}), EPB-K is still economically competitive with the commercial carbon. The cost sensibility analysis indicates that the cost of the salt medium and the electricity are the primary contributors to the COM. This work demonstrates that the EP could be employed as an abundant biowaste that produces bio-carbons with high performances, and the molten salt strategy could be further optimized by adopting low-cost salt to reduce the COM.

Supplementary Materials: The following supporting information can be downloaded at: https://www.mdpi.com/article/10.3390/nano12234289/s1, Figure S1: The Si distribution of EBC-K by HAADF-STEM and EDS mapping. Figure S2: XRD curves of the EBC-Na and EBC-K. Figure S3: The XPS survey curves of EBC-K (black) and EBC-Na (red). Figure S4: SMX degradation in pure water and practical water treatment (local running water as background). Table S1: the XPS results on the distribution of the C, N, O and O elements. References [51–56] are cited in the Supplementary Materials.

Author Contributions: Conceptualization, M.Z. and D.W.; methodology, K.H., Y.Z. (Yi Zhou) and X.W.; validation, Y.D. and Y.G.; formal analysis, D.W., M.Z. and Z.G.; investigation, K.H., M.Z. and X.W.; data curation, Y.D., K.H. and Y.Z. (Yi Zhang); writing—original draft preparation, M.Z.; writing—review and editing, D.W. and P.L.; supervision, D.W. and P.L.; funding acquisition, D.W. and Z.G.; All authors have read and agreed to the published version of the manuscript.

Funding: This research was funded by Thousand Talent Project of Henan Province (ZYQR201912167), Outstanding Youth Science Foundation of HTU (2021JQ03). College Students Innovative Entrepreneurial Training Plan Program of Higher Education of Henan Province (202210465017, 202210476020).

Institutional Review Board Statement: Not applicable.

Informed Consent Statement: Not applicable.

Data Availability Statement: The detailed data in the study are available from the corresponding authors by request. (Dapeng Wu and Pengfei Li).

Conflicts of Interest: The authors declare no conflict of interest.

References

1. Li, Y.; Li, J.; Pan, Y.; Xiong, Z.; Yao, G.; Xie, R.; Lai, B. Peroxymonosulfate activation on $FeCo_2S_4$ modified g-C_3N_4 ($FeCo_2S_4$-CN): Mechanism of singlet oxygen evolution for nonradical efficient degradation of sulfamethoxazole. *Chem. Eng. J.* **2020**, *384*, 123361. [CrossRef]
2. Sun, W.; Pang, K.; Ye, F.; Pu, M.; Zhou, C.; Yang, C.; Zhang, Q. Efficient persulfate activation catalyzed by pyridinic N, C-OH, and thiophene -S on N,S-co-doped carbon for nonradical sulfamethoxazole degradation: Identification of active sites and mechanisms. *Sep. Purif. Technol.* **2022**, *284*, 120197. [CrossRef]
3. Chen, Y.; Liang, Y.; Deng, Y.; Bin, Y.; Wang, T.; Luo, H. Peroxymonosulfate activation by concave porous S/N co-doped carbon: Singlet oxygen-dominated non-radical efficient oxidation of organics. *J. Environ. Chem. Eng.* **2022**, *10*, 107933. [CrossRef]
4. Sun, P.; Liu, H.; Feng, M.; Guo, L.; Zhai, Z.; Fang, Y.; Zhang, X.; Sharma, V.K. Nitrogen-sulfur co-doped industrial graphene as an efficient peroxymonosulfate activator: Singlet oxygen-dominated catalytic degradation of organic contaminants. *Appl. Catal. B Environ.* **2019**, *251*, 335–345. [CrossRef]
5. Ye, S.; Zeng, G.; Tan, X.; Wu, H.; Liang, J.; Song, B.; Tang, N.; Zhang, P.; Yang, Y.; Chen, Q.; et al. Nitrogen-doped biochar fiber with graphitization from Boehmeria nivea for promoted peroxymonosulfate activation and non-radical degradation pathways with enhancing electron transfer. *Appl. Catal. B* **2020**, *269*, 118850. [CrossRef]
6. Li, X.; Jia, Y.; Zhou, M.; Su, X.; Sun, J. High-efficiency degradation of organic pollutants with Fe, N co-doped biochar catalysts via persulfate activation. *J. Hazard. Mater.* **2022**, *397*, 122764. [CrossRef]
7. Li, X.; Jia, Y.; Zhang, J.; Qin, Y.; Wu, Y.; Zhou, M.; Sun, J. Efficient removal of tetracycline by H_2O_2 activated with iron-doped biochar: Performance, mechanism, and degradation pathways. *Chin. Chem. Lett.* **2022**, *33*, 2105–2110. [CrossRef]
8. Zeng, T.; Yu, M.; Zhang, H.; He, Z.; Chen, J.; Song, S. Fe/Fe_3C@N-doped porous carbon hybrids derived from nano-scale MOFs: Robust and enhanced heterogeneous catalyst for peroxymonosulfate activation. *Catal. Sci. Technol.* **2017**, *7*, 396–404. [CrossRef]
9. Wang, C.; Wu, D.; Wang, H.; Gao, Z.; Xu, F.; Jiang, K. Nitrogen-doped two-dimensional porous carbon sheets derived from clover biomass for high performance supercapacitors. *J. Power Sources* **2017**, *363*, 375–383. [CrossRef]
10. Xue, S.; Xie, Z.; Ma, X.; Xu, Y.; Zhang, S.; Jia, Q.; Wan, R.; Tao, H. Insights into the singlet oxygen mechanism of Fe-doped activated carbon for Rhodamine B advanced oxidation. *Microporous Mesoporous Mater.* **2022**, *337*, 111948. [CrossRef]
11. Inga, B.; Christina, G.; Alysson, D.R.; Silvia, P.; Peter, H.; Andreas, T. Carbon Adsorbents from Spent Coffee for Removal of Methylene Blue and Methyl from Water. *Materials* **2021**, *14*, 3996. [CrossRef]
12. Dibyashree, S. Efficiency of Wood-Dust of Dalbergia sisoo as Low-Cost Adsorbent for Rhodamine-B Dye Removal. *Nanomaterials* **2021**, *11*, 2217. [CrossRef]
13. Ding, J.; Xu, W.; Liu, Y.; Tan, X.; Li, X.; Li, Z.; Zhang, P.; Du, L.; Li, M. Activation of persulfate by nanoscale zero-valent iron loaded porous graphitized biochar for the removal of 17β-estradiol: Synthesis, performance and mechanism. *J. Colloid Interface Sci.* **2021**, *588*, 776–786. [CrossRef]
14. Patel, A.K.; Katiyar, R.; Chen, C.W.; Singhania, R.R.; Awasthi, M.K.; Bhatia, S.; Bhaskar, T.; Dong, C.D. Antibiotic bioremediation by new generation biochar: Recent updates. *Bioresour. Technol.* **2022**, *358*, 127348. [CrossRef]
15. He, L.; Liu, Z.; Hu, J.; Qin, C.; Yao, L.; Zhang, Y.; Piao, Y. Sugarcane biochar as novel catalyst for highly efficient oxidative removal of organic compounds in water. *Chem. Eng. J.* **2021**, *405*, 126895. [CrossRef]
16. Zhu, M.; Kong, L.; Xie, M.; Lu, H.; Li, N.; Feng, Z.; Zhan, J. Carbon aerogel from forestry biomass as a peroxymonosulfate activator for organic contaminants degradation. *J. Hazard. Mater.* **2021**, *413*, 125438. [CrossRef]
17. Liu, J.J.; Dickson, R.; Niaz, H.; Hal, J.W.V.; Dijkstra, J.W. Production of fuels and chemicals from macroalgal biomass: Current status, potentials, challenges, and prospects. *Renew. Sustain. Energy Rev.* **2022**, *169*, 112954. [CrossRef]
18. Wu, M.; Li, P.; Li, Y.; Liu, J.; Wang, Y. Enteromorpha based porous carbons activated by zinc chloride for supercapacitors with high capacity retention. *RSC Adv.* **2015**, *5*, 16575–16581. [CrossRef]

19. Yang, W.; Wang, Z.; Song, S.; Han, J.; Chen, H.; Wang, X.; Sun, R.; Cheng, J. Adsorption of copper(II) and lead(II) from seawater using hydrothermal biochar derived from Enteromorpha. *Mar. Pollut. Bull.* **2019**, *149*, 110586. [CrossRef]
20. Chen, C.; Ma, T.; Shang, Y.; Gao, B.; Jin, B.; Dan, H.; Li, Q.; Yue, Q.; Li, Y.; Wang, Y.; et al. In-situ pyrolysis of Enteromorpha as carbocatalyst for catalytic removal of organic contaminants: Considering the intrinsic N/Fe in Enteromorpha and non-radical reaction. *Appl. Catal. B* **2019**, *250*, 382–395. [CrossRef]
21. Huang, W.; Wang, F.; Qiu, N.; Wu, X.; Zang, C.; Li, A.; Xu, L. Enteromorpha prolifera-derived Fe_3C/C composite as advanced catalyst for hydroxyl radical generation and efficient removal for organic dye and antibiotic. *J. Hazard. Mater.* **2019**, *378*, 120728. [CrossRef]
22. Li, J.; Liu, Y.; Ren, X.; Dong, W.; Cheng, W.; Cai, T.; Zeng, W.; Li, W.; Tang, L. Soybean residue based biochar prepared by ball milling assisted alkali activation to activate peroxydisulfate for the degradation of tetracycline. *J. Colloid Interface Sci.* **2021**, *599*, 631–641. [CrossRef]
23. Hao, H.; Zhang, Q.; Qiu, Y.; Li, M.; Wei, X.; Sang, W.; Tao, J. Insight into the degradation of Orange G by persulfate activated with biochar modified by iron and manganese oxides: Synergism between Fe and Mn. *J. Water Process. Eng.* **2020**, *37*, 101470. [CrossRef]
24. Su, X.; Guo, Y.; Yan, L.; Wang, Q.; Zhang, W.; Li, X.; Song, W.; Li, Y.; Liu, G. MoS_2 nanosheets vertically aligned on biochar as a robust peroxymonosulf ateactivator for removal of tetracycline. *Sep. Purif. Technol.* **2022**, *282*, 120118. [CrossRef]
25. Chen, Z.; Jing, Y.; Wang, Y.; Meng, X.; Zhang, C.; Chen, Z.; Zhou, J.; Qiu, R.; Zhang, X. Enhanced removal of aqueous Cd (II) by a biochar derived from salt-sealing pyrolysis coupled with NaOH treatment. *Appl. Surf. Sci.* **2020**, *511*, 145619. [CrossRef]
26. Dai, S.; Zhao, Y.; Niu, D.; Li, Q.; Chen, Y. Preparation and reactivation of magnetic biochar by molten salt method: Relevant performance for chlorine-containing pesticides abatement. *J. Air Waste Manag.* **2019**, *69*, 58–70. [CrossRef]
27. Guo, T.; Ma, N.; Pan, Y.; Bedane, A.H.; Xiao, H.; Eić, M.; Du, Y. Characteristics of CO_2 adsorption on biochar derived from biomass pyrolysis in molten salt. *Can. J. Chem. Eng.* **2018**, *96*, 2352–2360. [CrossRef]
28. Yu, D.; Wang, L.; Wu, M. Simultaneous removal of dye and heavy metal by banana peels derived hierarchically porous carbons. *J. Taiwan Inst. Chem. E* **2018**, *93*, 543–553. [CrossRef]
29. Wu, D.; Chen, J.; Zhang, W.; Liu, W.; Li, J.; Cao, K.; Gao, Z.; Xu, F.; Jiang, K. Sealed pre-carbonization to regulate the porosity and heteroatom sites of biomass derived carbons for lithium-sulfur batteries. *J. Colloid Interface Sci.* **2020**, *579*, 667–679. [CrossRef]
30. Abdelhafez, A.A.; Li, J. Removal of Pb(II) from aqueous solution by using biochars derived from sugar cane bagasse and orange peel. *J. Taiwan Inst. Chem. Eng.* **2016**, *61*, 367–375. [CrossRef]
31. Wang, C.; Wu, D.; Wang, H.; Gao, Z.; Xu, F.; Jiang, K. Biomass derived nitrogen-doped hierarchical porous carbon sheets for supercapacitors with high performance. *J. Colloid Interface Sci.* **2018**, *523*, 133–143. [CrossRef] [PubMed]
32. Wang, C.; Wu, D.; Wang, H.; Gao, Z.; Xu, F.; Jiang, K. A green and scalable route to yield porous carbon sheets from biomass for supercapacitors with high capacity. *J. Mater. Chem. A* **2018**, *6*, 1244–1254. [CrossRef]
33. Bai, X.; Xing, L.; Liu, N.; Ma, N.; Huang, K.; Wu, D.; Yin, M.; Jiang, K. Humulus scandens-Derived Biochars for the Effective Removal of Heavy Metal Ions: Isotherm/Kinetic Study, Column Adsorption and Mechanism Investigation. *Nanomaterials* **2021**, *11*, 3255. [CrossRef] [PubMed]
34. Wu, D.; Wang, Y.; Wang, F.; Wang, H.; An, Y.; Gao, Z.; Xu, F.; Jiang, K. Oxygen-incorporated few-layer MoS_2 vertically aligned on three-dimensional graphene matrix for enhanced catalytic performances in quantum dot sensitized solar cells. *Carbon* **2017**, *123*, 756–766. [CrossRef]
35. Li, H.; Wang, Y.; Chen, H.; Niu, B.; Zhang, W.; Wu, D. Synergistic mediation of polysulfide immobilization and conversion by a catalytic and dual-adsorptive system for high performance lithium-sulfur batteries. *Chem. Eng. J.* **2021**, *406*, 126802. [CrossRef]
36. Wang, R.; Wang, H.; Zhou, Y.; Gao, Z.; Han, Y.; Jiang, K.; Zhang, W.; Wu, D. Green synthesis of N-doped porous carbon/carbon dot composites as metal-free catalytic electrode materials for iodide-mediated quasi-solid flexible supercapacitors. *Inorg. Chem. Front.* **2022**, *9*, 2530. [CrossRef]
37. Long, Y.; Ye, F.; Shi, L.; Lin, X.; Paul, R.; Liu, D.; Hu, C. N, P, and S tri-doped holey carbon as an efficient electrocatalyst for oxygen reduction in whole pH range for fuel cell and zinc-air batteries. *Carbon* **2021**, *179*, 365–376. [CrossRef]
38. Zhang, S.; Zhi, S.; Wang, H.; Guo, J.; Sun, W.; Zhang, L.; Jiang, Y.; Zhang, X.; Jiang, K.; Wu, D. Laser-assisted rapid synthesis of anatase/rutile TiO_2 heterojunction with Function-specified micro-zones for the effective photo-oxidation of sulfamethoxazole. *Chem. Eng. J.* **2023**, *453*, 139702. [CrossRef]
39. Wu, D.; Wang, X.; Wang, H.; Wang, F.; Wan, D.; Gao, Z.; Wang, X.; Xu, F.; Jiang, K. Ultrasonic-assisted synthesis of two dimensional $BiOCl/MoS_2$ with tunable band gap and fast charge separation for enhanced photocatalytic performance under visible light. *J. Colloid Interface Sci.* **2019**, *533*, 539–547. [CrossRef]
40. Piotr, K.; Malgorzata, S.; Izabela, K.; Maciej, L.; Victor, K.; Jerzy, P. Obtaining N-Enriched Mesoporous Carbon-Based by Means of Gamma Radiation. *Nanomaterials* **2022**, *12*, 3156. [CrossRef]
41. Zou, X.; Wu, D.; Mu, Y.; Xing, L.; Zhang, W.; Gao, Z.; Xu, F.; Jiang, K. Boron and Nitrogen Co-doped Holey Graphene Aerogels with Rich B-N Motifs for Flexible Supercapacitors. *Carbon* **2020**, *159*, 94–101. [CrossRef]
42. Bai, X.; Zhang, M.; Niu, B.; Zhang, W.; Wang, X.; Wang, J.; Wu, D.; Wang, L.; Jiang, K. Rotten sugarcane bagasse derived biochars with rich mineral residues for effective Pb (II) removal in wastewater and the tech-economic analysis. *J. Taiwan Inst. Chem. Eng.* **2022**, *132*, 104231. [CrossRef]

43. Zhang, P.; Chen, C.; Zhang, X.; Jiang, Z.; Huang, J.; Chen, J. Fe and S co-doped N-enriched hierarchical porous carbon polyhedron as efficient non-noble-metal electrocatalyst toward oxygen reduction reaction in both alkaline and acidic medium. *Electrochimica Acta* **2019**, *298*, 570–579. [CrossRef]
44. Meng, F.L.; Wang, Z.L.; Zhong, H.X.; Wang, J.; Yan, J.M.; Zhang, X.B. Reactive Multifunctional Template-Induced Preparation of Fe-N-Doped Mesoporous Carbon Microspheres Towards Highly Efficient Electrocatalysts for Oxygen Reduction. *Adv. Mater.* **2016**, *28*, 7948. [CrossRef] [PubMed]
45. Yin, M.; Bai, X.; Wu, D.; Li, F.; Jiang, K.; Ma, N.; Chen, Z.; Zhang, X.; Fang, L. Sulfur-functional group tunning on biochar through sodium thiosulfate modified molten salt process for efficient heavy metal adsorption. *Chem. Eng. J.* **2022**, *433*, 134441. [CrossRef]
46. Long, Y.; Li, S.; Su, Y.; Wang, S.; Zhao, S.; Wang, S.; Zhang, Z.; Huang, W.; Liu, Y.; Zhang, Z. Sulfur-containing iron nanocomposites confined in S/N co-doped carbon for catalytic peroxymonosulfate oxidation of organic pollutants: Low iron leaching, degradation mechanism and intermediates. *Chem. Eng. J.* **2021**, *404*, 126499. [CrossRef]
47. Zhou, Y.; Jiang, J.; Gao, Y.; Ma, J.; Pang, S.; Li, J.; Lu, X.; Yuan, L. Activation of Peroxymonosulfate by Benzoquinone: A Novel Nonradical Oxidation Process. *Environ. Sci. Technol.* **2015**, *49*, 12941. [CrossRef]
48. Ren, W.; Gao, J.; Lei, C.; Xie, Y.; Cai, Y.; Ni, Q.; Yao, J. Recyclable metal-organic framework/cellulose aerogels for activating peroxymonosulfate to degrade organic pollutants. *Chem. Eng. J.* **2018**, *349*, 766–774. [CrossRef]
49. Ye, J.; Wang, Y.; Li, Z.; Yang, D.; Li, C.; Yan, Y.; Dai, J. 2D confinement freestanding graphene oxide composite membranes with enriched oxygen vacancies for enhanced organic contaminants removal via peroxymonosulfate activation. *J. Hazard. Mater.* **2021**, *417*, 126028. [CrossRef]
50. Yun, E.T.; Lee, J.H.; Kim, J.; Park, H.D.; Lee, J. Identifying the Nonradical Mechanism in the Peroxymonosulfate Activation Process: Singlet Oxygenation Versus Mediated Electron Transfer. *Environ. Sci. Technol.* **2018**, *52*, 7032–7042. [CrossRef]
51. Sangon, S.; Hunt, A.J.; Attard, T.M.; Mengchang, P.; Ngernyen, Y.; Supanchaiyamat, N. Valorisation of waste rice straw for the production of highly effective carbon based adsorbents for dyes removal. *J. Clean. Prod.* **2018**, *172*, 1128–1139. [CrossRef]
52. National Bureau of Statistics of China. 2021. Available online: http://www.stats.gov.cn/tjsj/ndsj/2021/indexch.htm (accessed on 15 October 2022).
53. Lalamove. Available online: https://m.huolala.cn (accessed on 1 November 2022).
54. Zouping Changshan Town Zefeng Fertilizer Factory. 2020. Available online: https://www.alibaba.com/product-detail/food-grade-99-99-purity-potassium_60225947884.html?spm=a2700.galleryofferlist.normal_offer.d_title.f9be2c9fkqJoD6 (accessed on 1 February 2020).
55. Shouguang Hengyi Chemical Technology Co., Ltd. 2020. Available online: https://www.alibaba.com/product-detail/Sodium-Chloride-NaCL-99-3-Industrial_60571951782.html?spm=a2700.galleryofferlist.normal_offer.d_title.48187c45LVfvGC (accessed on 1 February 2020).
56. State Grid Shandong Electric Power Company. 2022. Available online: http://www.sd.sgcc.com.cn/html/main/col2752/2022-10/09/20221009083047270522735_1.html (accessed on 15 October 2022).

Article

Humulus scandens-Derived Biochars for the Effective Removal of Heavy Metal Ions: Isotherm/Kinetic Study, Column Adsorption and Mechanism Investigation

Xingang Bai [1], Luyang Xing [2], Ning Liu [1,*], Nana Ma [2], Kexin Huang [1], Dapeng Wu [1,2,*], Mengmeng Yin [1] and Kai Jiang [1,*]

[1] Key Laboratory for Yellow River and Huai River Water Environmental and Pollution Control, Henan Key Laboratory for Environmental Pollution Control, Ministry of Education, School of Environment, Henan Normal University, Xinxiang 453007, China; baixghtu@126.com (X.B.); ke1919124086@163.com (K.H.); ymmdyx0125@163.com (M.Y.)

[2] Key Laboratory of Green Chemistry Medias and Reactions, Ministry of Education, School of Chemistry and Chemical Engineering, Henan Normal University, Xinxiang 453007, China; xingluyang0517@163.com (L.X.); mann076@htu.edu.cn (N.M.)

* Correspondence: lnywzq@163.com (N.L.); dapengwu@htu.edu.cn (D.W.); jiangkai6898@126.com (K.J.)

Abstract: *Humulus scandens* was first adopted as a biomass precursor to prepare biochars by means of a facile molten salt method. The optimized biochar exhibits a high specific surface area of ~450 m^2/g, a rich porous structure and abundant oxygen functional groups, which demonstrate excellent adsorption performance for heavy metal ions. The isotherm curves fit well with the Langmuir models, indicating that the process is governed by the chemical adsorption, and that the maximum adsorption capacity can reach 748 and 221 mg/g for Pb^{2+} and Cu^{2+}, respectively. In addition, the optimized biochar demonstrates good anti-interference ability and outstanding removal efficiency for Cu^{2+} and Pb^{2+} in simulated wastewater. The mechanism investigation and DFT calculation suggest that the oxygen functional groups play dominant roles in the adsorption process by enhancing the binding energy towards the heavy metal ions. Meanwhile, ion exchange also serves as the main reason for the effective removal.

Keywords: biochar; adsorption; molten salt; heavy metal ion; water treatment

Citation: Bai, X.; Xing, L.; Liu, N.; Ma, N.; Huang, K.; Wu, D.; Yin, M.; Jiang, K. *Humulus scandens*-Derived Biochars for the Effective Removal of Heavy Metal Ions: Isotherm/Kinetic Study, Column Adsorption and Mechanism Investigation. *Nanomaterials* **2021**, *11*, 3255. https://doi.org/10.3390/nano11123255

Academic Editor: Abdelhamid Elaissari

Received: 16 October 2021
Accepted: 17 November 2021
Published: 30 November 2021

Publisher's Note: MDPI stays neutral with regard to jurisdictional claims in published maps and institutional affiliations.

Copyright: © 2021 by the authors. Licensee MDPI, Basel, Switzerland. This article is an open access article distributed under the terms and conditions of the Creative Commons Attribution (CC BY) license (https://creativecommons.org/licenses/by/4.0/).

1. Introduction

With the rapid development of China's economy, a large volume of wastewater containing heavy metal ions is being discharged by the mining, metallurgy, chemical, electronics and other production industries [1,2]. Heavy metal ion pollution remaining in different water sources seriously threatens people's safety and health [3,4]. Harmful heavy metal ions such as cadmium, chromium, lead, arsenic, copper, mercury and so on are commonly detected in different water bodies [5,6]. At present, chemical, physical and biological methods including electrochemical treatment, ion exchange, chemical precipitation and adsorption are the mostly commonly employed strategies to remove heavy metal ions from wastewater [7]. Among these methods, adsorption is considered an effective and economic method because of its facile operation and low cost [8,9]. At present, biomass-derived activated carbons with a large specific surface area and resource abundance have been widely used in wastewater treatment [10–12]. A large number of studies have proven that the outstanding adsorption performance of biochar originates from the large surface area, stable porous structure, abundant surface oxygen functional groups such as carboxyl and hydroxyl groups, and rich mineral species which can remove the harmful ions through the ion exchange process [13]. Therefore, it is of great significance to develop high-efficiency and low-cost biochar materials to remove heavy metals from polluted water environments [14].

Hitherto, many biomasses, such as agricultural wastes, woody materials, algae and so on [15–18], are widely employed to prepare biochar for the removal of heavy metal ions from the water system. For example, *Undaria* was adopted as a raw material to prepare biochar through rapid pyrolysis and physical steam activation. The as-obtained biochar was able to effectively remove Cu^{2+} from the aqueous solution [19]. In order to achieve the fast removal of Pb^{2+}, biomasses such as palm, soybean meal, shells and so on were also adopted to prepare biochars with more adsorption sites as well as rich ion diffusion channels [20–24]. Based on these studies, it was found that although they are prepared through similar treatment methods, the adsorption performance of biochars varies widely, which implies that the unique tissue structures as well as the composition of the biomass exerts great influence on the adsorption performance of the biochar. Therefore, it is of great significance to explore the family of biomass for high-performance biochars.

In addition, in order to improve the adsorption capacity of biochars, various approaches have been developed to achieve the chemical modification of amino, alkali, sulfur and phosphoric function groups on the biochar surface. For example, Zhang et al. modified the surface of a rice straw-derived biochar with an amino group through nitrification and amination to prepare high-efficiency Pb^{2+} adsorbent. The as-introduced amino group shows strong complexation with Pb^{2+} to improve the adsorption strength of biochars [23]. Ming et al. adopted EDTA to modify the activated carbon derived from shells, and the mechanism studies indicate that the electrostatic action and complexation of surface active groups play important roles in enhancing the Pb^{2+} adsorption capacity [24]. Zhu et al. prepared surface-oxidized biochar from porous biochar with a high Cd^{2+}-adsorption capacity, and they found that the adsorption capacity is mainly affected by the surface functional groups [25]. In addition to the oxygen and amino function groups, Cl active sites and sulfur groups were also deliberately introduced on the surface of the biochars to enhance the removal efficiency of Hg^+ and Cu^{2+}, taking advantage of the enhanced binding energy between the active sites and the target heavy metal ions [26,27]. Although many schemes have been developed to achieve the surface modification of biochars, these strategies still suffer from many drawbacks, such as a tedious preparation procedure, high cost, toxic modification reagents and potential secondary contamination risk. In addition to the modification, the current production of biochars usually requires inert gas protection and high-temperature treatment with corrosive activators, which often have many disadvantages such as a high cost and a complicated preparation process. Therefore, it is necessary to develop a facile preparation process to produce biochar with rich surface function groups through a one-step pyrolysis process.

The molten salt method represents a commonly adopted route to prepare porous carbon materials with inert salt as the sealing and activating bi-functional medium. The biochars prepared through the molten salt method are widely used in the fields of environmental adsorption, energy storage devices, air purification and so on [28]. Due to the etching effect of molten salt and the template effect of salt crystals, mesopores and macropores can be introduced into the products [29]. At the same time, oxygen in the atmosphere can further react with highly active carbon atoms, which could introduce a large number of micropores into the products [30]. Therefore, the biochar prepared by the molten salt method has a high specific surface area and rich oxygen functional groups, which is conducive to the adsorption performance [31].

Humulus scandens (HS) is a perennial climbing herb which is widely distributed in the northern region of China. However, the massive growth of HS causes negative impacts on agricultural production and human activities. Moreover, the tissues of HS have regular porous structure and rich cellulose, which allows it to be promising precursor to prepare porous carbon materials. In this study, HS was firstly used as a precursor to prepare biochar by means of the NaOH modified molten salt method. The removal efficiency of Cu^{2+} and Pb^{2+} ions from an aqueous solution was greatly enhanced due to the well-developed porous structure, rich oxygen function groups as well as the embedded mineral species. The maximum Q_m can reach 748 and 221 mg/g for Pb^{2+} and Cu^{2+}, respectively.

In addition, the effects of adsorption conditions including initial pH, contact time, and initial concentration on the adsorption capacities were systematically investigated. The adsorption mechanism was studied, which indicates that oxygen function groups and the imbedded mineral species play dominant roles in the adsorption process by enhancing the binding energy towards the heavy metal ions and facilitating the ion exchange process. In addition, the optimized biochar shows outstanding anti-interference ability and good column adsorption performance, permitting its potential application in industrial-scale utilization. Considering that HS is a harmful biomass for agricultural production, the effective conversion of HS into biochar with excellent adsorption capacity can not only enrich the biomass family to produce carbon materials, but also reduce the adverse effects of the massive growth of HS in the north of China.

2. Materials and Methods

2.1. Materials

Humulus scandens, collected at Henan Normal University in Xinxiang City (35°19′54.49″ N, 113°54′51.75″ E) (Supplementary Materials Figure S1), Henan Province, China, was washed and dried in an oven at 50 °C for 12 h. The dried and pulverized HS was screened through a 100-mesh sieve for later use. $Pb(NO_3)_2$, $Cu(NO_3)_2·5H_2O$, NaOH, NaCl and KCl were purchased from Sinopharm Chemical Reagent Co. Ltd. (Beijing, China), were of analytic grade and were used without further purification.

2.2. Preparation of the Biochar

NaCl and KCl were mixed in an agate mortar at a mass ratio of 1:1. Then, 3 g of HS powder was mixed with 6 g of mixed salt and placed in a 50 mL ceramic crucible, covered with about 1 cm of mixed salt. The mixture was heated to 400, 500, 600, 700 and 800 °C in a muffle furnace at a rate of 10 °C/min, respectively. After continuous pyrolysis for 3 h, it was naturally cooled and cleaned with DI water to neutral. After drying in a 60 °C vacuum oven for 12 h, a series of biochars were obtained, and the biochar material obtained at a pyrolysis temperature of 600 °C was named HSC-M.

Then, 1, 2 and 3 g of NaOH were added to the mixture of 3 g HS powder and 6 g salt. After grinding evenly, the mixture was transferred into a ceramic pot, and covered with salt mixture with a thickness of about 1 cm. It was heated in a muffle oven at a rate of 10 °C/min to 600 °C and maintained for 3 h. The as-prepared biochars were named HSC-MA-1, HSC-MA-2 and HSC-MA-3 according to the NaOH dosage.

In addition, the washed, dried and crushed HS powder was placed in a ceramic crucible and placed in a tubular furnace. Under N_2 protection, the powder was sintered for 3 h at 600 °C with a heating rate of 10 °C/min. After having cooled naturally, the biochar was demoted as HSC-N_2.

2.3. Characterization of the Biochar

The pyrolysis process of the biomass in the molten salt medium was studied using TG-DSC in the air. The surface morphology and element distribution of biochar were analyzed by means of field emission scanning electron microscopy equipped with an energy dispersion spectrometer (SEM, Hitachi SU8010, Tokyo, Japan). The internal structure of the samples was observed using transmission electron microscopy (TEM, TF20, JEOL 2100F, Tokyo, Japan). X-ray photoelectron spectroscopy (XPS, Escalab250XI, Thermo Fisher Scientific, Waltham, MA, USA) was used to record the surface elemental composition and chemical bonds of the samples. X-ray powder diffraction (XRD, Rigaku DMAX-2000, Rigaku, Tokyo, Japan) was used to demonstrate the lattice space structure of the sample. The contents of Ca, Mg, Fe and Cu in the samples were analyzed by inductively coupled plasma mass spectrometry (ICP-MS, ELAN DRC-e, PerkinElmer Inc, LAS, USA). A Fourier transform infrared spectrometer (FTIR, Nexus 470, GMI Inc., Lebanon, OH, USA) and Raman spectrometer (Labramhr EVO, Horiba France SAS, Palaiseau, France) were used to analyze the surface functional groups of the biochar in the wave number ranges of

4000~400 cm^{-1} and 3000~50 cm^{-1}, respectively. The specific surface area and pore diameter distribution of the samples were determined by means of the N_2 adsorption method using a surface area analyzer (BET, MicrotracBEL, Osaka, Japan).

2.4. Heavy Metal Sorption Experiments

2.4.1. Adsorption Experiment

$Cu(NO_3)_2 \cdot 5H_2O$ and $Pb(NO_3)_2$ were added into two 500 mL volumetric flasks, respectively, and were prepared with 1000 mg/L Cu^{2+} and Pb^{2+} reserve solutions. Ten milligrams of the as-prepared samples were added into a 250 mL conical flask containing 100 mL, 100 mg/L Pb^{2+} or Cu^{2+}, respectively. The suspension was oscillated in a constant temperature water bath oscillator at 180 r/min for 4 h at 25 °C. After the reaction, the suspension was filtered using a 0.45 µm cellulose acetate membrane. Finally, the concentration of Pb^{2+} and Cu^{2+} after adsorption equilibrium was determined using a flame atomic absorption spectrometer.

2.4.2. Influence of Initial pH Value

The initial pH value of 100 mg/L Cu^{2+} and Pb^{2+} solution was adjusted to 2, 3, 4, and 5 with 0.1 mol/L HNO_3 or 0.1 mol/L NaOH solution. The effect of pH value on the adsorption of Pb^{2+} and Cu^{2+} by biochar in 100 mL of the Pb^{2+} and Cu^{2+} solution was determined by using the above method with a HSC-MA-2 dosage of 10 mg and a concentration of 100 mg/L.

2.4.3. Influence of Adsorbent Dosage

To analyze the influence of adsorbent dosage, 10, 20, 50, 80, 100, and 200 mg of the biochar HSC-MA-2 were added into a conical flask containing 100 mL of 200 mg/L Pb^{2+}. Similarly, 10, 20, 50, 80, 100 and 200 mg of the biochar HSC-MA-2 were also introduced into a conical flask containing 100 mL of 100 mg/L Cu^{2+}. Different adsorbent concentrations (0.1–2.0 g/L) were studied to determine their influence on the removal of Pb^{2+} and Cu^{2+} ions.

2.4.4. Adsorption Kinetics and Isothermal

For the kinetics tests, the initial concentration of Pb^{2+} was set as 200 mg/L and the biochar amount was set as 0.2 g/L. Meanwhile, the initial solubility of Cu^{2+} was 100 mg/L, and the amount of biochar was 0.5 g/L. The contact time between the biochars and the metal ion solution was 5–300 min, and the adsorption capacities of Pb^{2+} and Cu^{2+} on biochar were measured at 5, 10, 20, 30, 60, 120, 180, 240 and 300 min, respectively.

For the isothermal tests, the initial concentration of Pb^{2+} was set as 20, 50, 100, 150, 200, 250, and 300 mg/L, and the addition amount of biochar was 0.2 g L^{-1}, and the initial concentration of Cu^{2+} was 20, 40, 80, 120, 160, 200, 240 mg/L, and the addition amount of biochar was 0.5 g/L. At a constant temperature of 25 °C, the water bath was oscillated at a speed of 180 RPM for 4 h, and the samples were sampled at adsorption equilibrium. After filtration, the concentrations of Pb^{2+} and Cu^{2+} in the filtrate were determined, and the adsorption capacities of the biochar HSC-MA-2 on Pb^{2+} and Cu^{2+} at corresponding initial concentrations were calculated.

The amount of heavy metal ions adsorbed per unit mass of the adsorbent (Q_e) was calculated from Equation (1):

$$Q_e = (c_0 - c_e) \times \frac{V}{W} \qquad (1)$$

The formula for calculating the removal rate of metal ions is:

$$\eta\% = \frac{c_0 - c_e}{c_0} \times 100\% \qquad (2)$$

c_0 is the initial concentration of the Cu^{2+} and Pb^{2+} ions (mg L^{-1}); c_e is the concentration of the Cu^{2+} and Pb^{2+} ions at adsorption equilibrium (mg L^{-1}); Q_e is the amount of heavy metal ions adsorbed per unit mass of the adsorbent (mg g^{-1}); V is the total volume of the solution (L); and W is the mass of the biochar (g).

In order to study adsorption mechanism and kinetic parameters, quasi-first order, quasi-second order and intra-molecular diffusion models were employed to fit the adsorption curves, respectively.

Quasi first order dynamic equation:

$$\ln(q_e - q_t) = \ln q_e - \frac{k_1}{2.303}t \tag{3}$$

Quasi second order dynamic equation:

$$\frac{t}{q_t} = \frac{1}{k_2 q_e^2} + \frac{t}{q_e}$$

The equation of the intramedullary diffusion model:

$$q_t = K_{id} t^{1/2}$$

In the equation, q_e and q_t are the adsorption capacity at adsorption equilibrium and the adsorption capacity at time t, respectively (mg/g); t is the adsorption time (min); k_1 is the primary kinetic adsorption rate constant (min^{-1}), and k_2 is the secondary kinetic adsorption constant (g (mg/min)$^{-1}$). The correlation coefficient obtained by the fitting calculation was used to determine the adsorption process.

Adsorption isotherms of adsorbents describe the relationship between solid and liquid phases. Langmuir and Freundlich models are two commonly used adsorption isotherm models. Langmuir and Freundlich adsorption isotherm models are defined as follows:

Freundlich equation:

$$\ln q_e = \frac{1}{n} \ln C + \ln k_f$$

Langmuir equation:

$$\frac{C}{q_e} = \frac{1}{q_m k_l} + \frac{C}{q_e}$$

In the equation, C is the adsorption concentration (mg/L), q_e is the equilibrium adsorption capacity (mg/g), q_m is the maximum adsorption capacity (mg/g), k_l is the Langmuir constant (L/mg), k_f is the characteristic constant proportional to the equilibrium constant (L/mg), and n is the characteristic constant representing the properties of adsorption force.

2.5. The Effect of Background Ions

The effect of background ions on the adsorption of Pb^{2+} and Cu^{2+} ions by biochar was studied; 10 and 20 mg of HSC-MA-2 were added to 100 mL of 100 mg/L Pb^{2+} and 100 mg/L Cu^{2+} containing sodium, magnesium and calcium ions (with concentration of 1, 3, 5, 7 and 10 mmol/L), respectively, and then oscillated for 4 h at pH 5.0 and 25 °C.

2.6. Multi-Metal Adsorption

The interaction of coexisting heavy metals in the adsorption process has more practical significance than the interaction of a single system. A mixture of Pb^{2+}, Cu^{2+}, and Cr^{3+} was prepared at pH 5.0, and the initial concentrations of all metal ions were set as 50, 100, 150 and 200 mg/L. Different doses (0.5–3.0 g/L) of HSC-MS-2 were added into 100 mL of mixed heavy metal solution and reacted at 25 °C for 4 h.

2.7. Column Sorption

The actual wastewater simulation was estimated by the column sorption trials. The column adsorptions were carried out in a glass column (11.0 mm in diameter and 150 mm in length). The bottom of column was firstly packed with quartz sand (100 meshes). Then, 1.0 g of HSC-MA-2 was inserted on top of the quartz sand layer. The actual wastewater was pumped into the BC-ST packing bed in an up-flow type. The flow rate (Q, mL/min) of effluents was controlled by a peristaltic pump.

The simulated wastewater was fetched and prepared from the Yellow River in Xinxiang city, Henan province (34°54′17.78″ N, 113°46′7.95″ E) (Supplementary Materials Figure S2). The Cu^{2+}, Pb^{2+} and Cr^{3+} concentration were controlled at 100 mg/L with a pH of 5 to determine the effect of wastewater treatment on column sorption.

2.8. Gaussian Computations

All computations were performed by the Gaussian 09 Program (Revision D.01, Gaussian, Inc., Wallingford, CT, USA). All the geometries of the carbon domains with different functional groups were optimized at the B3LYP/6-31+G(d) level. The electrostatic potential (ESP) maps describe the electron density distribution, and the red area represents electron-rich areas, while the green area represents electron-deficient areas. According to the ESP maps, the geometries were further optimized at the B3LYP/6-31+G(d)/SDD level. The interaction energy between M and xxx were calculated by considering basis set superposition error (BSSE) correction.

3. Results and Discussion

3.1. Material Characterization

The pyrolysis process of HS biomass in the air or in a nitrogen atmosphere, and the pyrolysis process of HS in mixed salt with and without NaOH in the air were studied by using a synchronous thermal analyzer. Figure 1 shows the corresponding TG, DSC and DTG curves. For the pyrolysis process of HS in the air, Figure 1a shows that the first ~10% of weight loss ends around ~170 °C, which corresponds to the evaporation of moisture and decomposition of organic matter in HS. The second weight loss stage occurs in the range of 170–420 °C, with a total weight loss of ~55% due to the pyrolysis of organic compounds and the burning of carbonized products. When the temperature rose to ~600 °C, the weight loss of HS reached the maximum of ~76.8%. When the temperature was elevated to 730 °C, the weight of HS did not change, which indicates that the organic components in HS are completely removed at high temperature without a salt seal. The residue corresponds to ~23.2% of the initial biomass. Based on Figure 1b, the weight loss under N_2 is controlled at ~67.6%, indicating that ~32.4% of the biomass was retained after the calcination. Compared with that of the sample test in the air, it could be concluded that many mineral species are present in the biochars. For the mixture of HS and salt at a mass ratio of 1:2, it is found in Figure 1c that the weight loss of HS in the first three stages is basically the same as that in Figure 1a. Based on the calculation, the biomass weight loss at 600 °C is at ~60%, which is much lower than that without the salt protection (~75.0%), indicating the salt sealing partially reduces the hydrolysis of the biomass. As depicted in Figure 1d, after the addition of NaOH, it can be observed that the weight loss experiences a slight increase from ~60% to ~64%, which indicates that the NaOH helps to cave the carbon structure during the high-temperature reaction. Based on these observations, a possible formation mechanism is proposed: due to the etching effect of the molten salt ions as well as the penetrated O_2 atoms at high temperature, marco- and meso-porous structures form in the biochar. In addition, the NaOH not only further increases the pore diameters, but also introduces abundant oxygen functional groups on the biochar surface.

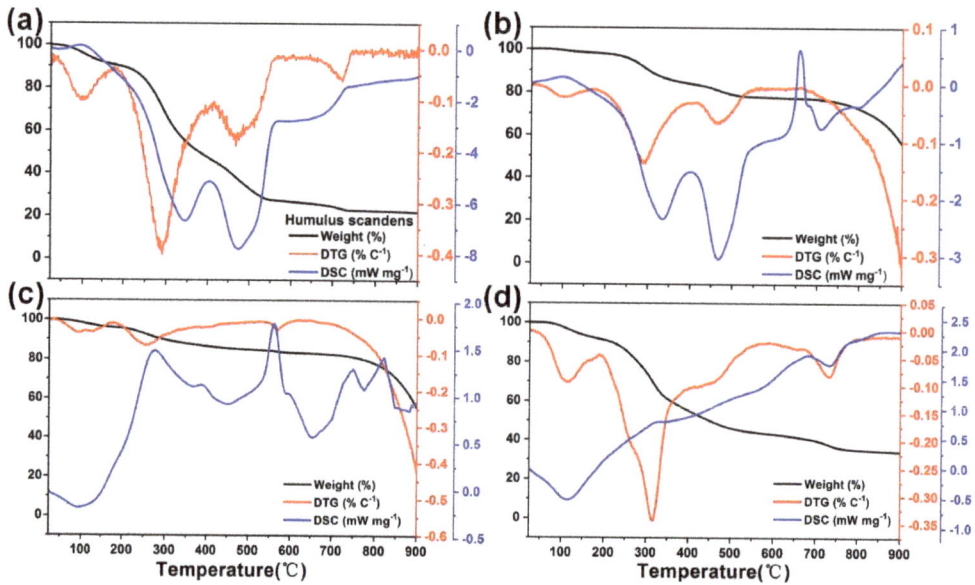

Figure 1. TG, DTG and DSC pyrolysis curves for (**a**) pure HS, (**b**) HS and salt and (**c**) HS with salt and NaOH in the air; (**d**) the pyrolysis of HS under N_2 protection.

Figure 2a shows an illustration of the preparation process of the HS-derived biochars. After the biomass was ground into a green powder, it was mixed with NaOH and the KCl/NaCl salt mix, and then sealed with KCl/NaCl in a crucible. Afterwards, the crucible was covered and heated to a given temperature for the pyrolysis. Finally, the product was washed with water to remove the excessive NaOH and salt. The biochars was dried for the adsorption of the heavy metal ions. The morphology and microstructure of the HS-derived biochars were studied using a field emission scanning electron microscope (FE-SEM). Figure 2b illustrates the excessive growth of HS in the north of China, which causes negative impacts on both agricultural production as well as the human activities. Figure 2c shows the FE-SEM image of the original biomass, indicating that the surface is smooth, with almost no pore structure detected. As depicted in Figure 2d, the surface of the biochar prepared in a nitrogen atmosphere is rougher, demonstrating that carbonation resulted in the formation of a porous structure in the as-yielded biochars (Supplementary Materials Figure S3). However, the porosities are closely packed together. However, as seen in Figure 2e, under the protection of molten salt, the HSC-MS exhibits an irregular sheet-like structure with a rougher surface. As shown in Supplementary Materials Figure S4, after the dual activation of molten salt and NaOH, the surfaces of the HSC-MAs all become rougher. In addition, the overall particle size of the HSC-MA-2 becomes smaller and the surface shows greater roughness, forming a rich porous structure (Figure 2f,g). Such porous structures allow the biochar to have large specific surface area, which can facilitate the mass diffusion in the pore channels of the biochar. The above results show that the formation of the porous structure of biomass results from the etching effects of both the molten salt and NaOH at high temperature. Meanwhile, the abundant porous structure and large specific surface area can promote the adsorption of heavy metal pollutants on the HS-derived biochars. Figure 2h depicts the TEM image of HSC-MA-2. The sample is composed of thin sheets with a large number of macropores with diameters of 50–100 nm. The TEM image with higher magnification (Figure 2i) shows a great number of mesoporous and micropores, due to the NaOH-assisted pyrolysis. The high-resolution TEM (HRTEM) image shown in Figure 2j indicates that many wrinkling lattice fringes can be detected at the surface of the carbon, indicating that the surface of the biochar was highly graphitized.

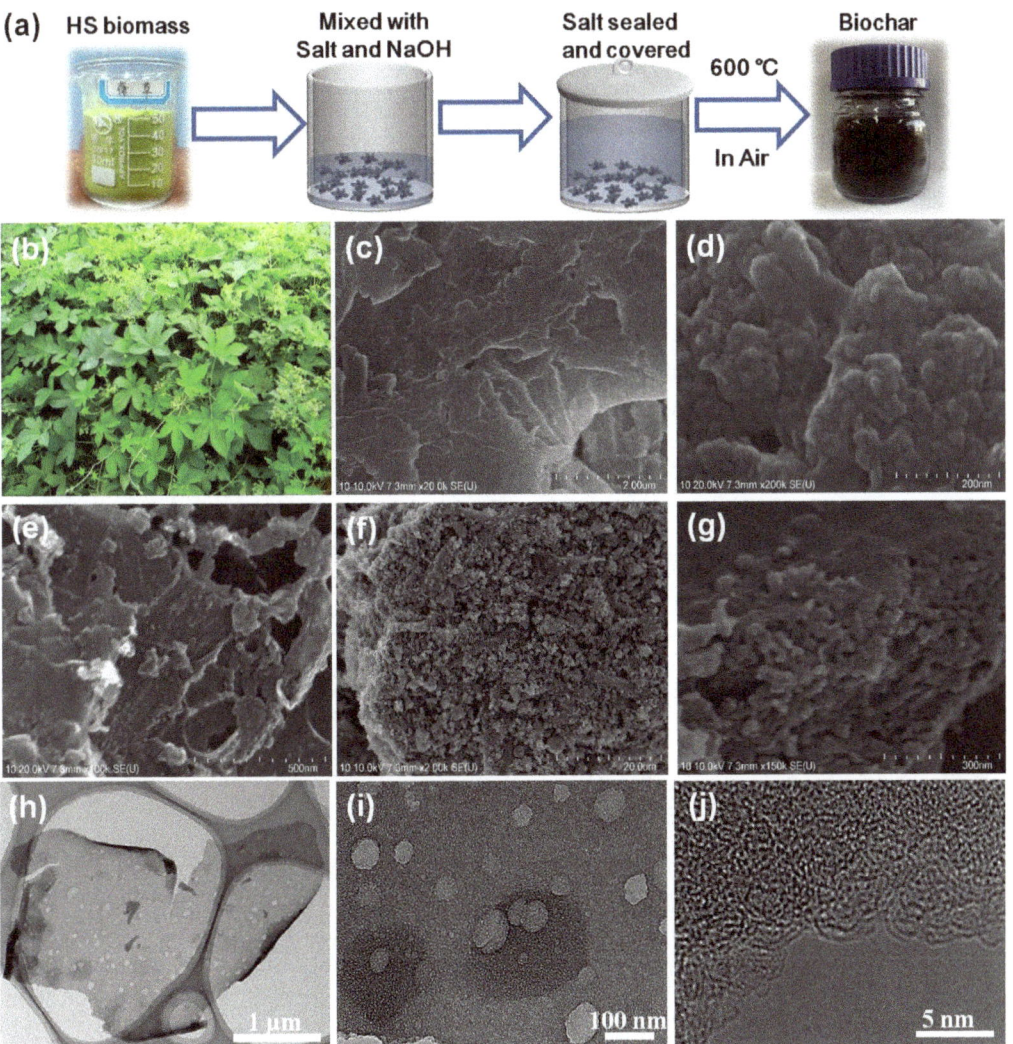

Figure 2. (**a**) Illustration of the preparation process of the biochar, (**b**) HS plant growth in the north of China, the inset is the corresponding ground powders. FE-SEM image of (**c**) the pristine HS biomass, (**d**) HSC-N$_2$, (**e**) HSC-M, and (**f,g**) HSC-MA-2 with different magnifications. (**h,i**) TEM images with different magnifications and the (**j**) HRTEM image.

X-ray photoelectron spectroscopy (XPS) was used to analyze the chemical elements of the biochars. Figure 3a shows the survey analysis of the HSC-N$_2$, HSC-M and HSC-MA-2. There are two distinct peaks corresponding to C1s and O1s at the binding energies of 284.6 and 532.4 eV in the spectrum. In addition, HSC-MA-2 shows a much enhanced content of mineral species such as Na, Mg and Ca. The detailed element compositions are summarized in Table 1.

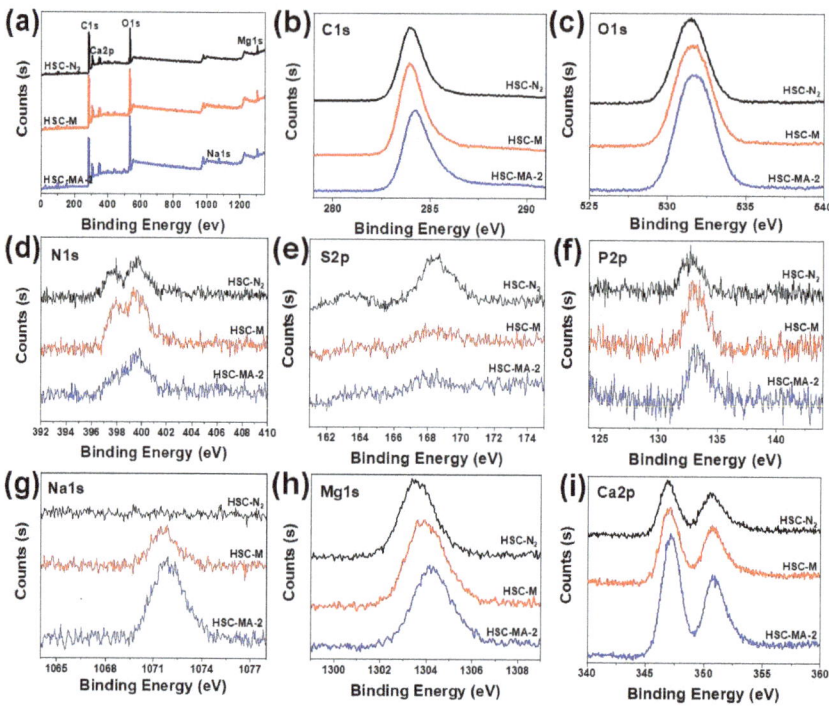

Figure 3. XPS image of the HSC-N$_2$, HSC-M and HSC-MA-2 (a) full spectra, (b) C1s, (c) O1s, (d) N2p, (e) S2p, (f) P2p, (g) Na1s, (h) Mg1s, (i) Ca2p.

Table 1. The element composition of the biochars measured by XPS.

Element	C (%)	O (%)	N (%)	P (%)	S (%)	Na (%)	Mg(%)	Ca (%)
HSC-N$_2$	60.23	24.53	3.69	0.79	1.28	0	2.82	3.28
HSC-M	60.67	24.45	3.65	1.12	0	0.71	2.93	3.36
HSC-MA-2	54.48	28.75	2.02	1.38	0	1.63	5.25	5.22

The C1s peak shown in Figure 3b indicates that HSC-N$_2$ and HSC-M have a similar relative carbon content, while HSC-MA-2 has a relatively low carbon content, suggesting that the etching of the molten salt and NaOH could substantially reduce the carbon sp^2 domain and introduce abundant oxygen motifs. Based on the fitting of the C1s signals shown in Supplementary Materials Figure S5, it can be also observed that the peaks belonging to C-C and C=C (at 284.8 eV and 286.5 eV) are reduced. Meanwhile, the peaks ascribed to C=O and COOH (at 288.7 eV and 290.9 eV) are slightly enhanced, and the increased binding energy suggests that carbon motifs with high valence forms are incorporated in the HSC-MA-2 [32,33].

Moreover, this change can be also detected in the O1s spectrum displayed in Figure 3c. It is obvious that the content of oxygen in HSC-MA-2 is the highest, which indicates that the oxygen functional groups are introduced into the biochar by the NaOH modification reagent. The deconvolution of O 1s spectra shown in Supplementary Materials Figure S6 can be fit into four types: C=O (531.3 eV), OH/RCOOR (532.6 eV), O-C=O (533.7 eV) and COOH (534.9 eV). Compared with that of HSC-N$_2$ and HSC-M, the peak ascribing to COOH is greatly enhanced in HSC-MA-2, indicating that carboxyl groups are successfully introduced into the biochars. This result is in good agreement with the C1s spectrum, and the increased carboxyl group population could in turn enhance the surface coordination effects on the heavy metal ions.

Figure 3d–f shows the contents of N, S and P in biochar. It can be seen that the biochars prepared from HS all contain a certain amount of N, P and S. Considering that no N-, P- and S-containing regents were used in the preparation, these heteroatoms originate from the pristine tissue of the HS biomass. Based on the previous reports, N, P and S with a lone pair of electrons show high affinity with the positively charged heavy metal ions [34]. Although HSC-MA-2 experience rigid MS and NaOH etching, the N and P could still be preserved in the biochar, which could further enhance the surface affinity towards the Pb and Cu ions. As depicted in Figure 3g–i, the modification of NaOH can greatly increase the content of Na, Mg and Ca in the biochar (the detail composition values are summarized in Table 1). These mineral ions are beneficial to the adsorption of heavy metals via the ion exchange process. For these biochars, no signal belonging to Pb and Cu could be detected, as shown in Supplementary Materials Figures S7 and S8.

In order to better study the element content in the biochars, we used inductively coupled plasma mass spectrometry (ICP-MS) to better reveal the metallic elements of the biochars. The detail results are shown in Table 2, and it is clear that the as-measured composition is much higher compared with that measured using XPS. Considering that XPS mainly detects the elements distributed on the surface of the biochars, the different composition results of ICP-MS demonstrate that the majority of the mineral species are embedded inside the biochars. Furthermore, HSC-MA-2 exhibits the highest contents in Na, Mg and Ca, which is in accordance with the XPS measurement.

Table 2. The element composition of the biochar measured by ICP-MS.

Element (mg/g)	Na_{23}	Mg_{24}	Ca_{43}
HS	0.89	8.45	9.26
HSC-N_2	3.48	22.60	17.00
HSC-M	3.19	26.40	23.70
HSC-MA-2	3.33	41.50	30.10

In order to analyze the graphitization degree and crystal phase of the biochars, the biochars were analyzed using laser Raman spectra, as shown in Figure 4a. There are characteristic diffraction peaks at ~1320 and ~1570 cm^{-1}, corresponding to the D band (disordered band) and G band (graphitized band) of the biochars [11]. The D-band is due to the defects and disorder induced by carbon atoms hybridizing, and the G-band is due to the stretching of carbon atoms. In general, the intensity ratio (I_D/I_G) of the D and G bands is used to describe the disorder degree of biochars. The calculated D and G band intensity ratios of HSC-N_2, HSC-M, HSC-MA-1, HSC-MA-2, HSC-MA-3 are 0.98, 0.99, 1.02, 1.04 and 1.01, respectively. The order of I_D/I_G was HSC-N_2 < HSC-M < HSC-MA-3 < HSC-MA-1 < HSC-MA-2, which indicated that the stable sp^2 domain of the carbon skeleton is etched and many surface defects are formed in HSC-MA-2. The increased disorder degree could permit the enhanced adsorption performance of the material due to the high surface energy.

Fourier transform infrared spectrometry (FTIR) is an important technique to identify the functional groups on the surface of adsorbents. As shown in Figure 4b, characteristic peaks belonging to C-O motifs are detected (C-O, 956 cm^{-1} (stretching tensile vibration), 1034 cm^{-1} (OH, hydroxyl stretching vibration), 1124 cm^{-1} (C-O-C, stretching vibration), 1301 cm^{-1} (COOH or CHO bending vibration) and 1560 cm^{-1} (C=O, stretching vibration)). Therefore, the three peaks centered at ~950, ~1000 and ~1300 cm^{-1} could be attributed to the C-O, OH and COOH or CHO groups, respectively. It can be seen that the intensity of oxygen functional group peaks, such as COOH or CHO, are greatly increased in HSC-MA-2 compared with other samples. The increased oxygen functional groups could provide abundant coordinating sites for the heavy metal ions on the biochars, which could greatly facilitate the surface adsorption process. [12,35]

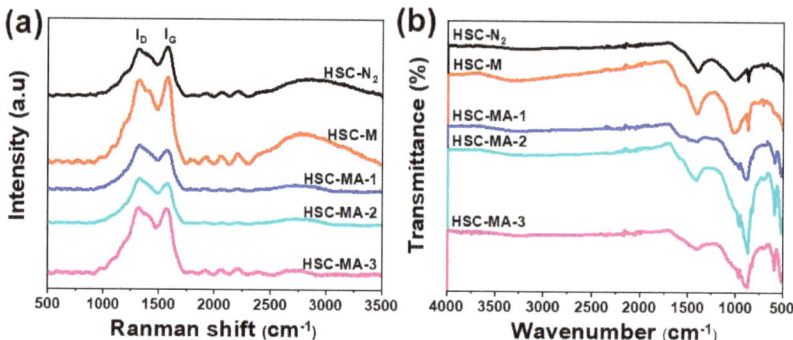

Figure 4. Raman (**a**) and FTIR (**b**) spectra of HSC-N$_2$, HSC-M, HSC-MA-1, HSC-MA-2, and HSC-MA-3.

Surface Area and Pore Structure

The pore structures of HSC-N$_2$, HSC-M and HSC-MA-2 were studied by means of the nitrogen adsorption/desorption method. Figure 5 shows the nitrogen adsorption/desorption isotherms of the biochars, which belongs to the mixture of the type I and type IV isotherms, indicating the co-existence of the micro- and meso-porous structures. The micropore adsorption initially occurs at a low pressure. The amount of nitrogen adsorbed is between 0.4 and 0.9, corresponding to the capillary condensation of meso-pores, which indicates that HSC-MA-2 has a hierarchical porous structure [36]. In the range of relative pressure greater than 0.9, the nitrogen desorption isotherm of the HSC-MA-2 shows a continuous rising trend compared with that of the other two curves, indicating that there are still large pores in the biochar. The results of nitrogen adsorption and desorption show that a lot of micro- and meso-pores with different size distributions can be detected in the biochars prepared by molten salt with NaOH. As listed in Table 3, the BET specific surface area of the biochar experiences a slight reduction from 182.9 to 161.4 and 171.9 m^2/g. However, the overall pore volume and pore diameter increase with the addition of NaOH. The results indicate that the optimal HSC-MA-2 exhibits well-developed pore structures with a greater pore diameter and larger pore volume, which could benefit the mass diffusion and facilitate the adsorption process on the exposed inner surfaces.

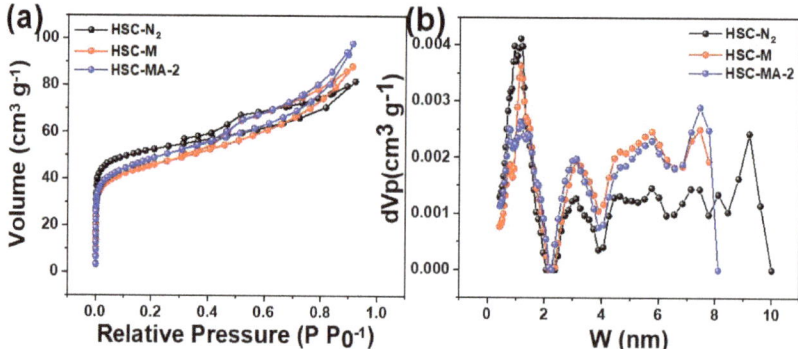

Figure 5. (**a**) Nitrogen adsorption/desorption analysis and the (**b**) pore size distribution of the HSC-N$_2$, HSC-M and HSC-MA-2.

Table 3. BET surface specific area (SSA) and pore structure of biochars.

Sample	SSA (m²/g)	Mean Pore Diameter (nm)	Total Pore Volume (cm³/g)
HSC-N$_2$	182.9	2.77	0.13
HSC-M	161.4	3.38	0.14
HSC-MA-2	171.9	3.52	0.15

3.2. Adsorption Tests

The adsorption of Pb^{2+} and Cu^{2+} on different biochars were studied. In general, the Pb^{2+} adsorption capacity of biochar prepared with HS (Figure 6a) is much higher than that of Cu^{2+}, which indicates that the as-prepared biochar has better affinity to Pb^{2+}. In addition, the adsorption capacity increases with the NaOH amount, and the maximum value was reached when the mass ratio of biomass to NaOH equaled 3/2. As shown in Figure 6b, the pyrolysis temperature was controlled at 400, 500, 600, 700 and 800 °C to study the temperature influences. Obviously, the biochar prepared at 600 °C demonstrated high adsorption capacity, indicating that 600 °C is the optimal pyrolysis temperature for HS biomass.

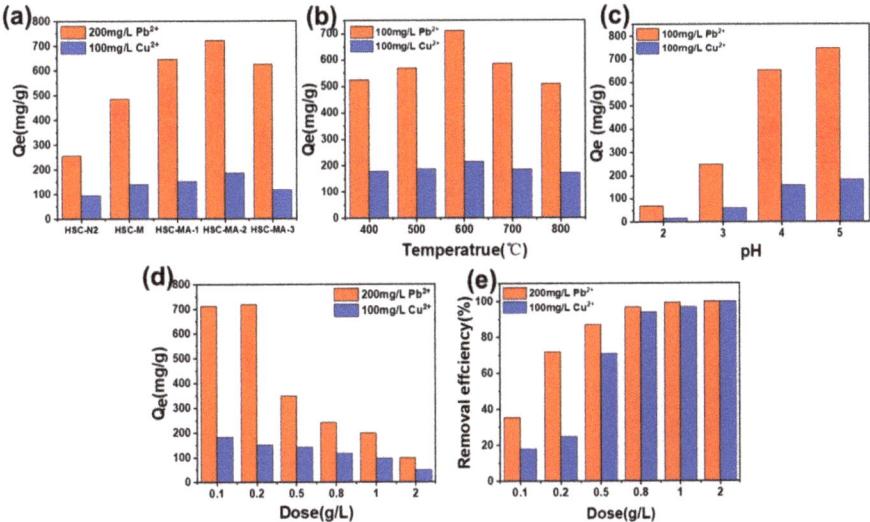

Figure 6. Influence of the modifying agents (a), temperatures (b), pH (c) and the dosage (d,e) on the adsorption capacity and removal efficiency of Pb^{2+} and Cu^{2+}.

It is generally believed that the pH value affects the states of metal ions, the ionization degree of the solution as well as the surface charge of biochars. It is of great significance to study the effects of pH on adsorption performances [37]. To avoid precipitation, the pH's effects on the Pb^{2+} and Cu^{2+} ions adsorption capacity of the biochars were investigated at the pH range of 2.0–5.0 (Figure 6c). The adsorption capacity of the Cu^{2+} and Pb^{2+} increased with increasing pH values. At a low pH, the competition adsorption of H^+ on the surface sites greatly limit the adsorption capacity of Pb^{2+} and Cu^{2+} of the biochar. To make it worse, some minerals in the biochar may also dissolve and release large amounts of cations, such as Ca^{2+} and Mg^{2+}. These released ions may compete with Pb^{2+} and Cu^{2+} for the adsorption sites, hindering the adsorption capacity of heavy metal ions. Thus, the sorption kinetics and sorption isotherm experiments were carried out at a pH of 5.

The effects of HSC-MA-2 with different dosages on the adsorption capacities (Figure 6d) and removal rates (Figure 6e) were studied to determine the optimal dosage. As expected, the adsorption capacity decreased with the increase in the biochar dosage. For Pb^{2+}, the removal efficiency was ~75% when the dosage was 0.2 g/L, and reached above ~90% when the dosage was elevated to 0.5 g/L. For Cu^{2+}, the removal efficiency reached over ~75% when the dosage was 0.5 g/L, and ~90% removal efficiency was achieved when the dosage was 1.0 g/L. Based on these analyses, the best dosages of Pb^{2+} and Cu^{2+} adsorbed by HSC-MA-2 are 0.2 g/L and 0.5 g/L, respectively. Considering that HSC-MA-2 has the best adsorption effect on Pb^{2+} and Cu^{2+}, the subsequent adsorption mechanism studies will focus on HSC-MA-2, adopting HSC-N_2 and HSC-M as references.

3.2.1. Adsorption Kinetics

The adsorption kinetics of Pb^{2+} and Cu^{2+} adsorption on different biochars are shown in Figure 7a–e. Table 4 depicts the kinetic parameters of the adsorption of Pb^{2+} and Cu^{2+}. The correlation coefficient (R^2) derived from the PSO model is much higher than that of the PFO model, which indicates that the adsorption of heavy metal ions is mainly dominated by the chemical adsorption, such as complexation and precipitation [4,5]. The adsorption rate is very fast at the beginning of the adsorption process, and then gradually reduces with the increase in contact time, until the adsorption equilibrium is reached after ~120 min.

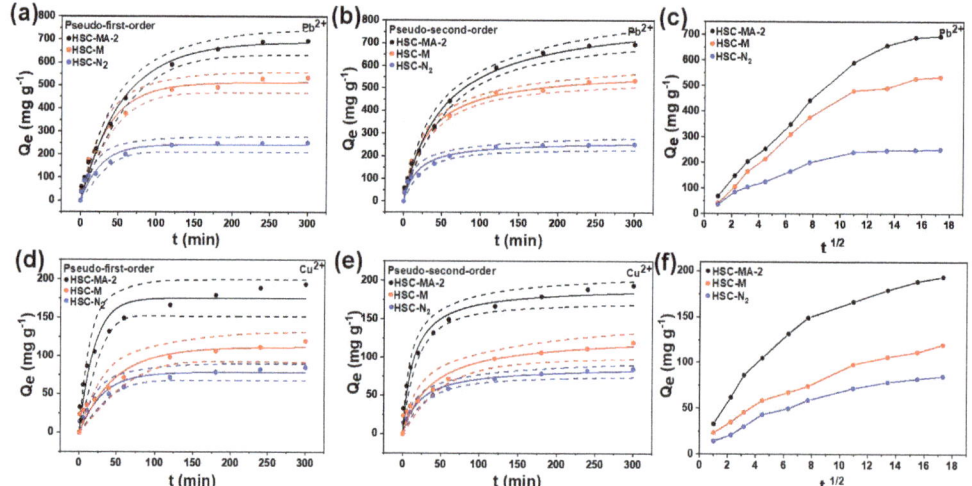

Figure 7. (**a,b**) Adsorption kinetics of Pb^{2+} on HSC-MA-2, HSC-M and HSC-N_2 (linear plot of pseudo-first-order and pseudo-second-order rate equation); (**d,e**) adsorption kinetics of Cu^{2+} on HSC-MA-2, HSC-M and HSC-N_2; intra-particle diffusion model for (**c**) Pb^{2+} and (**f**) Cu^{2+}.

Table 4. Kinetic parameters of Pb^{2+} and Cu^{2+} on HSC-N_2, HSC-M and HSC-MA-2.

	Sample	Pseudo-First-Order			Pseudo-Second-Order		
		R^2	Q_e	K_1	R^2	Q_e	K_2
Pb^{2+}	HSC-MS	0.9870	512.5411	0.2627	0.9943	578.1483	0.0100
	HSC-MS-2	0.9940	687.8283	0.2798	0.9946	729.8656	0.0100
	HSC-N_2	0.9625	242.9751	0.1138	0.9757	266.1626	0.0100
Cu^{2+}	HSC-MS	0.9525	111.7617	0.0208	0.9822	129.4631	0.0100
	HSC-MS-2	0.9609	161.0000	0.0655	0.9664	192.0009	0.0100
	HSC-N_2	0.9636	78.6980	0.0319	0.9819	88.1899	0.0100

The fitted curves of the intra-particle diffusion model are shown in Figure 7c,f. The intra-particle diffusion analysis shows that the adsorption of Pb^{2+} and Cu^{2+} by HSC-MA-2, HSC-M and HSC-N_2 can be divided into three stages. Rapid adsorption rates are observed at the first stage, which results from the presence of many adsorption sites on the biochar surface [24,37–39]. At the second stage, the adsorption is controlled by the intra-particle diffusion rate. Due to the decrease in heavy metal ion concentration and the very low mass transfer rate, the intra-particle diffusion process is slower than the first stage. The third stage is the equilibrium stage, and the adsorption rate is greatly reduced [38]. Based on these observations, the HSC-MA-2 shows the fastest adsorption rate among the samples, indicating that the well-developed porous structure could benefit the intra-particle diffusion, thus enhancing the adsorption kinetics.

3.2.2. Adsorption Isotherm

The effects of the initial concentration of Pb^{2+} and Cu^{2+} on adsorption capacity were investigated. Due to the concentration gradient between the solution and the adsorbent surface, the adsorption capacity increases with the initial concentration of Pb^{2+} and Cu^{2+}. It is generally believed that a higher initial concentration of Pb^{2+} and Cu^{2+} provides a greater driving force to overcome the mass transfer resistance in the solution, which results in a higher adsorption capacity. When the initial concentrations of Pb^{2+} and Cu^{2+} are higher, the adsorption capacity is prone to reach a balance due to the limitation of the maximum adsorption capacities [39]. Freundlich and Langmuir models were used to describe the relationship between adsorption capacity and equilibrium concentration, and the fitting curves are shown in Figure 8. The as-calculated Q_{max}, correlation coefficient (R^2), and other relevant constants are displayed in Table 5. The Langmuir isotherm has higher R^2 than the Freundlich model, which further proves that the chemical adsorption governs the removal process. The Q_{max} of HSC-MA-2 is 748.1 and 221.1 mg/g for Pb^{2+} and Cu^{2+}, respectively. These results are in good agreement with the above experimental results, indicating that the adsorption capacity of the modified biochar for Pb^{2+} and Cu^{2+} is significantly improved compared to the biochar prepared directly by the molten salt sealing method without the modification reagent (594.5 and 155.0 mg/g for Pb^{2+} and Cu^{2+}) or under N_2 protection (310.6 and 110.0 mg/g for Pb^{2+} and Cu^{2+}). These results further indicate that under the conditions of an air atmosphere and high temperature, the etching effects of molten salt and NaOH on carbon products can endow the biochar with high specific surface area and rich oxygen functional groups, which is conducive to the adsorption capacity of heavy metal ions. In addition, as shown in Table 6, the optimized HSC-MA-2 demonstrates much enhanced adsorption capacity compared with that of the previously reported biochars prepared from different biomasses [40–48].

Table 5. Langmuir and Freundlich constants and correlation coefficients (R^2) for Pb^{2+} and Cu^{2+} adsorption on HSC-N_2, HSC-M and HSC-MA-2.

	Sample	Langmuir			Freundlich		
		R^2	Q_m	K_L	R^2	N_f	K_f
Pb^{2+}	HSC-MS	0.9999	594.4691	1.0000	0.9916	20.2338	460.9877
	HSC-MS-2	0.9932	748.0998	1.0000	0.9900	13.0424	502.6869
	HSC-N_2	0.9913	310.6050	0.4741	0.9984	12.0606	210.2611
Cu^{2+}	HSC-MS	0.9814	155.0275	1.0000	0.9622	9.8353	96.2226
	HSC-MS-2	0.9879	221.0786	1.0000	0.9696	12.7665	152.2614
	HSC-N_2	0.9856	110.0282	1.0000	0.9656	8.4288	61.7383

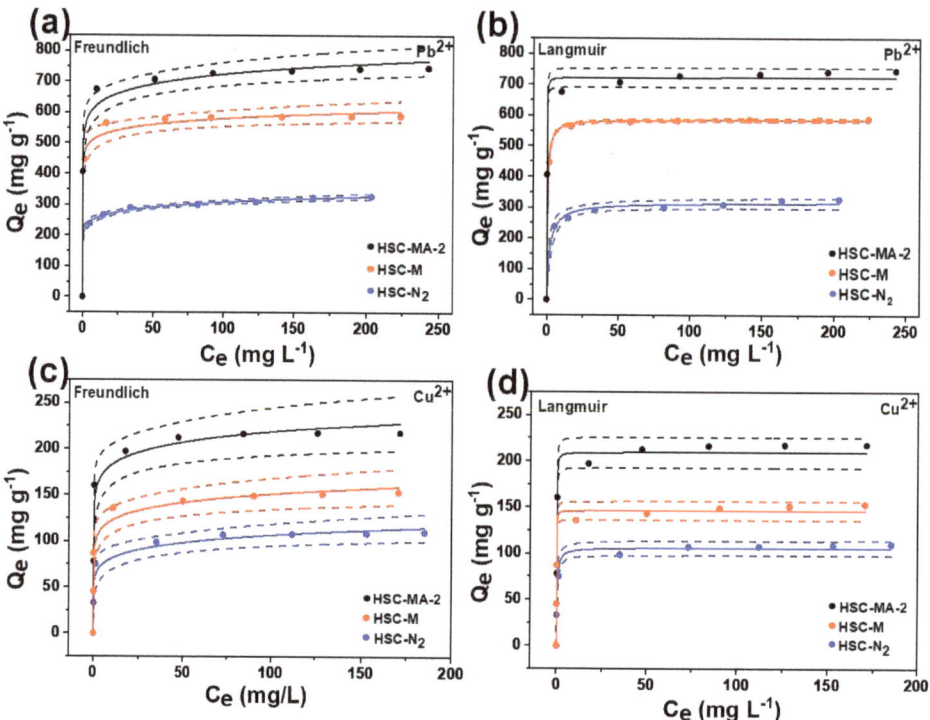

Figure 8. Freundlich sorption isotherms of (**a**) Pb^{2+} and (**c**) Cu^{2+}; and Langmuir sorption isotherms of (**b**) Pb^{2+} and (**d**) Cu^{2+} on the HSC-N$_2$, HSC-M and HSC-MA-2.

3.2.3. Effects of Competition Ions

In the adsorption process, the background ions (Na^+, Mg^{2+} and Ca^{2+}) affect the adsorption of target ions on the biochar. As shown in Figure 9, the results indicate that Na^+ has little effect on the removal of Pb^{2+} and Cu^{2+}, while the presence of Mg^{2+} and Ca^{2+} has a greater influence on the adsorption capacity of HSC-MA-2 compared with Na^+ (monovalent cation). As the concentration of background ions is increased, the removal efficiencies of Pb^{2+} and Cu^{2+} decrease due to the enhanced ion strength of Mg^{2+} and Ca^{2+}, which could be more competitive for the active sites on HSC-MA-2. In addition, the influence of Ca^{2+} on the adsorption of Pb^{2+} and Cu^{2+} is greater than that of Mg^{2+}. The higher valence sates of Mg^{2+} and Ca^{2+} may lead to stronger hydration, which makes Mg-OH more stable than Ca^{2+} in aqueous solutions and thus reduces its interaction with the surface active sites [39]. Therefore, the effect of Mg^{2+} on Pb^{2+} and Cu^{2+} adsorption by HSC-MA-2 is lower than that of Ca^{2+}, but greater than that of Na^+.

Table 6. The comparison on the adsorption performances.

Biomass	Target Ions	Q_{max}	SSA	References
soybean cake	Pb^{2+}	133.60 mg/g	32.7 m^2/g	[22]
Shell	Pb^{2+}	100.25 mg/g	499.2 m^2/g	[24]
Medulla tetrapanacis	Cu^{2+}, Pb^{2+}	Cu^{2+} 458.72 mg/g Pb^{2+} 1031.23 mg/g	246.85 m^2/g	[39]
Enteromorpha	Cu^{2+}, Pb^{2+}	Cu^{2+} 254 mg/g Pb^{2+} 98 mg/g	29.7 m^2/g	[40]
corn stalks	Cu^{2+}	152.61 mg/g	4.46 m^2/g	[41]
Rice husks	Cu^{2+}	265 mg/g	2330 m^2/g	[42]
Palm	Pb^{2+}	118.08 mg/g	Not given	[21]
sugar cane	Pb^{2+}	86.96 mg/g	92.30 m^2/g	[43]
water hyacinths	Pb^{2+}	128.95 mg/g	51.15 m^2/g	[44]
Enteromorpha compressa	Cu^{2+}	137 mg/g	52 m^2/g	[15]
sesame straw	Cu^{2+}, Pb^{2+}	Cu^{2+} 55 mg/kg Pb^{2+} 102 mg/kg	Not given	[45]
fresh banana peels	Cu^{2+}, Pb^{2+}	Cu^{2+} 24.27 mg/g Pb^{2+} 193 mg/g	82.4 m^2/g	[46,47]
corn stalk	Cu^2, Pb^{2+}	Cu^{2+} 161.9 mg/g Pb^{2+} 195.1 mg/g	603.4 m^2/g	[48]

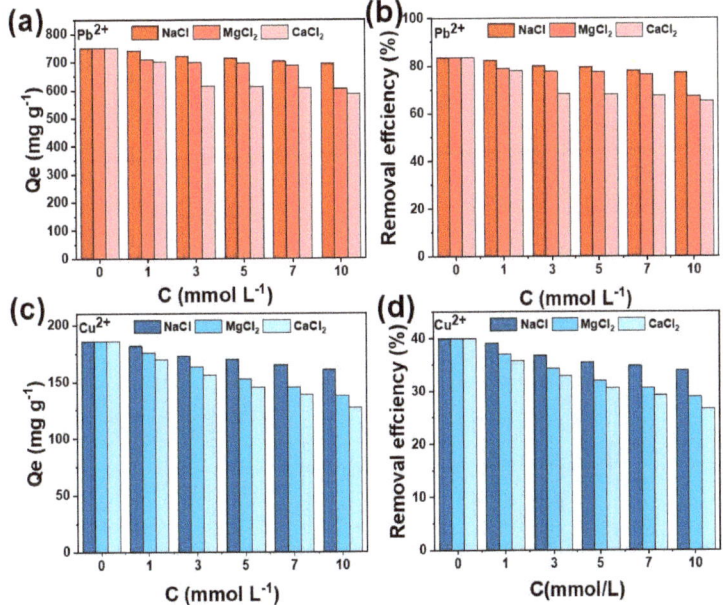

Figure 9. The effect of background ions on the sorption of (a) HSC-MA-2 and (b) HSC-N2 on Pb^{2+}, and (c) HSC-MA-2 and (d) HSC-N2 on Cu^{2+}.

3.2.4. Multi-Metal Adsorption and Column Test

The interaction of coexisting heavy metal ions in the adsorption process has more practical significance than in the single heavy metal ion system. Therefore, the adsorption capacity in the poly-metal ion system was studied (Figure 10a–c). As expected, the removal efficiencies of Pb^{2+}, Cu^{2+}, and Cr^{3+} increase with the increase in the biochar dosage. For HSC-MA-2, when the biochar dosage is increased from 50 to 300 mg, the removal efficiency of all heavy metal ions is enhanced significantly. The removal efficiency of all heavy metal ions decreases with the increase in the concentration of metal ions. As the dosage of HSC-MA-2 is increased to 300 mg, the removal efficiency of all heavy metal ions reaches ~100%. These results suggest that when the metal ions compete for the same adsorption sites, the metal ions with higher affinity replace the other ions with lower affinity, and the most intense competition takes place when there is a high concentration of heavy metal ions. For HSC-MA-2 (200 mg), the removal efficiencies of Pb^{2+}, Cu^{2+} and Cr^{3+} are 98.05%, 94.57% and 96.18%, respectively, when the concentration of all metal ions is set at 150 mg/L. It is evidenced that the presence of Cr^{3+} has a significant effect on the adsorption of Pb^{2+} and Cu^{2+} on the biochar, due to the strong affinity for Cr^{3+}. The experimental results prove that the HSC-MA-2 can effectively take up many types of heavy metal ions at the same time. Based on the abovementioned batch experiments, column adsorption experiments are also carried out in order to investigate the possible application of HSC-MA-2 in large-scale operations for the practical wastewater treatment. The column adsorption experiments were conducted using HSC-MA-2 as the filler (Figure 10d). The simulated wastewater was prepared from the Yellow River; 100 mg/L Cu^{2+}, Pb^{2+} and Cr^{3+} were added, and the average discharge of the column was adjusted to 0.4 mL/min. As the operation time increases, heavy metal ions adsorption gradually decreases. This phenomenon indicates that HSC-MA-2 is gradually saturated regarding heavy metal ions' adsorption capacity. Cu^{2+} and Cr^{3+} become saturated within 600 min, while the transition of Pb^{2+} occurs at 500 min and reaches saturation at 720 min. Generally, the longer the saturation time, the better the adsorption effect of the metal ions. Therefore, in the simulated column adsorption of practical wastewater, the affinity of HSC-MA-2 to different heavy metal ions follows the sequence of Pb^{2+} > Cu^{2+} > Cr^{3+}. Considering the outstanding removal effects for the heavy metal ions, the HSC-MA-2 provides a highly efficient removal rate for the practical wastewater treatment.

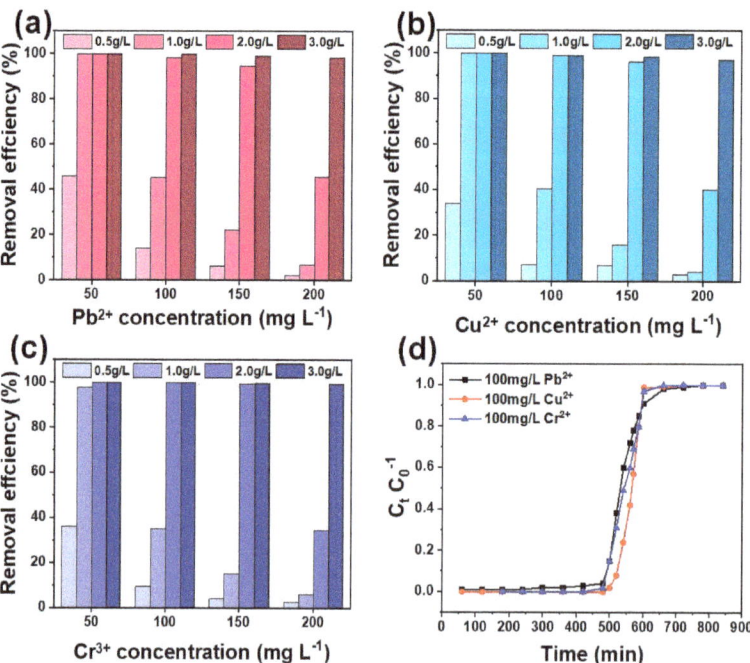

Figure 10. The effects of the absorbent dosages and the initial concentration of the solution coexisting with (**a**) Pb^{2+}, (**b**) Cu^{2+} and (**c**) Cr^{3+} on the removal efficiency of each metal ion, (**d**) kinetics of column adsorption of heavy metal ions in practical wastewater.

3.3. Adsorption Mechanism

As indicated in previous reports, the sorption mechanism probably involved surface complexation, metal precipitation, ion exchange, π–π interactions, and physical sorption [39,42]. In order to investigate the high sorption capacity of HSC-MA-2, the biochars before and after metal ion adsorption were investigated in detail. As shown in Figure 11a, the Raman signal is greatly subtracted after the adsorption of the heavy metal ions, especially for the case of the Pb^{2+} adsorption, which indicates that the signal is greatly affected by the interaction between the biochar and the positively charged heavy metal ions. The reduced I_G signal reflects the deviation of delocalized π electrons in the carbon domain, suggesting the strong π–π interaction between the biochar and the metal ions. In particular, the I_D intensity is much more reduced than that of the I_G, which indicates that the defects remaining in the carbon domain serve as the most effective sites for the adsorption. This phenomenon can be further proved by the FTIR spectra. As depicted in Figure 11b, the absorption peaks centered at ~950, ~1000 and ~1300 cm^{-1} are greatly abated, which is also attributed to the coordination of the oxygen group with the metal ions. Furthermore, the high-resolution XPS spectra on the C1s and O1s are disclosed in Figure 11c,d. The reduced C1s peaks can be explained by the strong interaction on the surface, which results from the heavy adsorption loading of the metal ions on the biochar. In addition, the O1s can be delineated into three peaks, corresponding to the OH, C=O and COOH, respectively. It is obvious that after the Pb^{2+} adsorption, the peaks ascribed to C=O and COOH are greatly reduced, which indicates that the C=O and COOH groups play significance roles in the heavy metal adsorption. Based on the Gauss simulation, COOH and C=O more effectively lead to the charge polarization on the carbon surface, which gives rise to the higher surface affinity towards the metal ions (362.8 kcal/mol) (Figure 11e) compared with the carbon domain (−137.6 eV) and other oxygen functional groups (C-O-C (−175.4 eV), C=O (−176.0 eV), OH (−140.8 eV) and H-C=O (−163.2 eV)). The computation results indicate that the COOH and C=O motifs demonstrate the most enhanced binding energy towards the Pb^{2+}, which agrees well with our experimental findings.

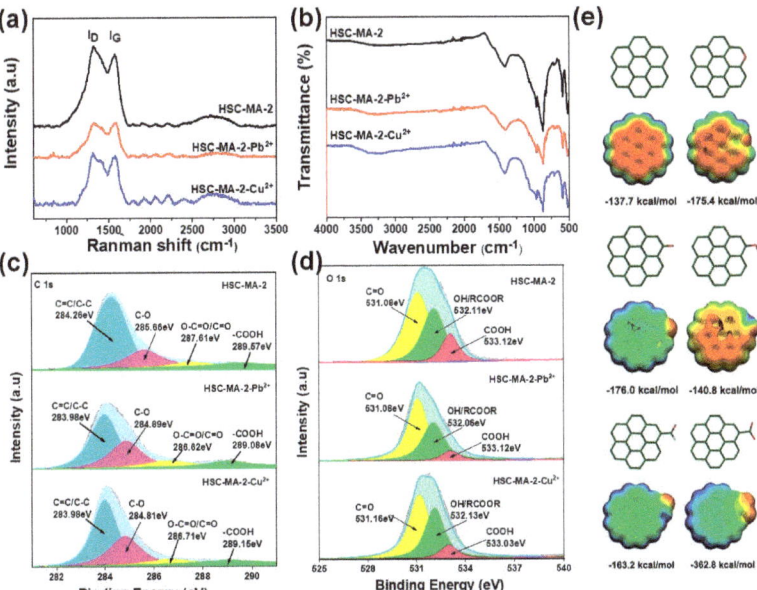

Figure 11. Raman (a), FTIR (b) spectra, high-resolution C1s (c) and O1s (d) spectra of HSC-MA-2 before and after Pb^{2+} and Cu^{2+} adsorption, (e) Gauss simulation on the Pb^{2+} binding energies with different types of oxygen functional groups on the carbon domain.

In order to investigate the ion exchange during the adsorption process, the release of Ca^{2+} and Mg^{2+} ions was monitored during the adsorption process (Figure 12a,b). The Ca^{2+} and Mg^{2+} concentrations experience continuous enhancement with prolonged adsorption times, which suggests that the release of Ca^{2+} and Mg^{2+} is accompanied by the heavy metal ion adsorption. In addition, the amount of Ca^{2+} released is much higher than that of the Mg^{2+}, implying that Ca^{2+} is the most effective mineral species helping the adsorption process. Moreover, the high-resolution XPS spectra shown in Figure 12c,d indicate that after the Pb^{2+} and Cu^{2+} adsorption, the Ca^{2+} and Mg^{2+} content in the spectra are greatly suppressed, especially for the Pb^{2+} adsorption case. Taking into account the reduced Ca^{2+} and Mg^{2+} remaining in the biochar, as well as the enhanced concentration of Ca^{2+} and Mg^{2+} ions in the final adsorption system, it can be concluded that ion exchange plays an important role in facilitating heavy metal ion adsorption.

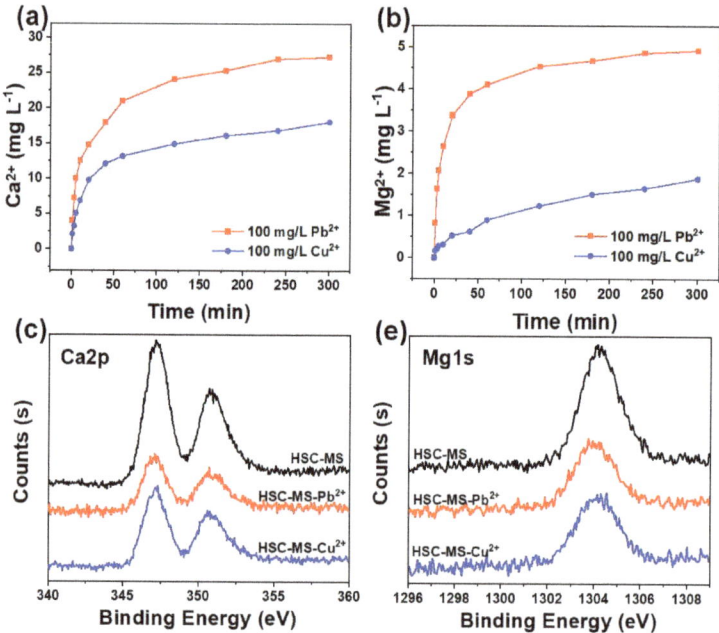

Figure 12. The concentration of Ca^{2+} (a) and Mg^{2+} (b) released in the adsorption process. High-resolution XPS spectra on the (c) Ca2p and (d) Mg1s of HSC-MA-2 before and after Pb^{2+} and Cu^{2+} adsorption.

4. Conclusions

The NaOH-modified molten salt method was adopted to prepare a series of biochars using *Humulus scandens* as a biomass precursor. The optimized biochar HSC-MA-2 exhibits excellent Pb^{2+} and Cu^{2+} adsorption performance, and the Q_m can reach 748 and 221 mg/g, respectively. In addition, HSC-MA-2 shows good anti-interference ability and high removal efficiency in the simulated wastewater using Yellow River water for practical application. The enhanced performance of the HSC-MA-2 can be ascribed to the following qualities of the proposed NaOH-modified molten salt process. Firstly, the synthetic etching effect of the molten salt and the NaOH media introduce a large surface area and a large porous structure to offer abundant adsorption sites as well as mass diffusion channels. Secondly, the penetrating O_2 molecules and the NaOH endow the carbon surface with a large population of the oxygen functional groups such as C=O and COOH, which substantially increase the coordinating strength with the positively charged metal ions. Finally, the

NaOH helps to preserve the pristine mineral species from the biomass in the biochar, which serve as the active centers to adsorb the foreign heavy metal ions through ion exchange. Therefore, this method could serve as a general protocol to produce high-performance biochar for water treatment.

Supplementary Materials: The following are available online at https://www.mdpi.com/article/10.3390/nano11123255/s1, Figure S1 Location of the site where the HS biomass is obtained. (The map is downloaded from an open source website: https://www.google.com/maps/); Figure S2 Location of the site where the Yellow river water is fetched. (The map is downloaded from an open source website: https://www.google.com/maps/); Figure S3 SEM images of HS, HSC-N2, HSC-M; Figure S4 SEM images of HSC-MA-1, HSC-MA-2, HSC-MA-3; Figure S5 detailed XPS analysis on the C1s of the biochars; Figure S6 detailed XPS analysis on the O1s of the biochars; Figure S7 XPS on the Cu2p of the biochar before adsorption; Figure S8 XPS on the Pb4f of the biochar before adsorption.

Author Contributions: Conceptualization, D.W. and X.B.; methodology, X.B.; software, N.M. and N.L.; validation, X.B. and L.X.; formal analysis, D.W., N.L. and K.J.; investigation, X.B.; data curation, N.L., K.H. and M.Y.; writing—original draft preparation, X.B.; writing—review and editing, D.W. and N.L.; supervision, D.W. and K.J.; funding acquisition, D.W. and K.J.; All authors have read and agreed to the published version of the manuscript.

Funding: This work is funded by National Natural Science Foundation of China (51772078), Thousand Talent Project of Henan Province (ZYQR201810115, ZYQR201912167), Outstanding Youth Science Foundation of HTU (2021JQ03).

Institutional Review Board Statement: Not applicable.

Informed Consent Statement: Not applicable.

Data Availability Statement: The detailed data in the study are available from the corresponding authors by request. (Ning Liu, Dapeng Wu and KaiJiang).

Conflicts of Interest: The authors declare no conflict of interest.

References

1. Zhang, J.; Chen, S.; Zhang, H.; Wang, X. Removal behaviors and mechanisms of hexavalent chromiun from aqueous solution by cephalosporin residue and derived chars. *Bioresour. Technol.* **2017**, *238*, 484–491. [CrossRef]
2. Liu, P.; Rao, D.; Zou, L.; Teng, Y.; Yu, H. Capacity and potential mechanisms of Cd(II) adsorption from aqueous solution by blue algae-derived biochars. *Sci. Total Environ.* **2021**, *767*, 145447. [CrossRef] [PubMed]
3. Rodriguez-Narvaez, O.M.; Peralta-Hernandez, J.M.; Goonetilleke, A.; Bandala, E.R. Biochar-supported nanomaterials for environmental applications. *J. Ind. Eng. Chem.* **2019**, *78*, 21–33. [CrossRef]
4. Meseldžija, S.; Petrovic, J.; Onjia, A.; Volkov-Husovic, T.; Nesic, A.; Vukelic, N. Utilization of agro-industrial waste for removal of copper ions from aqueous solutions and mining-wastewater. *J. Ind. Eng. Chem.* **2019**, *75*, 246–252. [CrossRef]
5. Li, H.; Dong, X.; da Silva, E.B.; de Oliveira, L.M.; Chen, Y.; Ma, L.Q. Mechanisms of metal sorption by biochars: Biochar characteristics and modifications. *Chemosphere* **2017**, *178*, 466–478. [CrossRef]
6. Hong, N.; Cheng, Q.; Goonetilleke, A.; Bandala, E.R.; Liu, A. Assessing the effect of surface hydrophobicity/hydrophilicity on pollutant leaching potential of biochar in water treatment. *J. Ind. Eng. Chem.* **2020**, *89*, 222–232. [CrossRef]
7. Hopkins, D.; Hawboldt, K. Biochar for the removal of metals from solution: A review of lignocellulosic and novel marine feedstocks. *J. Environ. Chem. Eng.* **2020**, *8*, 103975. [CrossRef]
8. Zhang, Y.P.; Adi, V.S.K.; Huang, H.-L.; Lin, H.-P.; Huang, Z.-H. Adsorption of metal ions with biochars derived from biomass wastes in a fixed column: Adsorption isotherm and process simulation. *J. Ind. Eng. Chem.* **2019**, *76*, 240–244. [CrossRef]
9. Zhang, L.X.; Tang, S.Y.; He, Y.; Liu, Y.; Mao, W.; Guan, Y.T. Highly efficient and selective capture of heavy metals by poly(acrylic acid) grafted chitosan and biochar composite for wastewater treatment. *Chem. Eng. J.* **2019**, *378*, 122215. [CrossRef]
10. Wang, J.; Lei, S.; Liang, L. Preparation of porous activated carbon from semi-coke by high temperature activation with KOH for the high-efficiency adsorption of aqueous tetracycline. *Appl. Surf. Sci.* **2020**, *530*, 147187. [CrossRef]
11. Takdastan, A.; Samarbaf, S.; Tahmasebi, Y.; Alavi, N.; Babaei, A.A. Alkali modified oak waste residues as a cost-effective adsorbent for enhanced removal of cadmium from water: Isotherm, kinetic, thermodynamic and artificial neural network modeling. *J. Ind. Eng. Chem.* **2019**, *78*, 352–363. [CrossRef]
12. Jia, X.X.; Zhang, Y.; He, Z.; Chang, F.; Zhang, H.; Wågberg, T.; Hu, G. Mesopore-rich badam-shell biochar for efficient adsorption of Cr(VI) from aqueous solution. *J. Environ. Chem. Eng.* **2021**, *9*, 105634. [CrossRef]
13. Tripathi, M.; Sahu, J.N.; Ganesan, P. Effect of process parameters on production of biochar from biomass waste through pyrolysis: A review. *Renew. Sustain. Energy Rev.* **2016**, *55*, 467–481. [CrossRef]

14. Lee, J.C.; Kim, H.J.; Kim, H.W.; Lim, H. Iron-impregnated spent coffee ground biochar for enhanced degradation of methylene blue during cold plasma application. *J. Ind. Eng. Chem.* **2021**, *98*, 383–388. [CrossRef]
15. Kim, B.-S.; Lee, H.W.; Park, S.H.; Baek, K.; Jeon, J.-K.; Cho, H.J.; Jung, S.-C.; Kim, S.C.; Park, Y.-K. Removal of Cu^{2+} by biochars derived from green macroalgae. *Environ. Sci. Pollut. Res.* **2015**, *23*, 985–994. [CrossRef]
16. Lang, J.; Matejova, L.; Cuentas-Gallegos, A.K.; Lobato-Peralta, D.R.; Ainassaari, K.; Gomez, M.M.; Solis, J.L.; Mondal, D.; Keiski, R.L.; Cruz, G.J.F. Evaluation and selection of biochars and hydrochars derived from agricultural wastes for the use as adsorbent and energy storage materials. *J. Environ. Chem. Eng.* **2021**, *9*, 105979. [CrossRef]
17. Wang, S.W.; Zhong, S.; Zheng, X.Y.; Xiao, D.; Zheng, L.L.; Yang, Y.; Zhang, H.D.; Ai, B.L.; Sheng, Z.W. Calcite modification of agricultural waste biochar highly improves the adsorption of Cu(II) from aqueous solutions. *J. Environ. Chem. Eng.* **2021**, *9*, 106215. [CrossRef]
18. Leng, L.J.; Xiong, Q.; Yang, L.H.; Li, H.; Zhou, Y.; Zhang, W.; Jiang, S.; Li, H.; Huang, H. An overview on engineering the surface area and porosity of biochar. *Sci. Total Environ.* **2020**, *763*, 144204. [CrossRef]
19. Cho, H.J.; Baek, K.; Jeon, J.-K.; Park, S.H.; Suh, D.J.; Park, Y.-K. Removal characteristics of copper by marine macro-algae-derived chars. *Chem. Eng. J.* **2013**, *217*, 205–211. [CrossRef]
20. Güzel, F.; Cumali, Y. Synthesis, characterization, and lead (II) sorption performance of a new magnetic separable composite: $MnFe_2O_4@$ wild plants-derived biochar. *J. Environ. Chem. Eng.* **2021**, *9*, 104567.
21. Zhao, H.-T.; Ma, S.; Zheng, S.-Y.; Han, S.-W.; Yao, F.-X.; Wang, X.-Z.; Wang, S.-S.; Feng, K. β–cyclodextrin functionalized biochars as novel sorbents for high-performance of Pb2+ removal. *J. Hazard. Mater.* **2018**, *362*, 206–213. [CrossRef] [PubMed]
22. Yin, W.; Dai, D.; Hou, J.; Wang, S.; Wu, X.; Wang, X. Hierarchical porous biochar-based functional materials derived from biowaste for Pb(II) removal. *Appl. Surf. Sci.* **2018**, *465*, 297–302. [CrossRef]
23. Cruz, G.J.F.; Mondal, D.; Rimaycuna, J.; Soukup, K.; Gómez, M.M.; Solis, J.L.; Lang, J. Agrowaste derived biochars impregnated with ZnO for removal of arsenic and lead in water. *J. Environ. Chem. Eng.* **2020**, *8*, 103800. [CrossRef]
24. Li, M.; Wei, D.; Liu, T.; Liu, Y.; Yan, L.; Wei, Q.; Du, B.; Xu, W. EDTA functionalized magnetic biochar for Pb(II) removal: Adsorption performance, mechanism and SVM model prediction. *Sep. Purif. Technol.* **2019**, *227*, 115696. [CrossRef]
25. Zhu, L.; Tong, L.; Zhao, N.; Wang, X.; Yang, X.; Lv, Y. Key factors and microscopic mechanisms controlling adsorption of cadmium by surface oxidized and aminated biochars. *J. Hazard. Mater.* **2019**, *382*, 121002. [CrossRef] [PubMed]
26. Luo, J.; Jin, M.; Ye, L.; Cao, Y.; Yan, Y.; Du, R.; Yoshiie, R.; Ueki, Y.; Naruse, I.; Lin, C.; et al. Removal of gaseous elemental mercury by hydrogen chloride non-thermal plasma modified biochar. *J. Hazard. Mater.* **2019**, *377*, 132–141. [CrossRef]
27. Liatsou, I.; Pashalidis, I.; Dosche, C. Cu(II) adsorption on 2-thiouracil-modified Luffa cylindrica biochar fibres from artificial and real samples, and competition reactions with U(VI). *J. Hazard. Mater.* **2020**, *383*, 120950. [CrossRef]
28. Li, J.; Tian, L.; Liang, F.; Wang, J.; Han, L.; Zhang, J.; Ge, S.; Dong, L.; Zhang, H.; Zhang, S. Molten salt synthesis of hierarchical porous N-doped carbon submicrospheres for multifunctional applications: High performance supercapacitor, dye removal and CO_2 capture. *Carbon* **2018**, *141*, 739–747. [CrossRef]
29. Li, K.; Tang, D.; Zhang, W.; Qiao, Z.; Liu, Y.; Huo, Q.; Liang, D.; Zhu, J.; Zhao, Z. Molten salt synthesis of Co-entrapped, N-doped porous carbon as efficient hydrogen evolving electrocatalysts. *Mater. Lett.* **2017**, *209*, 256–259. [CrossRef]
30. Wang, C.; Wu, D.; Wang, H.; Gao, Z.; Xu, F.; Jiang, K. A green and scalable route to yield porous carbon sheets from biomass for supercapacitors with high capacity. *J. Mater. Chem. A* **2017**, *6*, 1244–1254. [CrossRef]
31. Tian, L.; Li, J.; Liang, F.; Wang, J.; Li, S.; Zhang, H.; Zhang, S. Molten salt synthesis of tetragonal carbon nitride hollow tubes and their application for removal of pollutants from wastewater. *Appl. Catal. B Environ.* **2018**, *225*, 307–313. [CrossRef]
32. Wu, D.; Liu, J.; Chen, J.; Li, H.; Cao, R.; Zhang, W.; Gao, Z.; Jiang, K. Promoting sulphur conversion chemistry with tri-modal porous N, O-codoped carbon for stable Li-S batteries. *J. Mater. Chem. A* **2021**, *9*, 5497–5506. [CrossRef]
33. Li, H.; Wang, Y.; Chen, H.; Niu, B.; Zhang, W.; Wu, D. Synergistic mediation of polysulfide immobilization and conversion by a catalytic and dual-adsorptive system for high performance lithium-sulfur batteries. *Chem. Eng. J.* **2020**, *406*, 126802. [CrossRef]
34. Jiang, H.; Yang, Y.; Lin, Z.; Zhao, B.; Wang, J.; Xie, J.; Zhang, A. Preparation of a novel bio-adsorbent of sodium alginate grafted polyacrylamide/graphene oxide hydrogel for the adsorption of heavy metal ion. *Sci. Total. Environ.* **2020**, *744*, 140653. [CrossRef] [PubMed]
35. Wang, F.; Jin, L.; Guo, C.; Min, L.; Zhang, P.; Sun, H.; Zhu, H.; Zhang, C. Enhanced heavy metals sorption by modified biochars derived from pig manure. *Sci. Total. Environ.* **2021**, *786*, 147595. [CrossRef]
36. Kılıç, M.; Kırbıyık, C.; Çepelioğullar, Ö.; Pütün, A.E. Adsorption of heavy metal ions from aqueous solutions by bio-char, a by-product of pyrolysis. *Appl. Surf. Sci.* **2013**, *283*, 856–862. [CrossRef]
37. Zhou, X.; Zhou, J.; Liu, Y.; Guo, J.; Ren, J.; Zhou, F. Preparation of iminodiacetic acid-modified magnetic biochar by carbonization, magnetization and functional modification for Cd(II) removal in water. *Fuel* **2018**, *233*, 469–479. [CrossRef]
38. Mahdi, Z.; Yu, Q.J.; El Hanandeh, A. Investigation of the kinetics and mechanisms of nickel and copper ions adsorption from aqueous solutions by date seed derived biochar. *J. Environ. Chem. Eng.* **2018**, *6*, 1171–1181. [CrossRef]
39. Zhang, L.; Li, W.; Cao, H.; Hu, D.; Chen, X.; Guan, Y.; Tang, J.; Gao, H. Ultra-efficient sorption of Cu^{2+} and Pb^{2+} ions by light biochar derived from Medulla tetrapanacis. *Bioresour. Technol.* **2019**, *291*, 121818. [CrossRef]
40. Yang, W.; Wang, Z.; Song, S.; Han, J.; Chen, H.; Wang, X.; Sun, R.; Cheng, J. Adsorption of copper(II) and lead(II) from seawater using hydrothermal biochar derived from Enteromorpha. *Mar. Pollut. Bull.* **2019**, *149*, 110586. [CrossRef]

41. Liu, J.; Cheng, W.; Yang, X.; Bao, Y. Modification of biochar with silicon by one-step sintering and understanding of adsorption mechanism on copper ions. *Sci. Total. Environ.* **2019**, *704*, 135252. [CrossRef] [PubMed]
42. Dinh Viet, C.; Liu, N.-L.; Viet Anh, N.; Hou, C.-H. Meso/micropore-controlled hierarchical porous carbon derived from activated biochar as a high-performance adsorbent for copper removal. *Sci. Total Environ.* **2019**, *692*, 844–853.
43. Abdelhafez, A.A.; Li, J. Removal of Pb(II) from aqueous solution by using biochars derived from sugar cane bagasse and orange peel. *J. Taiwan Inst. Chem. Eng.* **2016**, *61*, 367–375. [CrossRef]
44. Ding, Y.; Liu, Y.; Liu, S.; Li, Z.; Tan, X.; Huang, X.; Zeng, G.; Zhou, Y.; Zheng, B.; Cai, X. Competitive removal of Cd(ii) and Pb(ii) by biochars produced from water hyacinths: Performance and mechanism. *RSC Adv.* **2016**, *6*, 5223–5232. [CrossRef]
45. Park, J.-H.; Ok, Y.S.; Kim, S.-H.; Cho, J.-S.; Heo, J.-S.; Delaune, R.D.; Seo, D.-C. Competitive adsorption of heavy metals onto sesame straw biochar in aqueous solutions. *Chemosphere* **2016**, *142*, 77–83. [CrossRef]
46. Zhou, B.; Wang, Z.; Shen, D.; Shen, F.; Wu, C.; Xiao, R. Low cost earthworm manure-derived carbon material for the adsorption of Cu^{2+} from aqueous solution: Impact of pyrolysis temperature. *Ecol. Eng.* **2016**, *98*, 189–195. [CrossRef]
47. Zhou, N.; Chen, H.; Xi, J.; Yao, D.; Zhou, Z.; Tian, Y.; Lu, X. Biochars with excellent Pb(II) adsorption property produced from fresh and dehydrated banana peels via hydrothermal carbonization. *Bioresour. Technol.* **2017**, *232*, 204–210. [CrossRef]
48. Yang, F.; Zhang, S.; Sun, Y.; Cheng, K.; Li, J.-S.; Tsang, D.C. Fabrication and characterization of hydrophilic corn stalk biochar-supported nanoscale zero-valent iron composites for efficient metal removal. *Bioresour. Technol.* **2018**, *265*, 490–497. [CrossRef]

Article

Co-Immobilization of Lactase and Glucose Isomerase on the Novel g-C$_3$N$_4$/CF Composite Carrier for Lactulose Production

Le Wang [1], Bingyu Jiao [1], Yan Shen [1], Rong Du [1], Qipeng Yuan [2,*] and Jinshui Wang [1,*]

- [1] National Engineering Laboratory for Wheat & Corn Further Processing, College of Biological Engineering, Henan University of Technology, Zhengzhou 450001, China
- [2] State Key Laboratory of Chemical Resource Engineering, Beijing University of Chemical Technology, Beijing 100029, China
- * Correspondence: y0100300@163.com (Q.Y.); tougaoyx123@126.com (J.W.); Tel./Fax.: +86-010-6445-1781 (Q.Y.); +86-0371-6775-6513 (J.W.)

Abstract: The g-C$_3$N$_4$/CF composite carrier was prepared by ultrasound-assisted maceration and high-temperature calcination. The enzyme immobilization using the g-C$_3$N$_4$/CF as the novel carrier to immobilize lactase and glucose isomerase was enhanced for lactulose production. The carbon fiber (CF) was mixed with melamine powder in the mass ratio of 1:8. The g-C$_3$N$_4$/CF composite carrier was obtained by calcination at 550 °C for 3 h. After the analysis of characteristics, the g-C$_3$N$_4$/CF was successfully composited with the carbon nitride and CF, displaying the improvement of co-immobilization efficiency with the positive effects on the stability of the enzyme. The immobilization efficiency of the co-immobilized enzyme was 37% by the novel carrier of g-C$_3$N$_4$/CF, with the enzyme activity of 13.89 U g^{-1} at 60 °C. The relative activities of co-immobilized enzymes maintained much more steadily at the wider pH and higher temperature than those of the free dual enzymes, respectively. In the multi-batches of lactulose production, the relative conversion rates in enzymes co-immobilized by the composite carrier were higher than that of the free enzymes during the first four batches, as well as maintaining about a 90% relative conversation rate after the sixth batch. This study provides a novel method for the application of g-C$_3$N$_4$/CF in the field of immobilizing enzymes for the production of lactulose.

Keywords: g-C$_3$N$_4$/CF; carbon fiber; co-immobilization; lactulose; enzyme stability

1. Introduction

Compared with traditional catalysts, biological enzymes have attracted more and more attention due to their high catalytic activity, high substrate specificity, and mild catalytic conditions, and have been widely used in biotechnology and industrial fields [1].

However, the poor stability of free enzymes and the difficulty of recovery bring some challenges to large-scale industrial production. The fragile nature of free enzymes and the narrow operating temperature and pH range limit their commercial use [2]. Immobilization has been reported to be an effective strategy for improving enzymes' stability and achieving low-cost recovery [3]. The choice of carrier material has a great influence on the catalytic effect of immobilized enzymes [4,5]. Adsorption is the fastest and most common method for immobilizing enzymes [6], and is used to immobilize laccase, resulting in improved preservation and pH stability, which maintained about 40% relative viability (4 °C, 30 d) and more than 50% relative viability in the pH range of 2.0–6.0 [7]. The co-immobilized dual-enzyme biocatalyst was successfully prepared by immobilizing horseradish peroxidase (HRP) and glucose oxidase (GOD) in a covalently bound manner on dopamine (DA)-modified cellulose (Ce)-chitosan (Cs) composite spheres, with the degradation rate of the co-immobilized dual enzyme biocatalyst remaining at 61.2% after 6 batches of degradation [8]. The old yellow enzyme (OYE3) was immobilized by acetal-agarose covalent binding (OYE3-GA) and affinity adsorption of EziGTM particles (OYE3-EziG). In the bioreduction reaction of α-methyl-trans-cinnamaldehyde, the OYE3-GA could be recycled for up to 12 reaction cycles, with a maximum conversion of 40% after 12 cycles [9]. However, immobilized enzymes also had some drawbacks [10]; for instance, the enzymatic activity and catalytic reaction rate of the immobilized enzymes were lower than the free enzymes. This is maybe due to the fact that the immobilization process changes the natural spatial structure of the enzymes' molecules [11] and reduces the freedom of movement of the immobilized enzymes' molecules, thus reducing the effective collision between the enzymes and the substrate. Moreover, the microenvironments around the enzymes' molecules change [12]. When multiple enzymes are immobilized, their active sites may be affected by various factors [13], such as the selection of the carrier [14,15], the immobilization method of the enzyme [6], changes in the microenvironment around the enzyme molecule [16], the reaction conditions [17], and others. In addition, the bioactivity of proteins may be affected by the stability of primary and higher structures, including the secondary, tertiary, and quaternary structures. Therefore, in the field of immobilized enzymes, various methods had been devised to preserve the conformation of proteins and thus avoid loss of activity. As the key determinant of immobilization technology, the selection of the immobilized carrier was very important.

Carbon fiber (CF) has become a common immobilized carrier material because of its high physical and chemical stability, good biocompatibility, and abundant active functional groups. Currently, CF is used for the immobilization of various enzymes, such as lipase [18], laccase [19], glucose oxidase, and so on [20]. However, the untreated CF had a smooth surface, inertia, and weak adsorption with enzymes. In order to further expand the application of CF in the immobilization field, researchers often modify it or prepare composite carriers. The application of composite carriers in immobilization technology has also been reported, such as CF/graphene [21], CF/carbon nanotubes, etc. [22].

Graphite phase carbon nitride (g-C_3N_4) is the most stable allotrope of carbon nitride, with excellent thermal and chemical stability [23]. In addition, the raw materials used in the preparation of g-C_3N_4, such as melamine, dicyandiamide, melamine, urea, etc., are inexpensive [10]. Therefore, a g-C_3N_4/CF composite carrier had the highest practical application value. The amino group at the edge of g-C_3N_4 can bind to enzymes by physical or covalent bonding to form a more stable complex [1]. In addition, g-C_3N_4 has high biocompatibility and a surface-active center, making it an ideal material for the immobilization of enzymes [24]. Glucose isomerase was immobilized on g-C_3N_4 nanosheets to prepare a novel sensitive biosensor for glucose detection under neutral conditions [25]. Glucose isomerase immobilized on a glass-carbon electrode was modified with an Au-g-C_3N_4 nanocomposite (Gox/Au-g-C_3N_4), establishing a sensing platform in the presence of luminal to detect glucose in samples [26].

g-C_3N_4/CF composite carriers have attracted extensive attention from researchers, but are currently mainly used in the field of photocatalysis [23,27,28], while there are few reports in the field of immobilization. It is noteworthy that the introduction of g-C_3N_4 could greatly improve the surface roughness, functional groups, and wettability of CF [29], which is conducive to the immobilization of enzymes.

Two or more enzymes were fixed on the same carrier to produce a cascade reaction that facilitates product separation and simplifies catalyst reuse [30]. In addition, due to the co-immobilization of different catalysts on the same carrier, the diffusion limitation of intermediates between active sites was reduced [31], which improves the overall reaction efficiency in the one-pot method [32]. Compared with the single immobilized enzyme, the overall performance of the immobilized double enzymes was improved [33]. The co-immobilization of enzymes could reduce mass transfer resistance and improve enzyme activity through effective substrate channels, and enhance reusability and stability [33]. The cellulose and glucose oxidase were co-immobilized on GO by covalent bonding, which showed that the co-immobilized double enzymes retained about 65% of their initial activity after seven repetitions [34].

Lactulose is a disaccharide synthesized from galactose and fructose, which does not exist in nature [35], and which has received increasing attention for its significant health-promoting effects [36]. Lactulose can be broken down and utilized by beneficial bacteria in the colon, and is widely added to processed foods as a probiotic [37]. It can also be used in the pharmaceutical industry for the treatment of hepatic encephalopathy and constipation [38]. As a high-value product, lactulose has many physiological functions, such as effectively promoting the proliferation of intestinal bacteria, inhibiting the growth of potential pathogenic bacteria, increasing the production of beneficial metabolites, and enhancing the absorption of minerals by the intestinal tract [39]. Lactulose has a wide range of applications in the food, pharmaceutical, and health food industries [40].

At present, the synthesis of lactulose mainly includes chemical and biological methods. Electrolysis is an effective technique to produce lactose from whey lactose [41]. The application of electroactivation (EA) technology for isomerization of lactose to lactofructose in an EA reactor is conditioned by anion and cation exchange membranes [42]. The glutaraldehyde-activated chitosan was used as a carrier to immobilize lactase, which resulted in an efficient and stable catalyst for the synthesis of lactofructose using cheese whey and fructose as substrates [43]. Lactulose was synthesized from fructose and lactose in continuous packed-bed reactor operation with glyoxyl-agarose immobilized *Aspergillus oryzae* β-galactosidase [44]. Lactulose synthesis with biological enzymes could overcome some disadvantages of industrial chemical synthesis, such as the degradation of lactulose, side reaction, purification of late products, and so on [45]. Lactose can be converted into galactose and glucose under the catalysis of lactase [46]. Glucose was isomerized to fructose by glucose isomerase, then fructose and galactose were linked by a β-1, 4 glycosidic bond to synthesize lactulose under the catalysis of lactase [47,48]. Glucose isomerase isomerized glucose to fructose, which reduced its inhibition on lactose hydrolysis and accelerated the process of lactose hydrolysis-glucose isomerization [49]. However, using lactose as a single substrate, the synthesis of lactulose by lactase and glucose isomerase will lead to a low lactulose yield [50]. In the initial stage of the reaction, the fructose substrate was insufficient. In the later stage, when the fructose content accumulated to the transglycoside reaction suitable for lactase, the lactose content decreased greatly and the transglycoside reaction was difficult to continue, resulting in a low lactulose content. Consequently, the appropriate amount of fructose could be added at the initial stage of the reaction to increase lactulose yield.

In our research, the novel carriers, using g-C_3N_4/CF and different amounts of melamine and CF, were prepared for the co-immobilization of lactose and isomeric glucose, as shown in Figure 1. The characterizations of composite carriers were performed to investigate the physicochemical properties of g-C_3N_4/CF for co-immobilization. Moreover, the effects of g-C_3N_4/CF on the co-immobilization of lactase and glucose isomerase were elucidated for the lactulose production.

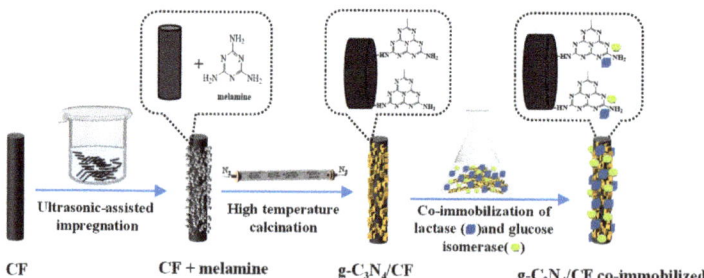

Figure 1. Preparation process diagram of the g-C_3N_4/CF composite carriers for co-immobilization. The acetone-treated CF was ultrasonically impregnated in melamine saturated solution for 1 h, dried, then their surfaces were covered with melamine powder uniformly according to different ratios. After that, they were calcined at 550 °C for 3 h under the nitrogen protection and cooled to produce a

g-C$_3$N$_4$/CF composite carrier. The g-C$_3$N$_4$/CF composite carrier was combined with enzyme molecules and finally the co-immobilization of enzymes with the g-C$_3$N$_4$/CF as the novel composite carrier was obtained.

2. Materials and Methods

2.1. Chemicals and Reagents

CF was purchased from Shanghai Yingjie Special Fiber Materials Co., Ltd. (Shanghai, China). Melamine (AR) was purchased from Tianjin Dingsheng Xin Chemical Co., Ltd. (Tianjin, China). Acetone (AR) and Coomassie Brilliant Blue (AR) were purchased from Sinopharm Group Chemical Reagent Co., Ltd., (Shanghai, China). Lactase was purchased from Harbin Meihua Biotechnology Co., Ltd., (Harbin, China). Glucose isomerase was purchased from Genenko Bioengineering Co., Ltd., (Wuxi, China). Bovine serum albumin (AR) was purchased from Shanghai Zhanyun Chemical Co., Ltd. (Shanghai, China). Lufluorescein isothiocyanate (FITC, $\geq$95%) was purchased from Biosharp Biotechnology Co., Ltd. (Hefei, China). O-nitrophenol (99%) was purchased from Shanghai Maclin Biochemical Technology Co., Ltd., (Shanghai, China). O-Nitrophenyl-β-D-galactopyranoside (ONPG AR) purchased from Shanghai Baoman Biotechnology Co., Ltd. (Shanghai, China). Fuctose (AR), lactose (AR), and rhodamine B, were purchased from Tianjin Comio Chemical Reagent Co., Ltd. (Tianjin, China). Lactulose (99%) was purchased from Sigma (Shanghai, China).

2.2. Preparation of g-C$_3$N$_4$/CF Composite Carriers

The g-C$_3$N$_4$/CF composite carriers were prepared by ultrasonic-assisted impregnation and high temperature calcination. The CF was impregnated in acetone solution for 5 h then dried at 50 °C to obtain the pretreated CF (all CF mentioned below refers to the pretreated CF). During the preparation of the typical g-C$_3$N$_4$/CF composite carrier, the CF was added to saturated melamine aqueous solution by ultrasonic impregnation for 1 h and melamine powder was coated evenly on the surface of the wet CF (m (CF): m (melamine) = 1:2, 1:4, 1:6, 1:8, 1:10) and this was placed in a crucible with the initial dried in an oven at 50 °C. The crucible was covered with a lid and completely wrapped in aluminum foil. The crucible was calcined in N$_2$ at 550 °C for 3 h, and the heating rate was 5 °C·min^{-1}. After cooling at room temperature, the prepared g-C$_3$N$_4$/CF composite carrier was washed in water to remove the loose g-C$_3$N$_4$/CF on the surface of the CF, then dried and reserved (the prepared samples were named g-C$_3$N$_4$/CF-1, g-C$_3$N$_4$/CF-2, g-C$_3$N$_4$/CF-3, g-C$_3$N$_4$/CF-4, and g-C$_3$N$_4$/CF-5).

In addition, the pure g-C$_3$N$_4$ was synthesized by calcining melamine directly in N$_2$ at 550 °C for 3 h at a heating rate of 5 °C min^{-1}.

2.3. Co-Immobilization and SDS-PAGE Analysis of Lactase and Glucose Isomerase

Lactase (2.5 g, 5.9 U g^{-1}) and glucose isomerase (1 g, 58.9 U g^{-1}), with an activity ratio of 1:4, were dissolved in 50 mL of 0.2 mol L^{-1} PBS solution at pH 6.0 and 0.5 g of immobilized carrier (Raw-CF, g-C$_3$N$_4$/CF-1, g-C$_3$N$_4$/CF-2, g-C$_3$N$_4$/CF-3, g-C$_3$N$_4$/CF-4, g-C$_3$N$_4$/CF-5) was added. The reaction was vibrated at 30 °C at 160 r min^{-1} for 5 h. The immobilized enzymes were obtained by filtration and washed 3 times with the PBS solution. Polyacrylamide gel electrophoresis (SDS-PAGE) was used to evaluate the protein profiles of the enzyme preparations used in this study.

2.4. Selection of the Composite Carrier with Enzyme Activity Ratio and the Kinetics of Co-Immobilized Enzymes

The optimal immobilized carrier was determined by the enzyme activity recovery and co-immobilization efficiency. Lactase and glucose isomerase were immobilized in the best immobilization carrier according to the activity ratios of 1:0, 1:1, 1:3, 1:4, 1:5, and 1:6 to obtain the best enzyme activity ratio.

The Kms of the immobilized enzymes were determined according to Micellis–Menton kinetics with different concentrations of lactose/fructose solutions (containing 10% lactose and 2% fructose) as substrates.

2.5. Enzyme Activity Assays

Control assays were performed to determine the effect of the immobilization experimental conditions on the stability of each enzyme. A solution of 1% lactase and 1% glucose isomerase was prepared and the enzyme activities of each enzyme before and after the immobilization experimental conditions (30 °C, 160 r min^{-1}, 5 h) were determined.

The enzyme activity recovery rate was calculated according to the following equation (Equation (1)).

With the bovine serum albumin as the standard protein, the amount of enzyme protein supported on an immobilized carrier was measured by using the Coomassie brilliant blue Bradford method by subtracting the residual protein content in the solution from the initial protein content [2]. The co-immobilization efficiency was calculated by Equation (2).

$$\text{The enzymes' activity recovery rate } (\%) = \frac{A_i}{A_0} \times 100 \qquad (1)$$

where A_0 is the total enzymes activity of free enzymes, A_i is the total enzyme activity of the immobilized enzymes.

$$\text{Co-immobilization efficiency } (\%) = \frac{m_0 - m_i}{m_0} \times 100 \qquad (2)$$

where m_0 is the total protein content of initial enzyme solution (mg), m_i is the total protein content of residual liquid after immobilization (mg).

Lactose can be hydrolyzed into galactose and glucose by lactase [51], the hydrolysis process is expressed in Equation (3). O-Nitrophenyl β-D-galactopyranoside (ONPG) is a chromogenic substrate of lactase, suitable for the detection of lactase activity. Its color development principle is that lactase can hydrolyze ONPG to produce galactose and alkaline conditions of the yellow product o-nitrophenol. O-nitrophenol detected absorbance at 420 nm [52]. The lactase activity was measured using ONPG as substrate and a pH of 4.5 at 45 °C for 30 min. The absorbance was measured at 420 nm and the amount of o-nitrophenol released was determined. Under the above conditions, the enzyme activity unit was defined as the amount of enzyme required to produce 1 μmoL of o-nitrophenol per minute of ONPG consumption.

$$\text{lactose} \xrightarrow{\text{lactase hydrolysis}} \text{galactose} + \text{glucose} \qquad (3)$$

Glucose could be isomerized by glucose isomerase [49], the isomerization process is represented by Equation (4). The activity of glucose isomerase was measured using glucose as a substrate and a pH of 7.5 at 70 °C for 30 min. The fructose content was determined by HPLC. The unit of activity of glucose isomerase was defined as the amount of enzyme required to consume glucose per minute to produce 1 μmoL of fructose at pH 7.5, 70 °C.

$$\text{glucose} \xrightarrow{\text{glucose isomerase isomerization}} \text{fructose} \qquad (4)$$

Galactose and fructose can be converted into lactulose by lactase [47]; the synthesis process of lactulose can be expressed by Equation (5). The free dual enzyme and co-immobilized enzyme activities were determined using a lactose/fructose solution (containing 10% lactose and 2% fructose) as a substrate, pH 6.0, at 55 °C for 4 h.

$$\text{galactose} + \text{fructose} \xrightarrow{\text{lactase undergo galactosylation}} \text{lactulose} \quad (5)$$

The concentrations of fructose and lactulose were determined by HPLC using a Hitachi HPLC system (Tokyo, Japan) and the N2000 software (Ejer Technol. Co., Ltd., Hangzhou, China). The specific method of HPLC was as follows. Column: sugarpakTM1, 6.5 × 309 mm (Milford, MA, USA). Mobile phase: ultra-pure water. Flow rate: 0.5 mL min^{-1}. Detector: refractive index detector (RID). Column temperature: 50 °C. Injection volume: 10 µL.

2.6. Characterization of Structure and Properties of Immobilized Enzyme with the g-C$_3$N$_4$/CF Composite Carriers

The carrier surface morphology and immobilized enzyme morphology were detected by scanning electron microscopy (SEM, Quanta 250FEG, FEI, Hillsboro, OR, USA). Lactase and glucose isomerase were labeled with rhodamine B and FITC, respectively, and immobilized on g-C$_3$N$_4$/CF composite carrier. The distribution of these two enzymes on the surface of g-C$_3$N$_4$/CF composite carrier was observed by a fluorescence microscope (Revolve FL, Echo Laboratories, San Diego, CA, USA). The specific surface area and pore size distribution of immobilized carrier were analyzed by the pore size and specific surface area tester (SSA, Biaode Electronic Technology Co., Ltd., Beijing, China). Fourier transform infrared spectroscopy (FTIR, WQF-530, Beijing Beifenruili Analytical Instrument Factory, Beijing, China) was used to detect the surface functional groups of the carrier. X-ray photoelectron spectroscopy (XPS, Nexsa, ThermoFisher, MA, USA) was used to detect the surface elemental composition and chemical states of the carrier. The chemical structure of the carrier was determined by X-ray diffractometer (XRD, Miniflex 600, Nippon Science Company, Tokyo, Japan)

2.7. Assay of Co-Immobilized Enzyme Activity

2.7.1. Effect of Temperature and pH on Enzymes' Activities

The enzymes and substrates reacted at 45–75 °C, and the enzymes' activities were determined. The enzyme activity was the relative activity, which was set as the highest enzyme activity of 100%, to detect the optimal temperature of the enzymes. The free double enzymes and co-immobilized enzymes were incubated in PBS buffer at 65, 70, and 75 °C for different times, respectively, and the enzyme activities were determined.

The enzymes and substrates reacted at pH 4.5–7.5 and the enzyme activities were determined. The enzyme activity was the relative activity, which was set as the highest enzyme activity of 100%, to detect the optimal pH of the enzymes. The free double enzymes and co-immobilized enzymes were incubated in PBS buffer at pH 4.5, 6.0, and 7.5 for different times, respectively, and the enzyme activities were determined.

The enzyme activity was the relative activity, which was set as the highest enzyme activity of 100% to detect the thermal stability and pH stability of the enzymes, respectively. Enzyme assays were according to Section 2.6.

2.7.2. Storage Stability of Co-Immobilized Enzymes

The free enzyme and co-immobilized enzyme were stored in PBS buffer with pH 6.0 at 4 °C for 30 days. The residual enzyme activity was measured every 5 days. The enzyme activity was the relative activity, which was set as the enzyme activity before storage of 100%. Enzyme assays were according to Section 2.6.

2.7.3. Operational Stability of Co-Immobilized Enzymes

Co-immobilized enzymes and free double enzymes were respectively added to the lactose/fructose (containing 10% lactose and 2% fructose) substrate solution and reacted for 4 h at 65 °C. After the reaction, the lactulose content was determined, and the co-immobilized enzyme was recovered.

The co-immobilized enzyme continued to the fresh substrate solution and circulated for six times. The concentration of lactulose generated from the free enzyme in the first reaction was set to 100%, and the lactulose content generated by the co-immobilized enzyme relative to the free dual enzyme was recorded as the relative conversion rate of the co-immobilized enzyme. Enzyme assays were according to Section 2.6.

The relative conversion of co-immobilized to pectose was calculated by Equation (6).

$$\text{The lactofructose yield rate } (\%) = \frac{C_i}{C_0} \times 100 \tag{6}$$

where C_i was the content of lactulose produced by the co-immobilized enzyme, C_0 was the content of lactulose produced by the free enzyme.

3. Results and Discussion

3.1. Structural Performance Characterization of the g-C$_3$N$_4$/CF Composite Carrier

The XPS C1s spectra and the XRD profiles of the CF, g-C$_3$N$_4$, and g-C$_3$N$_4$/CF are shown in Figure 2. The elemental compositions and chemical states of the surfaces of CF, g-C$_3$N$_4$, and g-C$_3$N$_4$/CF were determined by XPS. As shown in Table S1, the elements C, N, and O were present on the surfaces of CF, g-C$_3$N$_4$, and g-C$_3$N$_4$/CF. The introduction of g-C$_3$N$_4$ added more N elements to the surface of g-C$_3$N$_4$/CF compared to CF (10.31-fold increase).

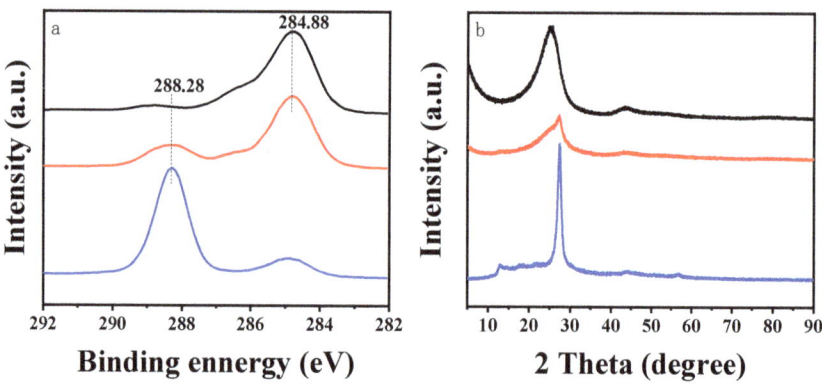

Figure 2. The XPS C1s spectra and the XRD profiles of the CF, g-C$_3$N$_4$, and g-C$_3$N$_4$/CF (CF: black line, g-C$_3$N$_4$/CF: red line, g-C$_3$N$_4$: blue line). (a) The XPS C1s spectra of CF, g-C$_3$N$_4$, and g-C$_3$N$_4$/CF. (b) The XRD profiles of CF, g-C$_3$N$_4$, and g-C$_3$N$_4$/CF.

In addition, XPS C1s spectra are shown in Figure 2a, to further understand the chemical states of CF, g-C$_3$N$_4$, and g-C$_3$N$_4$/CF surface elements. The CF had a characteristic peak at 284.88 eV which belonged to the sp^2 hybrid C atom in the C–C group [53]. Compared with the CF, the characteristic peak located at 288.28 eV was increased in the XPS C1s spectrum of g-C$_3$N$_4$/CF, reflecting the sp^2 bond C in the N=C(–N)$_2$ group [29], which was caused by the introduction of g-C$_3$N$_4$. The above results indicate the composite with g-C$_3$N$_4$ on the CF surface.

The phases and structures of CF, g-C$_3$N$_4$, and g-C$_3$N$_4$/CF were determined by XRD. As shown in Figure 2b, the CF had only a wide diffraction peak at about 22°, which may be caused by the (002) plane of the graphite structure [27]. There were two characteristic diffraction peaks in g-C$_3$N$_4$, located at 13.1° and 27.4°, respectively. The peak at 13.1° was related to the in-plane ordered (100) plane of the trihomogeneous thiazine ring with a distance of 0.670 nm. The peak at 27.4° was related to the (002) plane of interlayer stacking reflection of conjugated aromatic system with an interlayer distance of 0.323 nm [28,29]. The diffraction peaks of the g-C$_3$N$_4$/CF composite carrier were similar to those of CF and g-C$_3$N$_4$, which were located at 22° (from CF), 13.1,° and 27.4° (from g-C$_3$N$_4$), respectively, indicating that the surface of CF was composited with g-C$_3$N$_4$.

Specific surface area and porosity have great influence on the immobilization effect of the carrier. The N$_2$ adsorption–desorption isotherms and pore size distributions of different carriers as shown in Figure S1. Adsorption–desorption isotherms and pore size distributions of CF and g-C$_3$N$_4$/CF were measured by the BET method. The isotherms of CF and g-C$_3$N$_4$/CF were consistent with typical type II isotherms with hysteresis rings and shown in Figure S1a, which indicates that there is a complex pore structure on the surface of carriers [21]. As shown in Figure S1b, the aperture of the g-C$_3$N$_4$/CF composite carrier was mainly around 0.82 nm and 2.86 nm, while the aperture of CF was mainly between 2.4 and 5.74 nm. The results showed that, compared with CF, the pore structure of the g-C$_3$N$_4$/CF composite carrier was mainly microporous. Therefore, the adsorption capacity of g-C$_3$N$_4$/CF composite carriers was significantly higher than that of CF. The g-C$_3$N$_4$/CF composite carriers adsorption immobilization of biosynthetic enzymes was associated with physical interactions, including van der Waals forces, electrostatic interactions, hydrogen bonding, and ionic interactions [1].

The specific surface area, pore volume, and pore size of CF and g-C$_3$N$_4$/CF are shown in Table 1. The specific surface area of g-C$_3$N$_4$/CF was 5.08 m^2 g^{-1}, which was 8.33 times than that of CF. In addition, the microporous volume, mesoporous volume, and total pore volume of g-C$_3$N$_4$/CF were increased. The g-C$_3$N$_4$/CF composite carrier achieved a larger specific surface area and pore volume [23]. Within a certain range, the larger specific surface area and the smaller pore size may be favorable for the carrier adsorption of enzymes [25].

Table 1. Specific surface area, pore volume, and pore size of different carriers.

	CF	g-C$_3$N$_4$/CF
S_{BET} (m^2 g^{-1})	0.61 ± 0.04	5.08 ± 0.04
Microporous volume (cm^3 g^{-1})	1.91 × 10^{-4} ± 1.14 × 10^{-5}	2.15 × 10^{-3} ± 1.39 × 10^{-5}
Mesoporous volume (cm^3 g^{-1})	1.76 × 10^{-3} ± 2.24 × 10^{-4}	2.75 × 10^{-2} ± 1.23 × 10^{-3}
Total pore volume (cm^3 g^{-1})	1.95 × 10^{-3} ± 2.27 × 10^{-4}	2.96 × 10^{-2} ± 1.23 × 10^{-3}
Average pore diameter (nm)	12.87 ± 1.65	23.35 ± 0.97

The above results indicate that compared with CF, a g-C$_3$N$_4$/CF composite carrier could achieve the adsorption immobilization of double enzymes by the virtue of excellent physical properties (such as rough surface structure, smaller pore size, larger specific surface area, and pore volume). Figure 2 indicates that the CF surface was loaded with g-C$_3$N$_4$. The g-C$_3$N$_4$ was the stable isomer in carbon nitride, with good biocompatibility, thermal stability, chemical inertia, electronic structure and so on. It is an ideal candidate for the preparation of immobilized enzymes [54,55].

3.2. The Apparent Morphology of the Co-Immobilized Enzyme and Its Distribution on the g-C$_3$N$_4$/CF

The surface morphologies of CF, g-C$_3$N$_4$/CF, immobilized double enzyme, and free enzymes were observed by SEM. The CF surface was relatively smooth and neat (Figure 3a) [56]. After the treatment, some micron g-C$_3$N$_4$ was uniformly loaded on the surface of the g-C$_3$N$_4$/CF composite carrier (Figure 3b), resulting in the changed surface morphologies and rough surface of the carriers [29]. As shown in Figure 3c,d, the amount of immobilized double enzyme on the g-C$_3$N$_4$/CF surface was much more than that of on the CF; the dual enzyme was evenly distributed on the composite carrier in Figure 4d, which may be because the rough g-C$_3$N$_4$/CF surface provided more adsorption sites for the enzyme and enhanced the immobilized effect of the double enzyme.

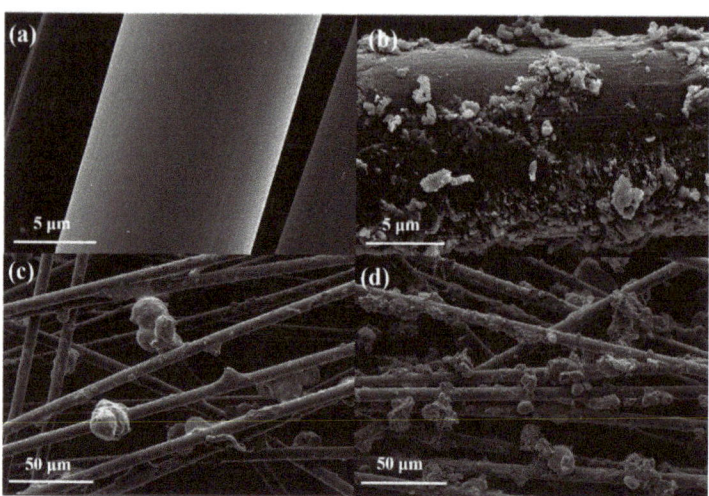

Figure 3. The SEM images of CF, g-C$_3$N$_4$/CF, and their immobilized enzymes. (**a**) CF; (**b**) g-C$_3$N$_4$/CF; (**c**) co-immobilized enzymes on CF; (**d**) co-immobilized enzymes on g-C$_3$N$_4$/CF.

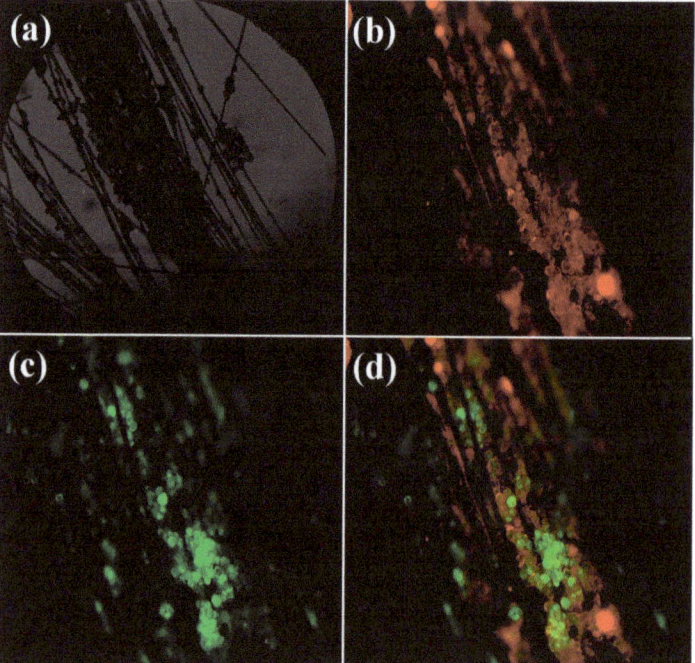

Figure 4. Distribution of double enzymes in g-C$_3$N$_4$/CF composite carrier. (**a**) Bright field image; (**b**) distribution of Rhoda mine B labeled lactase on g-C$_3$N$_4$/CF composite carrier; (**c**) distribution of FITC-labeled glucose isomerase on g-C$_3$N$_4$/CF composite carrier; (**d**) the distribution of lactase and glucose isomerase on g-C$_3$N$_4$/CF composite carrier. Herein, the g-C$_3$N$_4$/CF referred to the g-C$_3$N$_4$/CF composite carriers synthesized by being calcined with melamine and CF in 550 °C for 3 h.

The immobilized enzymes were tested by fluorescence microscopy to further analyze the distribution of lactase and glucose isomerase on the surface of the g-C$_4$N$_4$/CF composite carrier. As shown in Figure 4a, the surface of the composite carrier was obviously loaded with a lot of lactase and glucose isomerase. Figure 4b shows the red fluorescence of lactase labeled with Rhoda mine B. Figure 4c shows the green fluorescence of glucose isomerase labeled with FITC. Figure 4d shows that the two enzymes were uniformly distributed on the surface of the g-C$_3$N$_4$/CF composite carrier, which is beneficial for lactase to convert lactose into glucose and to then be utilized by glucose isomerase to fructose as soon as possible. The fructose and galactose are then used by nearby lactase to synthesize lactulose. The co-immobilization of lactase and glucose isomerase resulted in a multi-enzyme cascade reaction, which was beneficial to reduce the reactor volume, improved the production efficiency, and reduced waste generation [51].

The effect of the surface functional groups of the g-C$_3$N$_4$/CF composite carriers on dual-enzyme immobilization is shown in Figure 5. The surface functional groups of g-C$_3$N$_4$, g-C$_3$N$_4$/CF, and g-C$_3$N$_4$/CF co-immobilized enzymes were detected by FTIR.

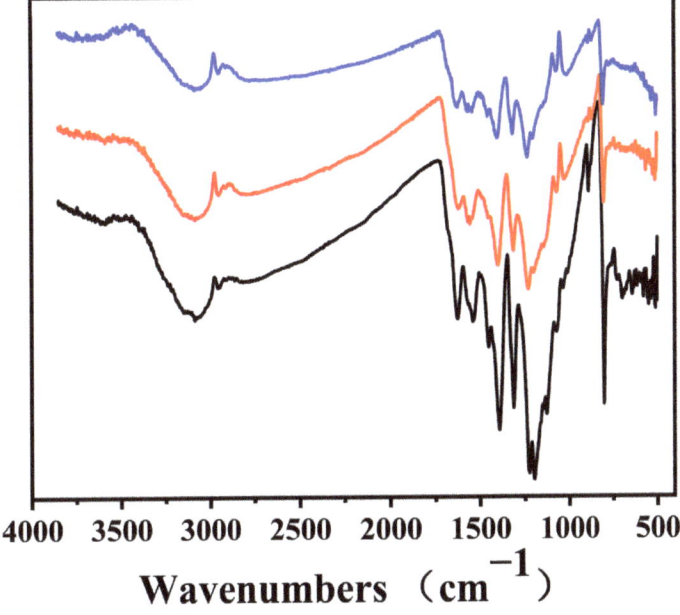

Figure 5. The FTIR spectra of g-C$_3$N$_4$, g-C$_3$N$_4$/CF, and g-C$_3$N$_4$/CF co-immobilized enzymes (g-C$_3$N$_4$/CF co-immobilized enzymes: blue line. g-C$_3$N$_4$/CF: red line. g-C$_3$N$_4$: black line).

The wide peak at 3500–3100 cm^{-1} was caused by the O–H stretching mode together with the vibrational mode of the N–H bond [28]. The characteristic peak at 1200–1700 cm^{-1} reflects the stretching vibration of the conjugated CN ring [29]. The characteristic peak at 804 cm^{-1} represents the respiratory vibration of the S-thiazine ring [57]. In addition, the characteristic peak of g-C$_3$N$_4$/CF was obviously weaker than that due to doping CF. Compared with g-C$_3$N$_4$/CF, the infrared spectrum intensity of g-C$_3$N$_4$/CF co-immobilized enzymes decreased significantly. The peaks of g-C$_3$N$_4$/CF co-immobilized enzymes at 1240, 1317, 1410, and 1625 cm^{-1} were attributed to aromatic C-N and C-N frameworks. These characteristic absorption peaks were highly overlapped with the absorption peaks of g-C$_3$N$_4$/CF, which was attributed to the physical interaction between the double enzyme and g-C$_3$N$_4$ [1]. The results showed that the basic surface structure of g-C$_3$N$_4$ was affected by the immobilization process.

In addition, some forces may exist between the enzyme and the surface of the composite carrier. The intensity of the amide band of the co-immobilized enzyme was reduced relative to the free double enzyme, from which it was inferred that the hydrophobic end of the enzyme interacts with the surface of the composite carrier. The immobilization process occurred due to hydrophobic forces, which contributed to the immobilization of the enzymes [58,59]. The N–H and H vibrational peaks of the g-C_3N_4/CF co-immobilized enzymes at 3500–3100 and 800 cm^{-1}. It is the possible that the mechanism of immobilization was the bind of the hydrogen atoms on the enzyme surface to the amino groups of g-C_3N_4 on the surface of the g-C_3N_4/CF composite carrier under physical conditions [1].

3.3. Enzymatic Properties Analysis of the g-C_3N_4/CF Co-Immobilized Double Enzyme

The SDS-PAGE analysis of glucose isomerase and lactase were performed and results are displayed in Figure S2. Figure S2A,B both show only one band, indicating that all protein in both crude extracts corresponded to these enzymes. Thus, it was feasible for us to calculate the immobilization efficiency by determining the total protein content of the initial enzyme solution [60,61]. In a further, SDS profiles indicated that the molecular weight of lactase was about 120 kDa and that of glucose isomerase was about 70 kDa. The relative molecular mass of glucose isomerase reported in the relevant literature was about 63 kDa [62] and the molecular weight of lactase was between about 60 and 120 kDa [61,63], similar to the results we obtained.

The test conditions for immobilized enzymes had some effects on enzyme viability [64]. As shown in Table S2, lactase activity decreased from 5.90 U g^{-1} to 5.13 U g^{-1}, with a decrease of 13%; glucose isomerase activity decreased from 58.94 U g^{-1} to 49.02 U g^{-1}, with a decrease of 16.4%. The change in enzyme activity was small. The lactase and glucose isomerase with immobilization show the good stability of the experimental condition.

The immobilization process affected the distribution of the substrate at the enzyme active site [65]. The Michaelis constant Km is considered the most important constant for studying the kinetics of the enzyme reaction, which represents the affinity between the enzyme and the substrate. Larger Km values indicated weaker affinity and vice versa [46]. The Km of the co-immobilized enzymes was determined by using a series of concentrations of the lactose/fructose mixture (10% lactose and 2% fructose) as the substrate according to the Michaelis–Menten kinetics. As shown in Figure S3, the Km value of the co-immobilized enzyme was 0.09, indicating that the co-immobilized enzyme had a favorable affinity for the substrate.

The optimal preparation of g-C_3N_4/CF composite carrier was selected by enzyme activity recovery and co-immobilization efficiency. As shown in Figure 6a, 100% enzyme activity recovery rate referred to the enzyme activity of the free dual enzyme prior to immobilization at 55 °C, pH 6.0, which was 19.07 U g^{-1}. During the preparation of the composite carriers, the recovery of enzyme activity and co-immobilization efficiency of the immobilized carriers gradually increased with the increase of melamine content. The enzyme activity recovery and co-immobilization efficiency of g-C_3N_4/CF-4 were 1.93 and 2.41 times that of CF, respectively. The g-C_3N_4/CF-4 had a higher co-immobilization efficiency of 37.31%. The enzyme activity recovery rate and co-immobilization efficiency of the composite carrier, prepared by continuously increasing the amount of melamine, did not obviously increase. This may be due to the fact that the g-C_3N_4 attached to the surface of CF had reached saturation in the modification process, leading to the properties of the composite carrier not being changed with the excessive melamine. Therefore, the g-C_3N_4/CF-4 was selected for subsequent experiments. For the convenience of description, the g-C_3N_4/CF composite carrier mentioned in the subsequent experiments only represents g-C_3N_4/CF-4.

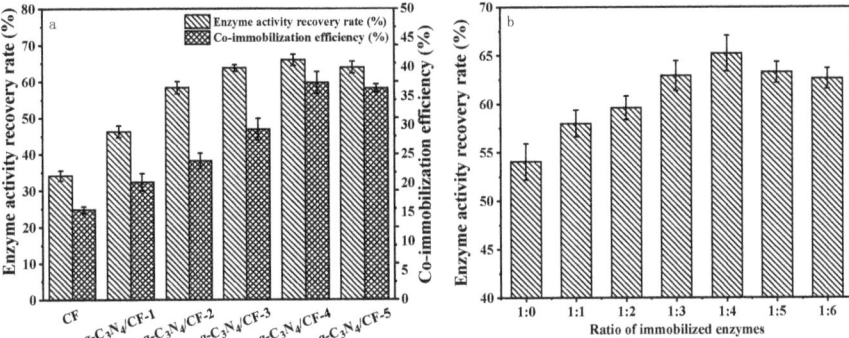

Figure 6. Effect of different carriers and enzyme activity ratios on co-immobilized enzyme activity. (**a**) Effect of different carriers on enzyme activity recovery and co-immobilization efficiency; (**b**) effects of double enzyme ratio on enzyme activity recovery of g-C_3N_4/CF co-immobilization enzyme.

The optimal ratio of immobilized lactase and glucose isomerase was determined by the enzyme activity recovery rate. As shown in Figure 6b, 100% enzyme activity recovery referred to the enzyme activity of the free double enzyme at each enzyme activity ratio, in which the enzyme activity of the free double enzyme was 19.5 U/g at the ratio of enzyme activity of 1:4. With the increase of the lactase and glucose isomerase activity ratio, the recovery rates of co-immobilized enzyme first increased and then reduced. When the glucose isomerase was insufficient, the glucose generated by lactose catalyzed by lactase inhibited the galactose conversion activity of lactase non-competitively, thus affecting the production of lactulose. This is similar to the reason for which Long et al. used glucose oxidase to relieve the inhibition of glucose to the fructose conversion activity of β-fructofuranosidase [3]. When the enzyme activity ratio was 1:4, the enzyme activity recovery rate was the highest, which may be related to the increase of glucose isomerase content promoting the conversion of glucose into fructose, thus promoting the increase of fructose concentration. In addition, the glucose conversion alleviated the inhibition of the galactose transfer activity of lactase, with the increased concentration of galactose. The increase of fructose and galactose promoted the production of lactulose. Subsequently, the enzyme activity recovery rate began to decrease, which may be due to the relative decrease of the active site binding lactase on the carrier surface with the excessive glucose isomerase. If glucose isomerase is excessive, the active sites of binding to lactase on the carrier surface would decrease relatively. Therefore, the enzyme activity ratio of lactase to glucose isomerase of 1:4 was selected for the subsequent experiments.

In the enzymatic catalytic process, the synthesis efficiency of the product depends not only on the preparation of the immobilized enzyme, but also on catalytic reaction conditions [66]. Temperature is one of the most important factors affecting enzyme catalysis.

The optimal temperature of the free and co-immobilized enzymes was determined at 45–75 °C with a pH of 6.0. In Figure 7a, the relative activity of 100% referred to the enzyme activity at the optimum temperature. The activity of the free double enzyme was 19.19 U g^{-1} at 55 °C. The activity of the co-immobilized enzyme was 13.89 U g^{-1} at 60 °C. As shown in Figure 7a, with the increase in temperature, the relative activities of immobilized and free double enzymes showed a trend of increasing and then decreasing. The decrease or loss of enzyme activity at a high temperature was due to the structural fragility of the enzyme and enzymatic dissociation [67]. The optimum temperature of the co-immobilized enzyme increased by 5 °C, indicating that the co-immobilized enzyme had higher thermal stability than the free enzyme. At the same time, the activity of the co-immobilized enzyme was significantly higher than that of the free enzyme at the temperature of 60–75 °C, indicating that the co-immobilized enzyme had the wider range of temperature application.

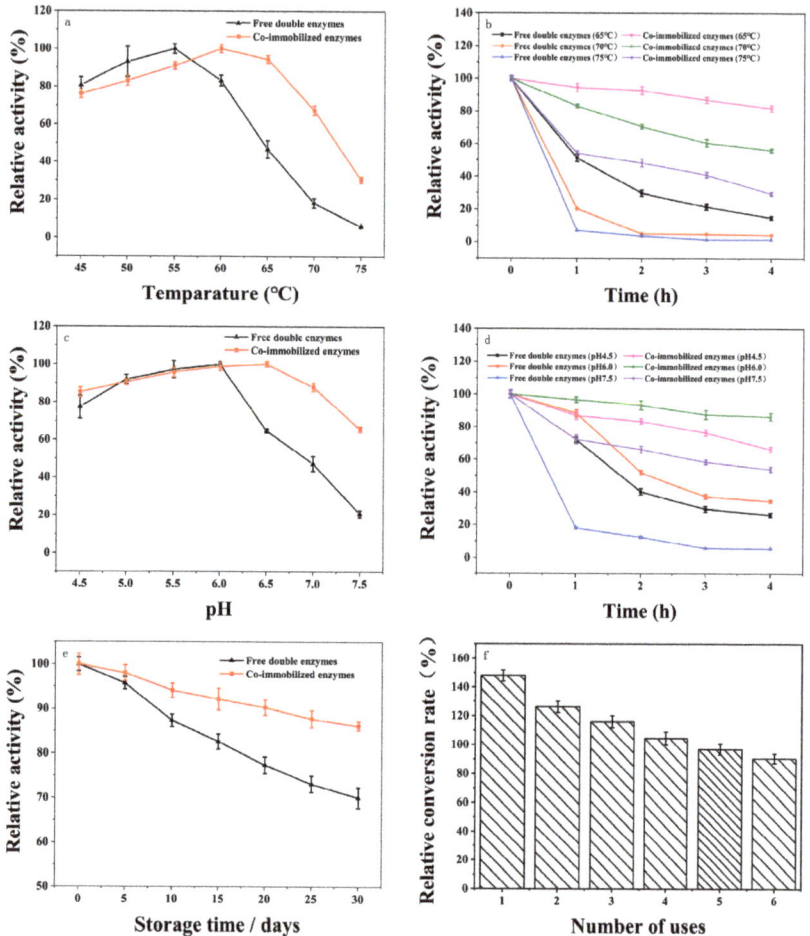

Figure 7. Enzymatic properties analysis of the g-C_3N_4/CF co-immobilized double enzyme. (**a**) The optimal reaction temperature changes of the free enzyme and the co-immobilized enzyme; (**b**) thermal stability of the free enzymes and the co-immobilized enzymes; (**c**) optimum pH change of the free enzyme and the co-immobilized enzyme; (**d**) the pH stability of the free enzymes and the co-immobilized enzymes; (**e**) storage stability of the free enzymes and co-immobilized enzymes; (**f**) lactulose relative conversion rate.

In order to further investigate the effect of immobilization on the thermal stability of the enzyme, the enzyme activities of free enzymes and co-immobilized enzymes were measured at 65 °C, 70 °C, and 75 °C with the reaction time. In Figure 7b, the relative activity of 100% referred to the initial enzyme activity with different temperatures at pH 6.0. The activity of the free double enzyme was 18.77 U g^{-1} and the activity of the co-immobilized double enzyme was 13.80 U g^{-1}. As shown in Figure 7b, the inactivation rate of free enzymes at the same high temperature over time was much higher than that of the co-immobilized enzyme, indicating that the co-immobilized enzyme had higher thermal stability, which was attributed to the increased rigidity of enzyme molecules with immobilization, preventing enzymatic subunit dissociation due to thermal denouement [68]. The chemical bonds formed during physical bonding could protect enzyme molecules from conformational changes at high temperatures [1].

pH is also one of the most important factors affecting enzyme activity, because the H^+ concentration in the environment not only affects the quaternary structure of the enzyme, but also affects the degree of ionization of the substrate, product, and active site residue [67]. With the increase of pH, the relative activities of co-immobilized and free double enzymes showed a trend of increasing and then decreasing. The free double enzymes and co-immobilized enzymes had the maximum relative activities at pH 6.0 and pH 6.5, respectively.

The changes in the activities of free and co-immobilized enzymes may be due to the effect of pH on the structural stability of the enzyme molecule, where very high or low pH levels denatured the protein deactivating the enzyme. Secondly, the change of pH value may also affect the dissociation state of enzymes and substrates [3]. The optimum pH values of free enzymes and co-immobilized enzymes were determined at 55 °C and a pH of 4.5–7.5.

In Figure 7c, the relative activity of 100% referred to the enzyme activity of at the optimum pH at 55 °C. The activity of the free double enzyme was 19.55 U g^{-1} at pH 6.0. The activity of the co-immobilized enzyme was 13.75 U g^{-1} at pH 6.5. As shown in Figure 7c, the optimal pH values of the free dual enzymes and the co-immobilized enzymes were 6.0 and 6.5, respectively. When the pH value was higher than 6.5, the wider pH stability was displayed in the co-immobilized enzymes, rather than the free dual enzymes. When the pH value was 7.5, the relative activity of the co-immobilized enzymes remained at 65.65% of the original enzyme activity, while the relative activity of the free enzymes decreased to 20.88%. The shift of optimal pH to the alkaline range may be related to the conformational changes of the enzyme on the carrier along with the protonation of the microenvironment. Electrostatic interactions between the pH and surface of the enzyme molecule leads to changes in the enzyme active center and conformational changes of the enzyme [24].

To further investigate the effect of immobilization on the pH stability of the enzyme, the activities of free enzymes and co-immobilized enzymes were measured after the incubation at pH of 4.5, 6.0, and 7.5 for different times. In Figure 7d, the relative activity of 100% referred to the initial enzyme activity with different pH values at 55 °C. The activity of the free double enzyme was 18.64 U g^{-1}. The activity of the co-immobilized double enzyme was 13.13 U g^{-1}. As shown in Figure 7d, the co-immobilized enzymes had a lower inactivation rate and wider pH stability compared to the free enzymes, which may be due to the fact that immobilized carriers could act as ion exchangers, making the pH of the solution around the enzyme different from that of the substrate solution [3].

It is known that enzyme leakage is an important factor on the effect of enzyme immobilization technology [66]. Thus, the stability of co-immobilized enzymes during operation and storage was investigated, which was the key performance index for practical applications.

To further explain the decrease in enzyme reusability, the storage stabilities of the enzyme were tested. In Figure 7e, the relative activity of 100% referred to the initial enzyme activity for storage. The activity of the free double enzyme was 19.29 U g^{-1}. The activity of the co-immobilized double enzyme was 13.57 U g^{-1}. As shown in Figure 7e, after 30 days of storage at 4 °C, the activity of co-immobilized enzyme was retained at 86.08%, while the activity of the free enzyme was retained at 70.06%, indicating that the storage stability of co-immobilized enzymes was significantly higher than that of free enzymes. The operating stability of the free double enzyme and co-immobilized enzymes were measured at 65 °C and a pH of 6.0, as shown in Figure 7f, where the free enzyme was not easily recovered and reused. In Figure 7f, the relative conversion of 100% was the amount of lactulose produced by the breakdown of the substrate with the free enzyme at 65 °C. The concentration of lactulose at this point was 7.19 g L^{-1}.

The relative conversion rate of the co-immobilized enzyme was higher in the first four batches than that of the free enzyme, remaining at around a 90% relative conversion rate after six cycles. The main reason why the co-immobilized enzymes were able to achieve multi-batch production was the positive effects of the composite carrier on the enzyme molecules. It was shown that the composite carrier with the large specific surface area and appropriate pore sizes enhanced the immobilization efficiency of enzymes [69]. The composite carrier had a rough surface with more enzyme molecules' immobilization than that of the CF, which may be due to the fact that the rough surface of the composite carrier could provide the larger binding area to increase the enzymes' immobilization efficiency. Furthermore, the images from laser scanning confocal microscopy indicated that the double enzymes were uniformly distributed on the composite carrier, which improved the immobilized enzymes for the production of lactose from the substrate. The presence of these conditions altered the microenvironment surrounding the enzyme molecule, which showed that the conformation of immobilized enzyme did not change much [68]. The negative effects outside on the co-immobilized enzymes were reduced by the co-immobilization. Furthermore, the enzymes were distributed on the surface of the carrier, which reduced the accumulation of enzymes during the storage of free enzymes [66]. Consequently, the stability of co-immobilized enzymes was raised, to the benefit of lactulose production, in multiple batches.

4. Conclusions

In this paper, g-C_3N_4/CF composite carriers were prepared by ultrasonically assisted impregnation and high-temperature calcination for the co-immobilization of lactase and glucose isomerase. The results showed that the surface of g-C_3N_4/CF composite carriers were uniformly composited with some certain microns of g-C_3N_4, which made the surface rougher, increased the specific surface area, optimized the pore structure, and provided more adsorption sites for the enzymes. The immobilization efficiency of 37% and enzyme activity of 13.89 U g^{-1} were obtained in the co-immobilized enzyme by the carrier of g-C_3N_4/CF. The lactase and glucose isomerase co-immobilized with the g-C_3N_4/CF composite carrier displayed good stability. Moreover, compared to the free enzymes, the co-immobilized enzymes performed higher relative conversion rates of lactulose during the multi-batch production. Therefore, the g-C_3N_4/CF composite carrier has wide application prospects in the immobilization of lactase and glucose isomerase for lactose production.

Supplementary Materials: The following supporting information can be downloaded at: https://www.mdpi.com/article/10.3390/nano12234290/s1, Table S1: Surface element composition of CF, g-C_3N_4, and g-C_3N_4/CF.; Figure S1: N_2 adsorption-desorption isotherms and pore size distributions of different carriers. (a) N_2 adsorption-desorption isotherms for CF and g-C_3N_4/CF (Adsorption isotherm plot of CF, black line. Desorption isotherm plot of CF, red line. Adsorption isotherm plot of g-C_3N_4/CF, blue line. Desorption isotherm plot of g-C_3N_4/CF, purple line). (b) Pore size distribution of CF and g-C_3N_4/CF (The CF, black line. The g-C_3N_4/CF, red line).; Figure S2: SDS-PAGE analysis of glucose isomerase and lactase. In each sample lane, 10.0 μL of protein sample was added. Lane M: protein marker, lane A: glucose isomerase, lane B: lactase.; Table S2: Effect of immobilization conditions on enzyme activity.; Figure S3: Lineweaver–Burk plots of the co-immobilized enzyme. The Km of the immobilized and free enzymes was determined by using a series of concentrations of the lactose\fructose mixture (10% lactose and 2% fructose) as the substrate according to the Michaelis-Menten kinetics.

Author Contributions: Conceptualization, L.W. and Q.Y.; Methodology, L.W., B.J. and Y.S.; Software, R.D.; Validation, B.J. and Y.S.; Formal analysis, B.J. and Y.S.; Investigation, L.W., B.J. and R.D.; Resources, L.W., Q.Y. and J.W.; Data curation, B.J. and Y.S.; Writing—original draft, L.W., B.J. and Y.S.; Writing—review & editing, L.W., B.J. and Y.S.; Supervision, Q.Y. and J.W.; Project administration, L.W. and J.W.; Funding acquisition, L.W. and J.W. All authors have read and agreed to the published version of the manuscript.

Funding: This work was supported by [the National Natural Science Foundation of China], grant no. [21978070, 21306040]; [Natural Science Foundation of Henan], grant no. [212300410032]; [Program for Science & Technology Innovation Talents in Universities of Henan Province], grant no. [20HASTIT034]; [National Key Research and Development Program], grant no. [2016YFD0400200]; [Key Young Teachers Plan in Henan University of Technology], grant no. [21420056].

Data Availability Statement: Not applicable.

Conflicts of Interest: The authors declare no conflict of interest.

References

1. Simsek, E.B.; Saloglu, D. Exploring the structural and catalytic features of lipase enzymes immobilized on g-C3N4: A novel platform for biocatalytic and photocatalytic reactions. *J. Mol. Liq.* **2021**, *337*, 116612. [CrossRef]
2. Ahmad, R.; Shanahan, J.; Rizaldo, S.; Kissel, D.S.; Stone, K.L. Co-immobilization of an Enzyme System on a Metal-Organic Framework to Produce a More Effective Biocatalyst. *Catalysts* **2020**, *10*, 499. [CrossRef]
3. Long, J.; Pan, T.; Xie, Z.; Xu, X.; Jin, Z. Co-immobilization of β-fructofuranosidase and glucose oxidase improves the stability of Bi-enzymes and the production of lactosucrose. *LWT* **2020**, *128*, 109460. [CrossRef]
4. Zdarta, J.; Meyer, A.S.; Jesionowski, T.; Pinelo, M. A General Overview of Support Materials for Enzyme Immobilization: Characteristics, Properties, Practical Utility. *Catalysts* **2018**, *8*, 92. [CrossRef]
5. Zdarta, J.; Meyer, A.S.; Jesionowski, T.; Pinelo, M. Multi-faceted strategy based on enzyme immobilization with reactant adsorption and membrane technology for biocatalytic removal of pollutants: A critical review. *Biotechnol. Adv.* **2019**, *37*, 107401. [CrossRef]
6. Jesionowski, T.; Zdarta, J.; Krajewska, B. Enzyme immobilization by adsorption: A review. *Adsorption* **2014**, *20*, 801–821. [CrossRef]
7. Wang, Z.; Ren, D.; Jiang, S.; Yu, H.; Cheng, Y.; Zhang, S.; Zhang, X.; Chen, W. The study of laccase immobilization optimization and stability improvement on CTAB-KOH modified biochar. *BMC Biotechnol.* **2021**, *21*, 74. [CrossRef]
8. Gu, Y.; Yuan, L.; Li, M.; Wang, X.; Rao, D.; Bai, X.; Shi, K.; Xu, H.; Hou, S.; Yao, H. Co-immobilized bienzyme of horseradish peroxidase and glucose oxidase on dopamine-modified cellulose–chitosan composite beads as a high-efficiency biocatalyst for degradation of acridine. *RSC Adv.* **2022**, *12*, 23006–23016. [CrossRef]
9. Tentori, F.; Bavaro, T.; Brenna, E.; Colombo, D.; Monti, D.; Semproli, R.; Ubiali, D. Immobilization of Old Yellow Enzymes via Covalent or Coordination Bonds. *Catalysts* **2020**, *10*, 260. [CrossRef]
10. Li, X.; Yin, Z.; Cui, X.; Yang, L. Capillary electrophoresis-integrated immobilized enzyme microreactor with graphene oxide as support: Immobilization of negatively charged L-lactate dehydrogenase via hydrophobic interactions. *Electrophoresis* **2019**, *41*, 175–182. [CrossRef]
11. Guzik, U.; Hupert-Kocurek, K.; Wojcieszyńska, D. Immobilization as a Strategy for Improving Enzyme Properties-Application to Oxidoreductases. *Molecules* **2014**, *19*, 8995–9018. [CrossRef] [PubMed]
12. Wu, D.; Feng, Q.; Xu, T.; Wei, A.; Fong, H. Electrospun blend nanofiber membrane consisting of polyurethane, amidoxime polyarcylonitrile, and β-cyclodextrin as high-performance carrier/support for efficient and reusable immobilization of laccase. *Chem. Eng. J.* **2018**, *331*, 517–526. [CrossRef]
13. Wang, L.; Sun, P.; Yang, Y.; Qiao, H.; Tian, H.; Wu, D.; Yang, S.; Yuan, Q.; Wang, J. Preparation of ZIF@ADH/NAD-MSN/LDH Core Shell Nanocomposites for the Enhancement of Coenzyme Catalyzed Double Enzyme Cascade. *Nanomaterials* **2021**, *11*, 2171. [CrossRef] [PubMed]
14. Sulman, A.M.; Matveeva, V.G.; Bronstein, L.M. Cellulase Immobilization on Nanostructured Supports for Biomass Waste Processing. *Nanomaterials* **2022**, *12*, 3796. [CrossRef] [PubMed]
15. Muley, A.B.; Mulchandani, K.H.; Singhal, R.S. Immobilization of enzymes on iron oxide magnetic nanoparticles: Synthesis, characterization, kinetics and thermodynamics. *Methods Enzymol.* **2019**, *630*, 39–79. [CrossRef]
16. Homouz, D.; Stagg, L.; Wittung-Stafshede, P.; Cheung, M.S. Macromolecular Crowding Modulates Folding Mechanism of α/β Protein Apoflavodoxin. *Biophys. J.* **2009**, *96*, 671–680. [CrossRef]
17. Chouyyok, W.; Panpranot, J.; Thanachayanant, C.; Prichanont, S. Effects of pH and pore characters of mesoporous silicas on horseradish peroxidase immobilization. *J. Mol. Catal. B Enzym.* **2009**, *56*, 246–252. [CrossRef]
18. Romanovskaya, I.I.; Bondarenko, G.I.; Davidenko, T.I. Immobilization of Penicillium solitum lipase on the carbon fiber material "Dnepr-MN". *Pharm. Chem. J.* **2008**, *42*, 360–362. [CrossRef]
19. Garcia, L.F.; Siqueira, A.C.R.; Lobón, G.S.; Marcuzzo, J.S.; Pessela, B.C.; Mendez, E.; Garcia, T.A.; Gil, E.D.S. Bio-electro oxidation of indigo carmine by using microporous activated carbon fiber felt as anode and bioreactor support. *Chemosphere* **2017**, *186*, 519–526. [CrossRef]
20. Kim, J.-H.; Cho, S.; Bae, T.-S.; Lee, Y.-S. Enzyme biosensor based on an N-doped activated carbon fiber electrode prepared by a thermal solid-state reaction. *Sens. Actuators B Chem.* **2014**, *197*, 20–27. [CrossRef]
21. Wang, L.; Jia, F.; Wu, D.; Wei, Q.; Liang, Y.; Hu, Y.; Li, R.; Yu, G.; Yuan, Q.; Wang, J. In-situ growth of graphene on carbon fibers for enhanced cell immobilization and xylitol fermentation. *Appl. Surf. Sci.* **2020**, *527*, 146793. [CrossRef]
22. Liu, Q.; Dai, G.; Bao, Y. Carbon nanotubes/carbon fiber hybrid material: A super support material for sludge biofilms. *Environ. Technol.* **2017**, *39*, 2105–2116. [CrossRef]
23. Zhang, J.; Huang, F. Enhanced visible light photocatalytic H2 production activity of g-C3N4 via carbon fiber. *Appl. Surf. Sci.* **2015**, *358*, 287–295. [CrossRef]
24. Zhang, H.; Wu, J.; Han, J.; Wang, L.; Zhang, W.; Dong, H.; Li, C.; Wang, Y. Photocatalyst/enzyme heterojunction fabricated for high-efficiency photoenzyme synergic catalytic degrading Bisphenol A in water. *Chem. Eng. J.* **2019**, *385*, 123764. [CrossRef]

25. Tian, K.; Liu, H.; Dong, Y.; Chu, X.; Wang, S. Amperometric detection of glucose based on immobilizing glucose oxidase on g-C3N4 nanosheets. *Colloids Surfaces A Physicochem. Eng. Asp.* **2019**, *581*, 123808. [CrossRef]
26. Jiang, J.; Chen, D.; Du, X. Ratiometric electrochemiluminescence sensing platform for sensitive glucose detection based on in situ generation and conversion of coreactants. *Sens. Actuators B Chem.* **2017**, *251*, 256–263. [CrossRef]
27. Shen, X.; Zhang, T.; Xu, P.; Zhang, L.; Liu, J.; Chen, Z. Growth of C3N4 nanosheets on carbon-fiber cloth as flexible and macroscale filter-membrane-shaped photocatalyst for degrading the flowing wastewater. *Appl. Catal. B Environ.* **2017**, *219*, 425–431. [CrossRef]
28. Ma, T.; Bai, J.; Li, C. Facile synthesis of g-C 3 N 4 wrapping on one-dimensional carbon fiber as a composite photocatalyst to degrade organic pollutants. *Vacuum* **2017**, *145*, 47–54. [CrossRef]
29. Song, B.; Wang, T.; Wang, L.; Liu, H.; Mai, X.; Wang, X.; Wang, N.; Huang, Y.; Ma, Y.; Lu, Y.; et al. Interfacially reinforced carbon fiber/epoxy composite laminates via in-situ synthesized graphitic carbon nitride (g-C3N4). *Compos. Part B Eng.* **2018**, *158*, 259–268. [CrossRef]
30. Morellon-Sterling, R.; Carballares, D.; Arana-Peña, S.; Siar, E.-H.; Braham, S.A.; Fernandez-Lafuente, R. Advantages of Supports Activated with Divinyl Sulfone in Enzyme Coimmobilization: Possibility of Multipoint Covalent Immobilization of the Most Stable Enzyme and Immobilization via Ion Exchange of the Least Stable Enzyme. *ACS Sustain. Chem. Eng.* **2021**, *9*, 7508–7518. [CrossRef]
31. Arana-Peña, S.; Carballares, D.; Morellon-Sterlling, R.; Berenguer-Murcia, Á.; Alcántara, A.R.; Rodrigues, R.C.; Fernandez-Lafuente, R. Enzyme co-immobilization: Always the biocatalyst designers' choice . . . or not? *Biotechnol. Adv.* **2020**, *51*, 107584. [CrossRef] [PubMed]
32. Wang, Y.; Zhang, N.; Hübner, R.; Tan, D.; Löffler, M.; Facsko, S.; Zhang, E.; Ge, Y.; Qi, Z.; Wu, C. Enzymes Immobilized on Carbon Nitride (C3N4) Cooperating with Metal Nanoparticles for Cascade Catalysis. *Adv. Mater. Interfaces* **2019**, *6*, 1801664. [CrossRef]
33. Bilal, M.; Hussain, N.; Américo-Pinheiro, J.H.P.; Almulaiky, Y.Q.; Iqbal, H.M. Multi-enzyme co-immobilized nano-assemblies: Bringing enzymes together for expanding bio-catalysis scope to meet biotechnological challenges. *Int. J. Biol. Macromol.* **2021**, *186*, 735–749. [CrossRef] [PubMed]
34. Zhang, H.; Hua, S.; Zhang, L. Co-immobilization of cellulase and glucose oxidase on graphene oxide by covalent bonds: A biocatalytic system for one-pot conversion of gluconic acid from carboxymethyl cellulose. *J. Chem. Technol. Biotechnol.* **2019**, *95*, 1116–1125. [CrossRef]
35. Ubilla, C.; Ramírez, N.; Valdivia, F.; Vera, C.F.V.; Illanes, A.; Guerrero, C. Synthesis of Lactulose in Continuous Stirred Tank Reactor With β-Galactosidase of Apergillus oryzae Immobilized in Monofunctional Glyoxyl Agarose Support. *Front. Bioeng. Biotechnol.* **2020**, *8*, 699. [CrossRef] [PubMed]
36. Wang, M.; Wang, L.; Lyu, X.; Hua, X.; Goddard, J.M.; Yang, R. Lactulose production from lactose isomerization by chemo-catalysts and enzymes: Current status and future perspectives. *Biotechnol. Adv.* **2022**, *60*, 108021. [CrossRef]
37. Dong, L.; Yu, Z.; Zhao, R.; Peng, B.; Zhang, Y.; Wang, S. The effect of lactulose thermal degradation products on β-lactoglobulin: Linear-, loop-, and cross-link structural modifications and reduced digestibility. *Food Chem.* **2023**, *403*, 134333. [CrossRef]
38. Zhu, X.; Wang, M.; Hua, X.; Yao, C.; Yang, R. An innovative and sustainable adsorption-assisted isomerization strategy for the production and simultaneous purification of high-purity lactulose from lactose isomerization. *Chem. Eng. J.* **2020**, *406*, 126751. [CrossRef]
39. Karakan, T.; Tuohy, K.M.; Solingen, G.J.-V. Low-Dose Lactulose as a Prebiotic for Improved Gut Health and Enhanced Mineral Absorption. *Front. Nutr.* **2021**, *8*, 672925. [CrossRef]
40. Karim, A.; Aider, M. Production of prebiotic lactulose through isomerisation of lactose as a part of integrated approach through whey and whey permeate complete valorisation: A review. *Int. Dairy J.* **2021**, *126*, 105249. [CrossRef]
41. Karim, A.; Aider, M. Sustainable Valorization of Whey by Electroactivation Technology for In Situ Isomerization of Lactose into Lactulose: Comparison between Electroactivation and Chemical Processes at Equivalent Solution Alkalinity. *ACS Omega* **2020**, *5*, 8380–8392. [CrossRef] [PubMed]
42. Karim, A.; Aider, M. Sustainable Electroisomerization of Lactose into Lactulose and Comparison with the Chemical Isomerization at Equivalent Solution Alkalinity. *ACS Omega* **2020**, *5*, 2318–2333. [CrossRef] [PubMed]
43. de Albuquerque, T.L.; Gomes, S.D.L.; D'Almeida, A.P.; Fernandez-Lafuente, R.; Gonçalves, L.R.B.; Rocha, M.V.P. Immobilization of β-galactosidase in glutaraldehyde-chitosan and its application to the synthesis of lactulose using cheese whey as feedstock. *Process. Biochem.* **2018**, *73*, 65–73. [CrossRef]
44. Guerrero, C.; Valdivia, F.; Ubilla, C.; Ramírez, N.; Gómez, M.; Aburto, C.; Vera, C.; Illanes, A. Continuous enzymatic synthesis of lactulose in packed-bed reactor with immobilized Aspergillus oryzae β-galactosidase. *Bioresour. Technol.* **2018**, *278*, 296–302. [CrossRef]
45. Ramírez, N.; Ubilla, C.; Campos, J.; Valencia, F.; Aburto, C.; Vera, C.; Illanes, A.; Guerrero, C. Enzymatic production of lactulose by fed-batch and repeated fed-batch reactor. *Bioresour. Technol.* **2021**, *341*, 125769. [CrossRef]
46. Ke, P.; Zeng, D.; Xu, K.; Cui, J.; Li, X.; Wang, G. Synthesis and characterization of a novel magnetic chitosan microsphere for lactase immobilization. *Colloids Surfaces A Physicochem. Eng. Asp.* **2020**, *606*, 125522. [CrossRef]
47. Ureta, M.M.; Martins, G.N.; Figueira, O.; Pires, P.F.; Castilho, P.C.; Gomez-Zavaglia, A. Recent advances in β-galactosidase and fructosyltransferase immobilization technology. *Crit. Rev. Food Sci. Nutr.* **2020**, *61*, 2659–2690. [CrossRef]

48. Song, Y.S.; Lee, H.U.; Park, C.; Kim, S.W. Batch and continuous synthesis of lactulose from whey lactose by immobilized β-galactosidase. *Food Chem.* **2013**, *136*, 689–694. [CrossRef]
49. Majore, K.; Ciprovica, I. Bioconversion of Lactose into Glucose–Galactose Syrup by Two-Stage Enzymatic Hydrolysis. *Foods* **2022**, *11*, 400. [CrossRef]
50. Song, Y.-S.; Lee, H.-U.; Park, C.; Kim, S.-W. Optimization of lactulose synthesis from whey lactose by immobilized β-galactosidase and glucose isomerase. *Carbohydr. Res.* **2013**, *369*, 1–5. [CrossRef]
51. Araya, E.; Urrutia, P.; Romero, O.; Illanes, A.; Wilson, L. Design of combined crosslinked enzyme aggregates (combi-CLEAs) of β-galactosidase and glucose isomerase for the one-pot production of fructose syrup from lactose. *Food Chem.* **2019**, *288*, 102–107. [CrossRef] [PubMed]
52. Inanan, T. Cryogel disks for lactase immobilization and lactose-free milk production. *LWT* **2021**, *154*, 112608. [CrossRef]
53. Song, B.; Wang, T.; Sun, H.; Liu, H.; Mai, X.; Wang, X.; Wang, L.; Wang, N.; Huang, Y.; Guo, Z. Graphitic carbon nitride (g-C3N4) interfacially strengthened carbon fiber epoxy composites. *Compos. Sci. Technol.* **2018**, *167*, 515–521. [CrossRef]
54. Li, C.; Yu, S.; Gu, L.; Han, J.; Dong, H.; Wang, Y.; Chen, G. A New Graphitic Carbon Nitride/Horseradish Peroxidase Hybrid Nano-Bio Artificial Catalytic System for Unselective Degradation of Persistent Phenolic Pollutants. *Adv. Mater. Interfaces* **2018**, *5*, 1801297. [CrossRef]
55. Ajiboye, T.O.; Kuvarega, A.T.; Onwudiwe, D.C. Graphitic carbon nitride-based catalysts and their applications: A review. *Nano-Struct. Nano-Objects* **2020**, *24*, 100577. [CrossRef]
56. Wang, L.; Liu, N.; Guo, Z.; Wu, D.; Chen, W.; Chang, Z.; Yuan, Q.; Hui, M.; Wang, J. Nitric Acid-Treated Carbon Fibers with Enhanced Hydrophilicity for Candida tropicalis Immobilization in Xylitol Fermentation. *Materials* **2016**, *9*, 206. [CrossRef] [PubMed]
57. Rizi, N.S.; Shahzeydi, A.; Ghiaci, M.; Zhang, L. Photocatalytic degradation of cationic and anionic organic pollutants in water via Fe-g-C3N4/CF as a macroscopic photo-Fenton catalyst under visible light irradiation. *J. Environ. Chem. Eng.* **2020**, *8*, 104219. [CrossRef]
58. de Brito, A.R.; Tavares, I.M.D.C.; de Carvalho, M.S.; de Oliveira, A.J.; Salay, L.C.; Santos, A.S.; dos Anjos, P.N.M.; Oliveira, J.R.; Franco, M. Study of the interaction of the lactase enzyme immobilized in a carbon nanotube matrix for the development of the chemically modified carbon paste electrode. *Surf. Interfaces* **2020**, *20*, 100592. [CrossRef]
59. Costa, J.B.; Lima, M.J.; Sampaio, M.J.; Neves, M.C.; Faria, J.L.; Morales-Torres, S.; Tavares, A.P.; Silva, C.G. Enhanced biocatalytic sustainability of laccase by immobilization on functionalized carbon nanotubes/polysulfone membranes. *Chem. Eng. J.* **2018**, *355*, 974–985. [CrossRef]
60. Gennari, A.; Simon, R.; Sperotto, N.D.D.M.; Bizarro, C.V.; Basso, L.A.; Machado, P.; Benvenutti, E.V.; Renard, G.; Chies, J.M.; Volpato, G.; et al. Application of cellulosic materials as supports for single-step purification and immobilization of a recombinant β-galactosidase via cellulose-binding domain. *Int. J. Biol. Macromol.* **2022**, *199*, 307–317. [CrossRef]
61. DE Carvalho, C.T.; Júnior, S.D.D.O.; Lima, W.B.D.B.; DE Medeiros, F.G.M.; Leitão, A.L.O.D.S.; Dantas, J.M.; DOS Santos, E.S.; DE Macêdo, G.R.; Júnior, F.C.d.S. Recovery of β-galactosidase produced by Kluyveromyces lactis by ion-exchange chromatography: Influence of pH and ionic strength parameters. *An. Acad. Bras. Ciênc.* **2022**, *94*, e20200752. [CrossRef] [PubMed]
62. Sharma, H.K.; Xu, C.; Qin, W. Isolation of Bacterial Strain with Xylanase and Xylose/Glucose Isomerase (GI) Activity and Whole Cell Immobilization for Improved Enzyme Production. *Waste Biomass-Valoriz.* **2020**, *12*, 833–845. [CrossRef]
63. Xia, Y.; He, L.; Mao, J.; Fang, P.; Ma, X.; Wang, Z. Purification, characterization, and gene cloning of a new cold-adapted β-galactosidase from Erwinia sp. E602 isolated in northeast China. *J. Dairy Sci.* **2018**, *101*, 6946–6954. [CrossRef] [PubMed]
64. Boudrant, J.; Woodley, J.M.; Fernandez-Lafuente, R. Parameters necessary to define an immobilized enzyme preparation. *Process. Biochem.* **2019**, *90*, 66–80. [CrossRef]
65. Keshta, B.E.; Gemeay, A.H.; Khamis, A.A. Impacts of horseradish peroxidase immobilization onto functionalized superparamagnetic iron oxide nanoparticles as a biocatalyst for dye degradation. *Environ. Sci. Pollut. Res.* **2021**, *29*, 6633–6645. [CrossRef] [PubMed]
66. Cui, C.; Ming, H.; Li, L.; Li, M.; Gao, J.; Han, T.; Wang, Y. Fabrication of an in-situ co-immobilized enzyme in mesoporous silica for synthesizing GSH with ATP regeneration. *Mol. Catal.* **2020**, *486*, 110870. [CrossRef]
67. Souza, C.J.; Garcia-Rojas, E.E.; Favaro-Trindade, C.S. Lactase (β-galactosidase) immobilization by complex formation: Impact of biopolymers on enzyme activity. *Food Hydrocoll.* **2018**, *83*, 88–96. [CrossRef]
68. Zhang, J.; Dai, Y.; Jiang, B.; Zhang, T.; Chen, J. Dual-enzyme co-immobilization for the one-pot production of glucose 6-phosphate from maltodextrin. *Biochem. Eng. J.* **2020**, *161*, 107654. [CrossRef]
69. Mehnati-Najafabadi, V.; Taheri-Kafrani, A.; Bordbar, A.-K. Xylanase immobilization on modified superparamagnetic graphene oxide nanocomposite: Effect of PEGylation on activity and stability. *Int. J. Biol. Macromol.* **2018**, *107*, 418–425. [CrossRef]

Article

Supramolecular Self-Assembly Strategy towards Fabricating Mesoporous Nitrogen-Rich Carbon for Efficient Electro-Fenton Degradation of Persistent Organic Pollutants

Ye Chen [†], Miao Tian [*,†] and Xupo Liu

School of Materials Science and Engineering, Henan Engineering Research Center of Design and Recycle for Advanced Electrochemical Energy Storage Materials, Henan Normal University, Xinxiang 453007, China
* Correspondence: tianmiao@htu.edu.cn; Tel.: +86-156-9078-0795
† These authors contributed equally to this work.

Abstract: The electro-Fenton (EF) process is regarded as an efficient and promising sewage disposal technique for sustainable water environment protection. However, current developments in EF are largely restricted by cathode electrocatalysts. Herein, a supramolecular self-assembly strategy is adopted for synthetization, based on melamine–cyanuric acid (MCA) supramolecular aggregates integrated with carbon fixation using 5-aminosalicylic acid and zinc acetylacetonate hydrate. The prepared carbon materials characterize an ordered lamellar microstructure, high specific surface area (595 m^2 g^{-1}), broad mesoporous distribution (4~33 nm) and high N doping (19.62%). Such features result from the intrinsic superiority of hydrogen-bonded MCA supramolecular aggregates via the specific molecular assembly process. Accordingly, noteworthy activity and selectivity of H$_2$O$_2$ production (~190.0 mg L^{-1} with 2 h) are achieved. Excellent mineralization is declared for optimized carbon material in several organic pollutants, namely, basic fuchsin, chloramphenicol, phenol and several mixed triphenylmethane-type dyestuffs, with total organic carbon removal of 87.5%, 74.8%, 55.7% and 54.2% within 8 h, respectively. This work offers a valuable insight into facilitating the application of supramolecular-derived carbon materials for extensive EF degradation.

Keywords: supramolecule; mesoporous nitrogen-rich carbon; carbon fixation; electro-Fenton; organic pollutant degradation

1. Introduction

As a result of the rapid evolution of industry and agriculture, a huge amount of organic matter, such as dyeing, pharmaceutical and residual organic chemical wastewater and so forth, is discharged into the aquatic ecosystem, inducing serious environmental pollution issues. The electro-Fenton (EF) process, is a widely used electrochemical advanced oxidation process (EAOP) and is accepted as valid in sewage management to comply with measures to protect the water environment. The EF process undergoes a typical two-electron oxygen reduction reaction (2e$^-$ ORR) process, which relies heavily on the performance of cathodic catalytic materials [1–3]. In terms of appealing cathode catalysts, carbon materials with their advantages of cost-effective availability, eco-friendliness and good electrical conductivity were widely utilized in the EF degradation of various organic contaminants [4–6]. However, the current development of EF is hampered by the limited catalytic activity and stability of cathode carbon materials.

Extensive efforts were devoted to improving the performance of carbon materials, including heteroatom doping, functionalization, structural modification and molecular doping (molecular copolymerization), etc. However, common doping or modification of carbon materials have the disadvantages of tedious processing, large consumption of reagents, low doping contents and uncontrolled/non-uniform morphology. Fortunately, simple, feasible and low-toxic molecular copolymerization can eliminate the above shortcomings well. It can induce molecular assembly to generate supramolecular polymers with highly ordered microstructure/macroscopic structures via hydrogen bonding formation [7]; these are beneficial for creating efficient carbon catalysts for EF, following high-temperature treatment.

The self-assembled melamine–cyanuric acid (MCA) is a common supramolecular aggregate obtained by molecular copolymerization and regarded as a promising precursor for preparing high-performance carbon materials [8–11]. Compared with single nitrogen-containing precursors (such as urea, melamine or dicyandiamide), MCA has the following advantages: (i) MCA can solve, effectively, the problems, such as the small specific surface area and bits of exposed active sites, of carbon materials derived from single precursors. (ii) The supramolecular self-assembly process is beneficial in maintaining a specific ordered morphology at high temperatures due to the non-covalent interactions of hydrogen bonding with strong directionality [12]. Generally, most studies focused on the application of MCA supramolecular aggregates to prepare g-C_3N_4 at ~550 °C. Nevertheless, g-C_3N_4 with an ultra-high level of nitrogen doping and poor electrical conductivity is far from satisfactory for electrochemical employment. The shortcomings can be well circumvented by further converting g-C_3N_4 into N-doped carbon materials with higher conductivity at higher temperature. However, there is little or no residue when the carbonization temperature is further elevated. That is, the carbon yield can be negligible. It was found that the metal-assisted fixation of g-C_3N_4-derived carbon is valid in resolving the above issue [13–15]. Moreover, suitable metal compounds are not only able to fix the carbon, but they also tailor the pore structures and heteroatom doping of carbon materials, which, in turn, supply sufficient active sites in pursuit of the competent 2e^- ORR in the EF reaction [16,17]. Thus, exploring appropriate strategies for utilizing MCA supramolecular aggregates to fabricate efficient cathode carbon catalysts is indispensable for advancing the development of the EF process.

Herein, mesoporous nitrogen-rich carbon materials were established for EF degradation processes by carbonizing MCA supramolecular aggregates. The supramolecular precursor has a significant effect on the microstructure adjustment and N doping of the carbon matrix. The adopted 5-aminosalicylic acid and zinc acetylacetonate hydrate not only play a crucial role in carbon fixation, but also affect the specific surface area and N/O content of carbon materials. The synthesized carbon materials show a regular lamellar-like morphology, large specific surface area, rational structural defect and high nitrogen content, owing to the structural superiority of supramolecular aggregates. The optimized carbon catalysts exhibit a promising application in EF degradation. Numerous organic pollutants, namely, basic fuchsin (BF), mixed triphenylmethane-type dyes, phenolic substances and antibiotics containing chloramphenicol were highly degraded and mineralized with admirable stability and reusability. Therefore, this work supplies a simple approach for the preparation of advanced cathode carbon materials to facilitate the practical applications of EF in organic pollutant treatment.

2. Materials and Methods

2.1. Chemicals and Materials

Melamine ($C_3H_6N_6$, MA), cyanuric acid ($C_3H_3N_3O_3$, CA), 5-aminosalicylic acid ($C_7H_7NO_3$), the antibiotic chloramphenicol ($C_{11}H_{12}Cl_2N_2O_5$), phenol (C_6H_6O), and four triphenylmethane-type dyes of basic fuchsin ($C_{20}H_{20}ClN_3$, BF), malachite green ($C_{23}H_{26}N_2$, MG), victoria blue B ($C_{33}H_{32}ClN_3$, VB) and leucocrystal violet ($C_{25}H_{31}N_3$, LV) were all purchased from Aladdin (Shanghai, China). Zinc acetylacetonate hydrate ($C_{10}H_{14}ZnO_4 \cdot xH_2O$) was supplied by Macklin. Sodium hydroxide (NaOH), sulfuric acid (H_2SO_4), ethanol (C_2H_6O), hydrochloric acid (HCl), phosphoric acid (H_3PO_4), iron (II) sulfate heptahydrate ($FeSO_4 \cdot 7H_2O$), sodium sulfate (Na_2SO_4), isopropyl alcohol (C_3H_8O) and potassium titanyl oxalate ($K_2TiO \cdot C_4O_8 \cdot 2H_2O$) were acquired from Alfa Aesar (Shanghai, China). All chemicals were of analytical grade and employed without extra purification.

2.2. Preparation of Mesoporous Nitrogen-Rich Carbon Materials

For the typical preparation of carbon materials, 2 g melamine, 2 g cyanuric acid, 0.31 g 5-aminosalicylic acid and different contents of zinc acetylacetonate hydrate were dispersed in 200 mL ethanol under vigorous stirring and subsequent ultrasonic processing for 4 h [10,11]. The milky precipitates were gathered by centrifugation and rinsed with ethanol several times, and then vacuum dried at 60 °C overnight. Afterward, the ground powders were delivered to the tube furnace for calcining at 550 °C (at a ramp rate of 2.3 °C min^{-1}), subsequently rising to 900 °C at 5 °C min^{-1} under a continuous N$_2$ flow for 2 h. The products were washed in HCl solution and deionized water several times, followed by drying at 40 °C. The finally obtained samples were referred to as MCAN-x, in which x represented the mass of zinc acetylacetonate hydrate (x = 0.5, 1, 2 and 3 g).

Additionally, the control samples of MAN-1 and CAN-1 were prepared using the same synthetic process except for precipitating the single melamine and cyanuric acid, respectively. The samples without 5-aminosalicylic acid or zinc acetylacetonate hydrate participation were labelled MCA-1 and MCAN-0, respectively. The N-1 sample was also prepared using 5-aminosalicylic acid and zinc acetylacetonate hydrate only.

2.3. Characterizations

Field emission scanning electron microscopy (FE-SEM) for the microstructure and morphological characteristics was conducted at Hitachi High-Technologies Corporation SU8010 and allied to transmission electron microscopy (TEM) at JEOL JEM-2100. The textural properties and pore structures were acquired from Quantachrome Autosorb Station iQ2 apparatus. X-Ray Diffraction (XRD) patterns for crystalline structure were employed on the Bruker-D8 apparatus (Cu Kα radiation) in the scope of 5~70°. Raman spectra for the graphitization degree of the samples were measured using a LabRAM HR Evolution (532 nm laser source). The surface element information and functional groups were gathered by the X-ray photoelectron spectroscopy (XPS) spectra (Thermo Scientific ESCALAB250) and Fourier transform infrared spectroscopy (FTIR) spectra (Thermo Nicolet Corporation NEXUS equipment). The hydrophilicity was recorded on a KRüSS DSA25 contact angle meter. A TG/DTA thermal analyzer (NETZSCH STA449F3) was employed for weight variation of the samples.

2.4. Electrochemical Measurements

To investigate the electrochemical performance of the obtained catalysts, a cyclic voltammogram (CV) was firstly operated within a three-electrode system (3 mm-diameter glassy carbon coated by catalysts as working electrodes, platinum foil and saturated calomel electrode (SCE) as counter and reference electrodes) and measured by an electrochemical workstation (CHI 660E). The electrolyte was 0.05 mol·L^{-1} Na$_2$SO$_4$ solution (pH = 2.0) purified with N$_2$ or O$_2$. Subsequently, the experimental conditions of the linear-sweep voltammogram (LSV) were similar to those of the CV, except that the working electrode diameter was 4 mm. The H$_2$O$_2$ concentration was estimated by a UV–Vis spectrophotometer (TU-1810) using the titanium (IV) spectrophotometric method at an absorption wavelength of 400 nm.

The EF degradation systems were established for various pollutants and different conditions (cathode potentials with −0.8 V vs. SCE, Fe^{2+} contents, initial concentrations and pH values). The target organic pollutants were BF (C_0 = 10 mg·L^{-1}), the mixed dyes (C_0 = 10 mg·L^{-1}, containing BF, malachite green (MG), victoria blue B (VB) and crystal violet (LV)), the chloramphenicol (C_0 = 20 mg·L^{-1}) and the phenol (C_0 = 20 mg·L^{-1}). O$_2$ was supplied for 15 min in the EF systems with the precipitation of FeSO$_4$ (0.2 mmol·L^{-1}) and Na$_2$SO$_4$ (0.05 mol·L^{-1}). The pH values for all the electrolytes were 2.0 and were adjusted with 1 mol·L^{-1} H$_2$SO$_4$ and 0.1 mol·L^{-1} NaOH.

The removal efficiency of the pollutant was obtained from Equation (1).

$$\text{removal efficiency (\%)} = \frac{C_0 - C_t}{C_0} \times 100\% \qquad (1)$$

where C_t and C_0 represent the final and initial pollutant concentrations (mg L^{-1}), respectively. The pseudo-first-order kinetics were estimated by the following Equation (2).

$$\ln \frac{C_t}{C_0} = -kt \tag{2}$$

where the rate constant (k, min^{-1}) means the apparent rate constant, the representations of C_t and C_0 are the same as the above.

Furthermore, the mineralization of different pollutants was evaluated by the Vario TOC analyzer (Elementar Analysensysteme GmbH). The mineralization efficiency was obtained by Equation (3).

$$\text{TOC removal (\%)} = \frac{TOC_0 - TOC_t}{TOC_0} \times 100\% \tag{3}$$

where TOC_0 and TOC_t are the measured TOC values before and after the degradation treatments, respectively.

3. Results and Discussion

3.1. Catalyst Characterization

Supramolecular self-assembly strategy was proposed to synthesize the mesoporous nitrogen-rich carbon materials. The hydrogen-bonded MCA supramolecular precursor was generated through the molecular cooperative assembly process. MCA supramolecular aggregates were further converted into nitrogen-doped porous carbon after subsequent calcination and pickling treatments with the assistance of the carbon-fixation effect via 5-aminosalicylic acid and zinc acetylacetonate hydrate. The formation mechanism of carbon materials is shown in Figure 1a. MCA supramolecular aggregates are employed as a structure-directing agent for creating a mesoporous lamellar structure. As shown in Figure 1b, the preparative MCA precursor shows the morphology of stacked-sheet aggregates. The morphology becomes much thinner during the calcination process. The obtained MCAN-x carbon materials all exhibit a crimped and flexible lamellar-like architecture with uniformly dispersed pores (Figures 1c,d and S1), which is profitable for the exposure of active sites to improve electrocatalytic activity; however, the MAN-1 or CAN-1 carbon materials prepared only with melamine or cyanuric acid show an irregularly blocky structure (Figure S1e,f). The obvious difference in morphologies confirms the noteworthy structural superiority of the supramolecular aggregates via the self-assembly process for the formation of specific flaky carbon nanosheets in contrast with the single precursors. The TEM image of MCAN-1 in Figure 1e,f reveals an appealing nanosheet-assembly structure. The nanosheets possess a lattice space of 0.34 nm (Figure 1g), which corresponds well to the (002) plane of graphitic carbon [18]. The results illustrate that the porous carbon materials with ultrathin graphite nanosheets are successfully synthesized through the supramolecular self-assembly strategy. Porous carbon nanosheets can increase the specific surface areas and expose more active sites, promoting ORR electrocatalytic activity.

To investigate the formation mechanism of porous carbon materials, TG analysis of different precursors was conducted and the results are shown in Figure 2a,b. It is observed that the pure MCA supramolecular aggregates only exhibit a residual mass of 0.9 wt.% after high-temperature calcination. Residuals of only 4.9 wt.% and 1.5 wt.% are achieved for MCAN-0 and MCA-1, respectively. However, an ~10.0 wt.% residual is detected for MCAN-1 at the pyrolysis temperature of 900 °C. It can be inferred that the actively reactive groups of amino, hydroxyl and carboxyl in 5-aminosalicylic acid precipitate in the molecular-assembly process, thereby protecting the carbon skeletons well. The hydroxyl and carbonyl groups interact with the MCA carbon skeleton via hydrogen bonding for carbon matrix fixation. Meanwhile, the decomposition products from zinc acetylacetonate hydrate (such as zinc nitride and zinc oxide) can also prevent the complete decomposition of g-C_3N_4 during the calcination process. Together, 5-aminosalicylic acid and zinc acetylacetonate hydrate wield a synergistic impact on the reconstitution process of the gaseous nitrogen-containing carbonaceous species, which transforms the g-C_3N_4 intermediates to the MCA-derived N-doped carbon materials. The above carbon-fixation process is in competition with the pyrolysis of the MCA thermal polycondensation-produced g-C_3N_4 that both of which volatilize to induce the pore structures. In other words, the MCA precursor not only plays a role in N doping during the evolution process, but also serves as a soft template to create pores and increase the specific surface area by generating the N-containing molecular gas.

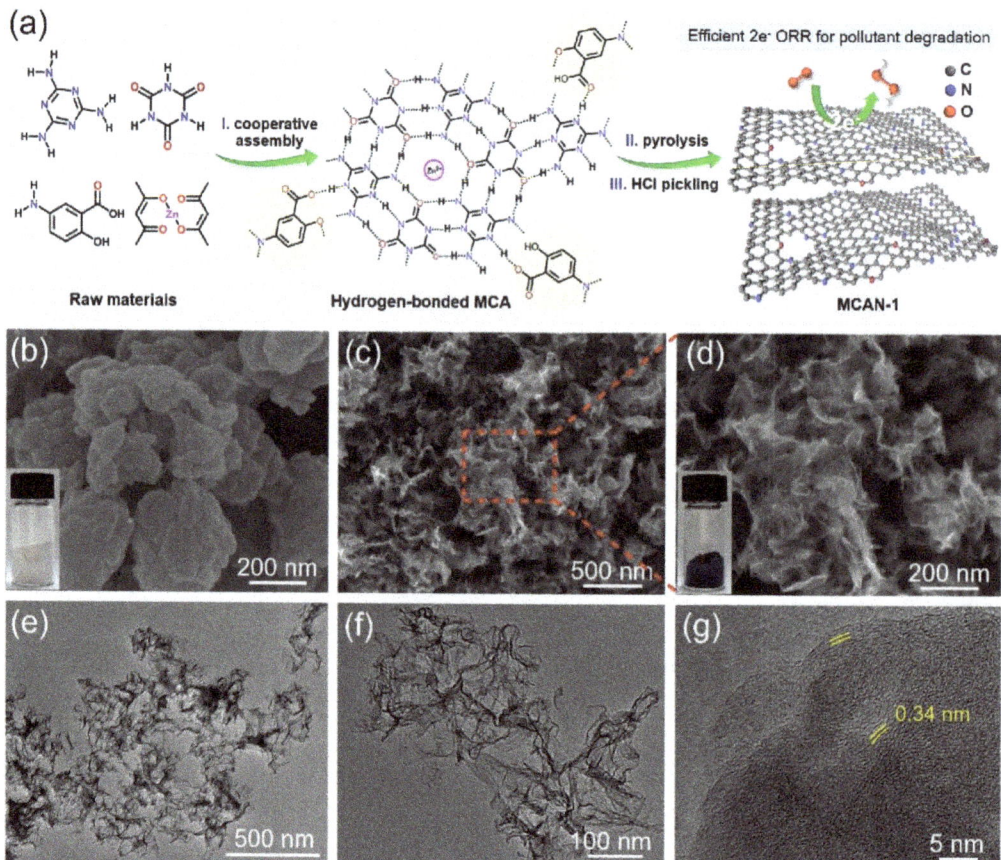

Figure 1. (a) MCAN-1 synthetic procedure; (b) SEM images of the MCAN-1 precursor (inset: the corresponding photograph); (c,d) SEM images with different magnifications of the obtained MCAN-1 product (inset: the corresponding photograph of MCAN-1); (e,f) TEM and (g) HR-TEM MCAN-1 images.

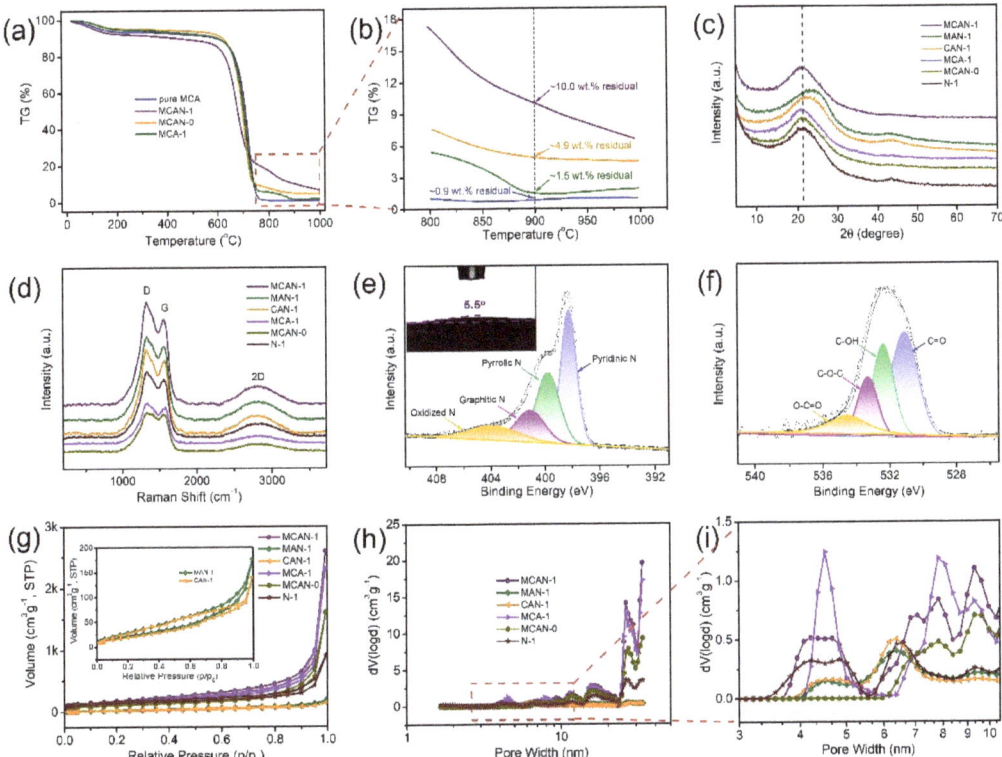

Figure 2. (a) Thermogravimetric analysis of intermediate products of pure MCA, MCAN-1, MCAN-0 and MCA-1; (b) enlarged figure of thermogravimetric curves; (c) XRD patterns and (d) Raman spectra for MCAN-1, MAN-1, CAN-1, MCA-1, MCAN-0 and N-1. High-resolution XPS spectra for (e) N 1s (inset is contact angle of representative MCAN-1) and (f) O 1s for MCAN-1; (g) N_2 adsorption/desorption isotherm (illustration: enlarged isotherms for MAN-1 and CAN-1); (h) pore size distribution for MCAN-1, MAN-1, CAN-1, MCA-1, MCAN-0 and N-1 and (i) enlarged figure of pore size distributions at about 3~10 nm.

In Figures 2c and S2, all the as-synthesized samples exhibit similar XRD patterns. The discernible peak at ~21° and the indistinct peak at ~43° are indexed to (002) and (100) crystal faces of graphitic carbon, respectively [19,20]. There is no metal-related peak observed as the zinc doping amounts gradually increase, signifying that the metal species are thoroughly removed. It is notable that the XRD peaks of MAN-1 and CAN-1 at ~23° exhibit a positive shift compared with the carbon materials derived from the MCA supramolecular precursor, revealing more defects and broader interlayer distances for MCAN-x [14,21]. Furthermore, the defect density was explored by Raman spectra. In Figures 2d and S3, the characteristic peaks at ~1347 cm^{-1} (D band) and ~1560 cm^{-1} (G band) point to the disordered carbon and sp^2-hybridized carbon, respectively [22,23]. The I_D/I_G values of all samples are in the range of 1.38~1.53 with massive defective sites. The defects originating from the N-doping and porous structures can efficiently promote the H_2O_2 transfer process from the cathode surface and further accumulate the pivotal OOH* intermediates to hinder the generation of H_2O in electro-Fenton degradation [24].

The surface C, N and O contents of MCAN-1 are estimated as 75.99 at.%, 19.62 at.% and 4.39 at.%, respectively, by XPS measurement (listed in Table 1), showing an obvious carbon-fixation effect from the self-assembly strategy based on 5-aminosalicylic acid and zinc acetylacetonate hydrate. The related hydrophilic groups contribute to a strong hydrophilicity with a contact angle of 5.5° for MCAN-1 (Figure 2e inset) [25]. The N-rich feature of catalysts causes a favorable electron-donating ability, which enhances electron delocalization and, consequently, creates sufficient active sites for ORR [26,27]. The MCAN-x samples are enriched with N functional groups, containing pyridinic N (~398.3 eV), pyrrolic N (~399.8 eV), graphitic N (~401.1 eV) and oxidized N (~403.8 eV) (Figure 2e) [28]. The deconvoluted O 1s spectra exhibit four peaks related to C=O (~531.3 eV), C–OH (~532.3 eV), C–O–C (~533.5 eV) and O–C=O (~534.5 eV) bonds (Figure 2f). The latter two groups are widely admitted as active sites to facilitate the reduction process from O_2 to H_2O_2 [29,30]. The N contents in the carbon matrix can be readily tailored via varying different precursors, such as MAN-1 (15.61 at.%), CAN-1 (13.6 at.%) and especially N-1 (without MCA precursor, 5.79 at.%), indicating that the self-assembly of MCA provides a large amount of nitrogen (Table 1 and Figures S4–S6). Differing from the MCAN-0 (13.42 at.%), the N contents of the samples with Zn salts are 15.41~19.62 at.%. Thus, it can be inferred that the partial N from MCA supramolecular aggregates can be retained in the carbon frameworks during the carbon-fixation process by Zn species, while the remaining MCA decomposes resulting in pore creation during high-temperature calcination.

Table 1. Summary of characteristics and performance for the synthesized samples.

Sample	(A) S_{BET} [$m^2 \cdot g^{-1}$]	I_D/I_G	N Content [at.%]	O Content [at.%]	(B) Removal Efficiency [%]	k [min^{-1}] (R^2)	(C) TOC Removal [%]
MCAN-0.5	626	1.47	15.42	8.66	95.7	0.050 (0.992)	81.8
MCAN-1	595	1.52	19.62	4.39	98.0	0.063 (0.992)	87.5
MCAN-2	492	1.51	19.34	4.94	93.2	0.044 (0.990)	73.3
MCAN-3	397	1.53	18.26	5.12	90.9	0.039 (0.993)	70.3
MAN-1	84	1.44	15.61	6.21	84.1	0.030 (0.995)	49.9
CAN-1	77	1.41	13.63	7.25	78.0	0.024 (0.994)	43.6
MCAN-0	374	1.38	13.42	4.03	88.3	0.034 (0.993)	52.5
MCA-1	494	1.43	14.39	4.24	93.6	0.043 (0.990)	57.3
N-1	467	1.44	5.79	10.38	81.1	0.026 (0.990)	57.0

(A) Specific surface area of the samples; (B) BF removal efficiency for different samples within 1 h; (C) TOC removal in EF degradation of BF dyestuff for different samples after 8 h.

Specific surface area and porosity were explored by N_2 adsorption–desorption experiments. All the samples feature similar Type IV isotherms with H3 hysteresis loops (Figures 2g and S7a), which correspond to widespread slit-shaped mesoporosity in pore size distributions (PSDs) (Figures 2h,i and S7b). The specific surface areas of MCAN-x are located in the range of 374~626 $m^2 \cdot g^{-1}$ with pore volumes of 2~4 $cm^3 \cdot g^{-1}$. As expected, the MCAN-x samples derived from MCA supramolecular aggregates exhibit much larger surface areas than those of MAN-1 (84 $m^2 \cdot g^{-1}$) and CAN-1 (77 $m^2\ g^{-1}$), demonstrating that the textural characteristics of carbon materials can be improved by the supramolecular self-assembly strategy [31]. In addition, the lower specific surface area of N-1 attains 467 $m^2\ g^{-1}$ compared with that of MCAN-1 with the addition of MCA precursor (595 $m^2\ g^{-1}$), which further attests to the pore-forming ability of MCA. The results match well with the former SEM/TEM analysis. The higher surface area can expose more active species and provide desirable access to the reactants and the cathode material surface, thereby facilitating, in a promising way, the 2e⁻ ORR catalytic activity. In addition, similar PSDs are acquired and distributed at about 4~33 nm for the synthesized samples. The widespread mesoporous structures can act as diffusion channels for the O_2 and penetration of electrolytes, consequently expediting the activity and selectivity of H_2O_2 generation via the 2e⁻ ORR pathway.

3.2. Electrochemical Measurements

To unravel ORR activity and selectivity in the representative sample of MCAN-1, CV and LSV measurements were conducted in both N_2/O_2-saturated Na_2SO_4 electrolytes (pH = 2.0). Upon O_2 saturating, MCAN-1 exhibits a distinct reduction peak at −0.064 V (vs. SCE), as shown in Figure 3a. In view of the diffusion-controlled measurements, the limiting current density gradually rises with the increased rotation rates (400–2025 rpm), as shown in Figure 3b. The corresponding Koutecky–Levich (K–L) curves are further estimated and feature a definite linearity (inset of Figure 3b). The electron transfer numbers (n) are calculated to be about 2.0~2.6 in the potential range of −0.8~−0.3 V from the K–L equation, indicating a two-electron ORR process (Figure 3c). Besides, the uninterrupted H_2O_2 accumulation in MCAN-1 was recorded to be ~190.0 mg·L^{-1} with 120 min, which is much higher than most of the reported H_2O_2 yields (Figure 3d) [32–37]. The depletion of H_2O_2 concentrations can be ignored after three cycles of continuous measurement by adopting the same MCAN-1 cathode, indicating its noteworthy durability. The H_2O_2 can be activated into •OH on cathode surfaces and bulk solution for further mineralization of the organic dye molecules to CO_2 and H_2O [38].

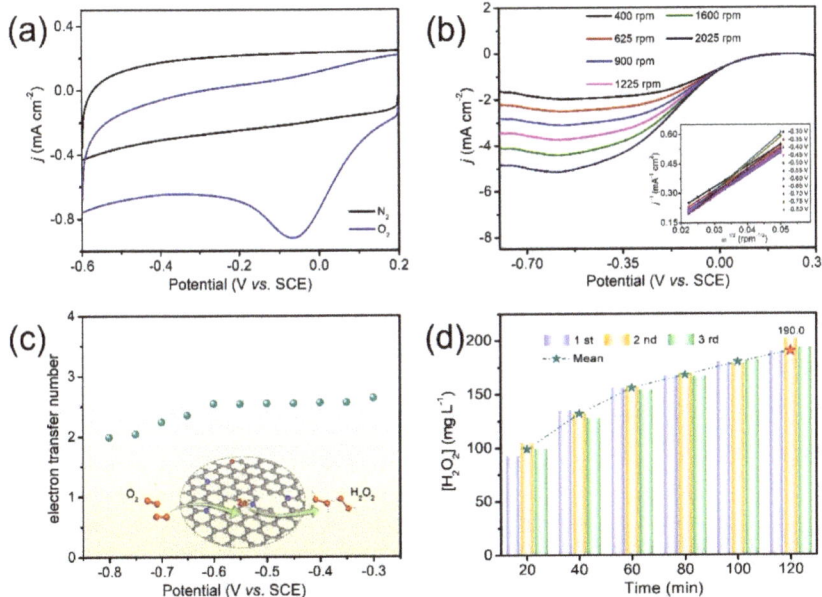

Figure 3. (**a**) CV curves and (**b**) RDE curves at 400~2025 rpm (illustration: the corresponding K–L plot) of MCAN-1; (**c**) electron transfer numbers acquired at diverse potentials of MCAN-1; (**d**) accumulated H_2O_2 yields.

3.3. Degradation of Organic Pollutants

The application of synthetic samples in the EF degradation of organic pollutants was further investigated. The schematic diagram of the EF degradation process is depicted in Figure 4a. The removal efficiency (C/C_0) of BF is employed to evaluate the performances of MCAN-x and control samples with −0.8 V. As shown in Figure 4b and Figure S8a, the C/C_0 value of MCAN-1 after 60 min was determined to be 0.021, which is the lowest among all the samples, demonstrating the superior degradation ability. The excellent performance is attributed to the large S_{BET} and porous characteristic for accelerating O_2 diffusion and providing plentiful active sites for H_2O_2 generation. The chemical kinetics of the oxidative degradation process was further investigated. As Figure 4c, Figure S8b and Table 1 show, the $-\ln(C/C_0)$ has a linear relationship to the reaction time (t) for all samples, indicating the typical pseudo-first-order kinetics of the BF degradation reaction. The k from the curve slope of $-\ln(C/C_0)$ vs. t directly reflects the degradation rate of BF, in which MCAN-1 with the highest k-value of 0.063 min^{-1} has the fastest degradation reaction process. The degradation efficacy of MCAN-1 can reach 98.0% in the short period of 1 h, evaluated by UV–Vis curves (Figure 4d). The absorption peaks of BF dyestuff are intuitively faded away accompanied by the magenta decoloration (Figure 4e,f). Thus, the as-prepared MCAN-1 delivers promising application in EF degradation of organic pollutants.

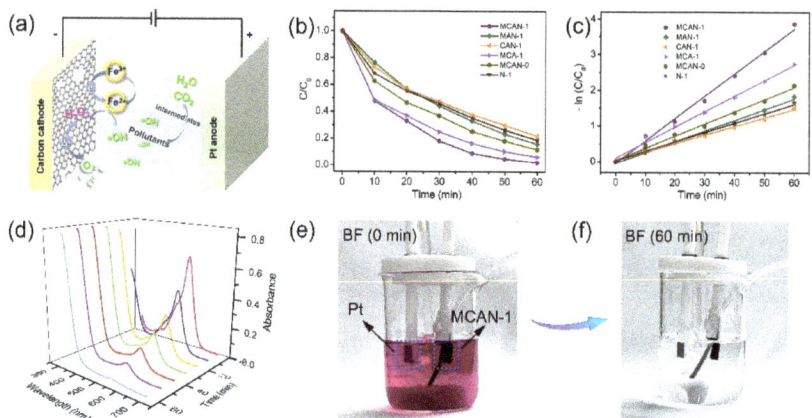

Figure 4. (a) Schematic diagram of EF degradation; (b) EF degradation of BF by MCAN-1 and the control samples and (c) the corresponding variation of $-\ln(C/C_0)$. (d) UV–Vis spectra of BF degradation for MCAN-1. Photographs of color changes for basic fuchsin solutions (e) before and (f) after EF degradation process.

In addition, isopropanol generally acts as an •OH scavenger and was added to the BF solution for the purpose of demonstrating the contribution of •OH during EF degradation. As a result, the BF degradation efficiency is apparently inhibited at ~51.8% (k = 0.011 min^{-1}) within 1 h (Figure 5a), confirming the significance of •OH species on the oxidation degradation of BF dyestuff in the EF process. To investigate the practicability of catalysts, their electrochemical stability and reusability for the oxidization of organic contaminants were further explored. The response of the current shows that they do not decay alongside the continuous mineralization process of BF within ~8 h (Figure 5b). The relatively steady current manifests a good stability in MCAN-1, which profits from the stable carbon skeletons and porous features for active site preservation. This also implies an indirect cathodic oxidation process (EF) instead of the direct electrochemical destruction during the dyestuff removal [39]. Furthermore, the MCAN-1 catalyst only yields about 6% removal efficiency recession (~98% to ~92%) within 10 successive runs (Figures 5c and S9 and Table S1) because of its well-preserved reusability and stability during the EF degradation process.

To achieve the optimized degradation conditions, the influencing parameters for EF degradation of BF were also systematically exploited for the MCAN-1 catalyst, including Fe^{2+} contents, pH values, initial concentrations and cathode potentials. Firstly, the influence of Fe^{2+} contents was investigated as shown in Figure 5d (top). Poor degradation efficiency results from a low concentration of 0.1 mM being supplied. An enhanced degradation efficiency is seen with the incremental Fe^{2+} content (0.2 mM), which supplies sufficient •OH species and strong oxidation ability [40]. However, as the concentrations increase further, a negative BF-oxidation efficiency is presented resulting from the •OH depletion by the redundant Fe^{2+} [24]. The k-values are calculated to be 0.026, 0.063, 0.043 and 0.030 min^{-1} from 0.1 mM to 0.8 mM (Figure S10a and Table S2). The highest k-value was obtained for 0.2 mM, indicating the fastest reaction mechanism. Thus, 0.2 mM is selected as the optimal Fe^{2+} concentration for the EF degradation of BF. Secondly, it is well known that the pH values of electrolytes have an important effect on ORR properties, owing to the proton-coupled electron transfer process [41]. The BF degradation efficiencies were explored in electrolytes with various pH values, with results shown in Figure 5d (bottom). Maximum efficiency was achieved at pH = 2.0. A distinct decrement in the BF degradation ability from 98.0% to 22.3% was obtained with the relevant k values from 0.063 min^{-1} to 0.004 min^{-1} when the pH values rose from 2.0 to 7.0 (Figure S10b and Table S2). The decreased acidity causes H_2O_2 decomposition and insoluble iron precipitation, thereby reducing catalytic activity and hindering sustainably oxidative degradation performance [24,42].

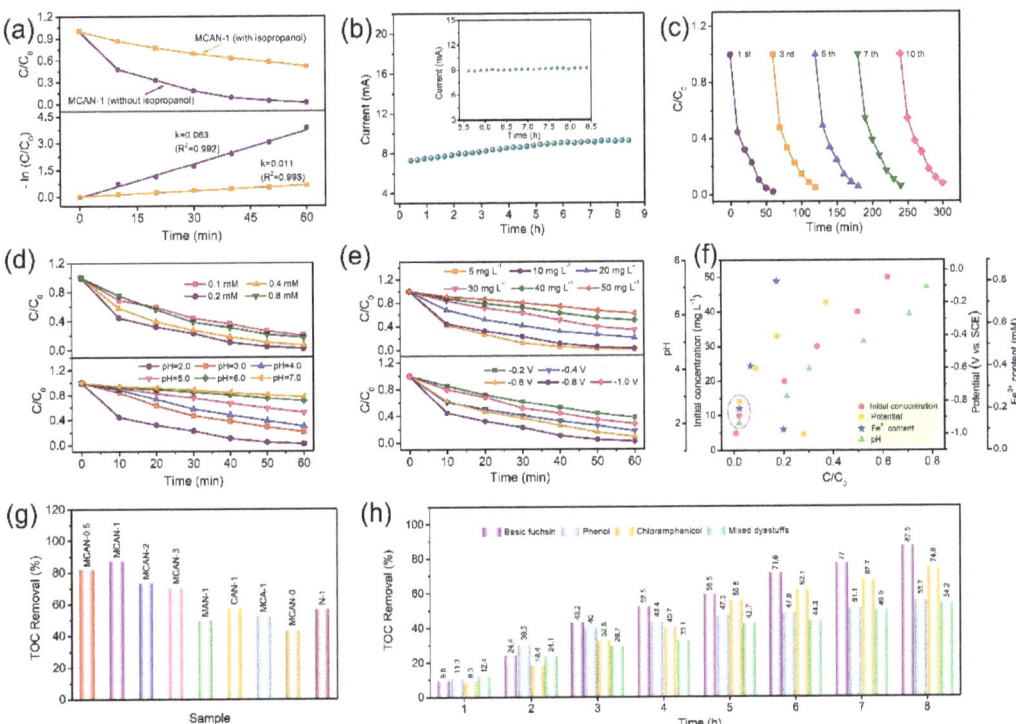

Figure 5. (a) The effect of radical scavenger (isopropanol); (b) chronoamperometry curve for MCAN-1 (illustration: enlarged curve at about 5.5~8 h); (c) reusability of MCAN-1 for BF degradation; (d) EF degradation for BF using MCAN-1 cathode with different Fe^{2+} contents (**top**) and pH values (**bottom**); (e) EF degradation for BF by using MCAN-1 cathode with different initial concentrations (**top**) and potentials (**bottom**); (f) analysis of the optimized degradation conditions for BF degradation process; (g) TOC removal of BF for all samples; (h) TOC removal of BF, phenol, chloramphenicol and mixed triphenylmethane-type dyes by MCAN-1.

Furthermore, the impact of initial BF concentrations on degradation performance was concretely explored in Figure 5e (top). As anticipated, intensive BF degradation efficiency is accomplished at 5 mg L^{-1} (the lowest concentration) due to the fact that the few dyestuff molecules must be degraded based on the interfacial reaction at the same •OH production ability. All the removal efficiencies decrease as the initial concentrations gradually increase. The decreasing k values of 0.079, 0.063, 0.026, 0.018, 0.012 and 0.008 min^{-1} are presented from 5 to 50 mg·L^{-1} initial BF concentrations, respectively (Figure S10c and Table S2). In addition, the effect of supplied potentials was studied. In general, the supplied potentials manage the electron transfer between two electrodes and have a significant effect on H_2O_2 production. Furthermore, the appropriate potential is also beneficial in reducing the energy consumption [24]. In Figure 5e (bottom), the different potentials are compared against BF degradation using the same MCAN-1 cathode. As the applied potentials increase from −0.2 V to −1.0 V, the degradation efficiency gradually rises and then declines, with the optimum performance at −0.8 V. The negative potentials are conducive to the high-efficiency electron transfer, promoting the reduction of O_2 to H_2O_2, while excessive potentials lead to the preferential H_2 evolution or H_2O_2 degradation with electron consumption, consequently reducing the current efficiency [34,43]. In Figure S10d and Table S2, the degradation process follows the pseudo-first-order kinetics at different potentials. The corresponding k-value of MCAN-1 under −0.8 V is also higher than those of −0.2 V (0.017 min^{-1}), −0.4 V (0.028 min^{-1}), −0.6 V (0.039 min^{-1}) and −1.0 V (0.021 min^{-1}), indicating that the fastest degradation rate can be reached when the potential is −0.8 V. As a result, the optimal efficiency for EF degradation based on MCAN-1 is achieved in Na_2SO_4 electrolyte (appropriate concentration, 0.2 mM of Fe^{2+}, pH = 2.0) at the reaction voltage of −0.8 V (Figure 5f). The as-prepared MCAN-1 catalyst shows a potential application for practical contaminant removal. Furthermore, the mineralization performances of all the prepared catalysts were assessed by TOC measurement. As can be seen from Figure 5g and Table 1, MCAN-1 exhibits the optimal mineralization efficiency with a TOC removal value of 87.5%, much higher than the efficiencies of MCAN-0.5 (81.8%), MCAN-2 (73.3%), MCAN-3 (70.3%), MCA-1 (57.3%), N-1 (57.0%), MCAN-0 (52.5%), MAN-1 (49.9%) and CAN-1 (43.6%). The consistency of the EF degradation capacity for the MCAN-1 cathode was also studied with the antibiotic chloramphenicol, phenol, and the mixed triphenylmethane-type dyes containing BF, MG, VB and CV as representative contaminants. The TOC removal efficiencies of MCAN-1 were determined to be 74.8%, 55.7% and 54.2% within 8 h for the above three target contaminants (Figure 5h), revealing its appealing degradation ability for various types of organic contaminants.

4. Conclusions

In summary, mesoporous nitrogen-rich carbon materials were fabricated relying on the MCA supramolecular aggregates with the assistance of 5-aminosalicylic acid and zinc acetylacetonate hydrate. The supramolecular precursor has a significant effect on the microstructure of the carbon material, owing to the non-covalent interactions of hydrogen bonding with strong directionality. The optimized MCAN-1 is composed of a lamellar morphology, high N content of 19.62 at.%, abundant mesoporosity and a large specific surface area, resulting in enhanced exposure of active sites for facilitating $2e^-$ ORR. The as-prepared carbon materials exhibit noteworthy potential application in the degradation and mineralization of different models of organic pollutants in EF systems. Serving as the cathode for EF degradation, the representative MCAN-1 delivers excellent mineralization ability for basic fuchsin (87.5%), mixed triphenylmethane-type dyes (54.2%), phenol (55.7%) and chloramphenicol (74.8%) in 8 h. The carbon materials containing desirable stability and reusability show a promising application prospect for contaminant removal. This work provides a simple and delicate approach to synthesizing the mesoporous N-rich carbon materials with adjustable microstructures to be used as promising electrocatalysts for aquatic environmental pollution improvement.

Supplementary Materials: The following supporting information can be downloaded at: https://www.mdpi.com/article/10.3390/nano12162821/s1, Figure S1: SEM images of (a) MCAN-0.5, (b) MCAN-2, (c) MCAN-3, (d) MCAN-0, (e) MAN-1, (f) CAN-1, (g) MCA-1 and (h) N-1; Figure S2: XRD patterns of MCAN-0.5, MCAN-2 and MCAN-3; Figure S3: Raman spectra of MCAN-0.5, MCAN-2 and MCAN-3; Figure S4: (a) Full−scan XPS spectra of all the samples and (b) the high-resolution XPS spectra of C 1s of MCAN-1; Figure S5: High-resolution XPS spectra of C 1s of (a) MCAN-0.5, (d) MCAN-2, (g) MCAN-3, (j) N-1. High-resolution XPS spectra of N 1s of (b) MCAN-0.5, (e) MCAN-2, (h) MCAN-3, (k) N-1. High-resolution XPS spectra of O 1s of (c) MCAN-0.5, (f) MCAN-2, (i) MCAN-3, (l) N-1; Figure S6: High-resolution XPS spectra of C 1s of (a) MAN-1, (d) CAN-1, (g) MCA-1, (j) MCAN-0. High-resolution XPS spectra of N 1s of (b) MAN-1, (e) CAN-1, (h) MCA-1, (k) MCAN-0. High-resolution XPS spectra of O 1s of (c) MAN-1, (f) CAN-1, (i) MCA-1, (l) MCAN-0; Figure S7: (a) N_2 adsorption/desorption isotherm and (b) pore size distribution (illustration: enlarged pore size distributions at about 3~10 nm) of MCAN-0.5, MCAN-2 and MCAN-3; Figure S8: (a) EF degradation of BF by MCAN-0.5, MCAN-2 and MCAN-3 and (b) the corresponding variation of

−ln(C/C$_0$); Figure S9: Variation of −ln(C/C$_0$) for BF degradation with MCAN-1 as cathode in 10 cycles; Figure S10: Variation of −ln(C/C$_0$) for BF degradation with MCAN-1 as cathode under different (a) Fe^{2+} contents, (b) pH values, (c) initial concentrations and (d) potentials; Table S1: The rate constant k in 10 cycles; Table S2: The rate constant k under different influencing parameters, i.e., cathode potential, Fe^{2+} content, initial concentration (C$_0$), and pH value.

Author Contributions: Conceptualization, Y.C. and M.T.; methodology, Y.C. and M.T.; validation, Y.C. and M.T.; formal analysis, M.T.; investigation, M.T.; resources, Y.C. and M.T.; data curation, M.T.; writing—original draft preparation, M.T.; writing—review and editing, Y.C. and X.L.; visualization, M.T.; supervision, X.L.; funding acquisition, Y.C. All authors have read and agreed to the published version of the manuscript.

Funding: This work was supported by the National Natural Science Foundation of China (Grant Nos. U1804255), the Key Research and Development and Promotion Projects in Henan Province (Grant no. 222102520038, 212102210651 and 222102320147), the Key Scientific Research Project of Henan Province Higher Education of China (No. 21A430020) and the Scientific and Technological Research Project of Xinxiang, China (No. GG2020020).

Data Availability Statement: The data presented in this study are available on request from the corresponding author.

Conflicts of Interest: The authors declare no conflict of interest.

References

1. Tian, Q.; Xiao, F.; Zhao, H.; Fei, X.; Shen, X.; Postole, G.; Zhao, G. Simultaneously accelerating the regeneration of FeII and the selectivity of 2e$^−$ oxygen reduction over sulfide iron-based carbon aerogel in electro-Fenton system. *Appl. Catal. B Environ.* **2020**, *272*, 119039. [CrossRef]
2. Huang, B.-C.; Jiang, J.; Wang, W.-K.; Li, W.-W.; Zhang, F.; Jiang, H.; Yu, H.-Q. Electrochemically catalytic degradation of phenol with hydrogen peroxide in situ generated and activated by a municipal sludge-derived catalyst. *ACS Sustain. Chem. Eng.* **2018**, *6*, 5540–5546. [CrossRef]
3. Ko, Y.-J.; Kim, H.-G.; Seid, M.G.; Cho, K.; Choi, J.-W.; Lee, W.-S.; Hong, S.W. Ionic-liquid-derived nitrogen-doped carbon electrocatalyst for peroxide generation and divalent iron regeneration: Its application for removal of aqueous organic compounds. *ACS Sustain. Chem. Eng.* **2018**, *6*, 14857–14865. [CrossRef]
4. Zhu, Y.; Deng, F.; Qiu, S.; Ma, F.; Zheng, Y.; Lian, R. Enhanced electro-Fenton degradation of sulfonamides using the N, S co-doped cathode: Mechanism for H$_2$O$_2$ formation and pollutants decay. *J. Hazard. Mater.* **2021**, *403*, 123950. [CrossRef] [PubMed]
5. Yang, Y.; Qiao, S.; Zhou, J.; Quan, X. A novel porous-carbon-based hollow fiber membrane with electrochemical reduction mediated by in-situ hydroxyl radical generation for fouling control and water treatment. *Appl. Catal. B Environ.* **2019**, *255*, 117772. [CrossRef]
6. Ren, G.; Zhou, M.; Su, P.; Yang, W.; Lu, X.; Zhang, Y. Simultaneous sulfadiazines degradation and disinfection from municipal secondary effluent by a flow-through electro-Fenton process with graphene-modified cathode. *J. Hazard. Mater.* **2019**, *368*, 830–839. [CrossRef]
7. Huang, J.; Lu, Y.; Zhang, H.; Shangguan, L.; Mou, Z.; Sun, J.; Sun, S.; He, J.; Lei, W. Template-free synthesis of mesh-like graphic carbon nitride with optimized electronic band structure for enhanced photocatalytic hydrogen evolution. *Chem. Eng. J.* **2021**, *405*, 126685. [CrossRef]
8. Wu, J.; Chen, J.; Huang, Y.; Feng, K.; Deng, J.; Huang, W.; Wu, Y.; Zhong, J.; Li, Y. Cobalt atoms dispersed on hierarchical carbon nitride support as the cathode electrocatalyst for high-performance lithium-polysulfide batteries. *Sci. Bull.* **2019**, *64*, 1875–1880. [CrossRef]
9. Yang, H.; Liu, Y.; Luo, Y.; Lu, S.; Su, B.; Ma, J. Achieving high activity and selectivity of nitrogen reduction via Fe–N$_3$ coordination on iron single-atom electrocatalysts at ambient conditions. *ACS Sustain. Chem. Eng.* **2020**, *8*, 12809–12816. [CrossRef]
10. Zhao, L.; Sui, X.-L.; Li, J.-Z.; Zhang, J.-J.; Zhang, L.-M.; Huang, G.-S.; Wang, Z.-B. Supramolecular assembly promoted synthesis of three-dimensional nitrogen doped graphene frameworks as efficient electrocatalyst for oxygen reduction reaction and methanol electrooxidation. *Appl. Catal. B Environ.* **2018**, *231*, 224–233. [CrossRef]
11. Yu, W.; Zhang, T.; Zhao, Z. Garland-like intercalated carbon nitride prepared by an oxalic acid-mediated assembly strategy for highly-efficient visible-light-driven photoredox catalysis. *Appl. Catal. B Environ.* **2020**, *278*, 119342. [CrossRef]
12. Guo, Y.; Li, J.; Yuan, Y.; Li, L.; Zhang, M.; Zhou, C.; Lin, Z. A rapid microwave-assisted thermolysis route to highly crystalline carbon nitrides for efficient hydrogen generation. *Angew. Chem. Int. Ed.* **2016**, *55*, 14693–14697. [CrossRef]
13. Chen, J.; Mao, Z.; Zhang, L.; Wang, D.; Xu, R.; Bie, L.; Fahlman, B.D. Nitrogen-deficient graphitic carbon nitride with enhanced performance for lithium ion battery anodes. *ACS Nano* **2017**, *11*, 12650–12657. [CrossRef]
14. Liu, J.; Zhang, Y.; Zhang, L.; Xie, F.; Vasileff, A.; Qiao, S.Z. Graphitic carbon nitride (g-C$_3$N$_4$)-derived N-rich graphene with tuneable interlayer distance as a high-rate anode for sodium-ion batteries. *Adv. Mater.* **2019**, *31*, e1901261. [CrossRef]

15. Ding, Y.; Yu, C.; Chang, J.; Yao, C.; Yu, J.; Guo, W.; Qiu, J. Effective fixation of carbon in g-C$_3$N$_4$ enabled by Mg-induced selective reconstruction. *Small* **2020**, *16*, e1907164. [CrossRef]
16. Li, X.; Guan, B.Y.; Gao, S.; Lou, X.W. A general dual-templating approach to biomass-derived hierarchically porous heteroatom-doped carbon materials for enhanced electrocatalytic oxygen reduction. *Energy Environ. Sci.* **2019**, *12*, 648–655. [CrossRef]
17. Tian, M.; Zhu, Y.; Chen, Y.; Liu, X.; Yang, Y.; Gao, S. Template-assisted self-activation of mesoporous carbon with active nitrogen/oxygen configurations for sustainable triboelectric nanogenerator powered electro-Fenton degradation. *Nano Energy* **2021**, *83*, 105825. [CrossRef]
18. Zhao, Y.; Liu, Y.; Chen, Y.; Liu, X.; Li, X.; Gao, S. A treasure map for nonmetallic catalysts: Optimal nitrogen and fluorine distribution of biomass-derived carbon materials for high-performance oxygen reduction catalysts. *J. Mater. Chem. A* **2021**, *9*, 18251–18259. [CrossRef]
19. Tian, M.; Zhu, Y.; Zhang, D.; Wang, M.; Chen, Y.; Yang, Y.; Gao, S. Pyrrolic-nitrogen-rich biomass-derived catalyst for sustainable degradation of organic pollutant via a self-powered electro-Fenton process. *Nano Energy* **2019**, *64*, 103940. [CrossRef]
20. Chen, C.; Zhu, Y.; Tian, M.; Chen, Y.; Yang, Y.; Jiang, K.; Gao, S. Sustainable self-powered electro-Fenton degradation using N, S co-doped porous carbon catalyst fabricated with adsorption-pyrolysis-doping strategy. *Nano Energy* **2021**, *81*, 105623. [CrossRef]
21. Yi, S.; Qin, X.; Liang, C.; Li, J.; Rajagopalan, R.; Zhang, Z.; Song, J.; Tang, Y.; Cheng, F.; Wang, H.; et al. Insights into KMnO$_4$ etched N-rich carbon nanotubes as advanced electrocatalysts for Zn-air batteries. *Appl. Catal. B Environ.* **2020**, *264*, 118537. [CrossRef]
22. Chen, Y.; Zhu, Y.; Tian, M.; Chen, C.; Jia, X.; Gao, S. Sustainable self-powered electro-Fenton degradation of organic pollutants in wastewater using carbon catalyst with controllable pore activated by EDTA-2Na. *Nano Energy* **2019**, *59*, 346–353. [CrossRef]
23. Zhang, J.; Chen, Y.; Liu, Y.; Liu, X.; Gao, S. Self-catalyzed growth of Zn/Co-N-C carbon nanotubes derived from metal-organic frameworks as efficient oxygen reduction catalysts for Zn-air battery. *Sci. China Mater.* **2022**, *65*, 653–662. [CrossRef]
24. Qi, H.; Sun, X.; Sun, Z. Porous graphite felt electrode with catalytic defects for enhanced degradation of pollutants by electro-Fenton process. *Chem. Eng. J.* **2021**, *403*, 126270. [CrossRef]
25. Tian, M.; Zhang, D.; Wang, M.; Zhu, Y.; Chen, C.; Chen, Y.; Jiang, T.; Gao, S. Engineering flexible 3D printed triboelectric nanogenerator to self-power electro-Fenton degradation of pollutants. *Nano Energy* **2020**, *74*, 104908. [CrossRef]
26. Han, G.F.; Li, F.; Zou, W.; Karamad, M.; Jeon, J.P.; Kim, S.W.; Kim, S.J.; Bu, Y.; Fu, Z.; Lu, Y.; et al. Building and identifying highly active oxygenated groups in carbon materials for oxygen reduction to H$_2$O$_2$. *Nat. Commun.* **2020**, *11*, 2209. [CrossRef]
27. Peng, L.; Hung, C.T.; Wang, S.; Zhang, X.; Zhu, X.; Zhao, Z.; Wang, C.; Tang, Y.; Li, W.; Zhao, D. Versatile nanoemulsion assembly approach to synthesize functional mesoporous carbon nanospheres with tunable pore sizes and architectures. *J. Am. Chem. Soc.* **2019**, *141*, 7073–7080. [CrossRef] [PubMed]
28. Gao, S.; Geng, K.; Liu, H.; Wei, X.; Zhang, M.; Wang, P.; Wang, J. Transforming organic-rich amaranthus waste into nitrogen-doped carbon with superior performance of the oxygen reduction reaction. *Energy Environ. Sci.* **2015**, *8*, 221–229. [CrossRef]
29. Chang, Q.; Zhang, P.; Mostaghimi, A.H.B.; Zhao, X.; Denny, S.R.; Lee, J.H.; Gao, H.; Zhang, Y.; Xin, H.L.; Siahrostami, S.; et al. Promoting H$_2$O$_2$ production via 2-electron oxygen reduction by coordinating partially oxidized Pd with defect carbon. *Nat. Commun.* **2020**, *11*, 2178. [CrossRef] [PubMed]
30. Wu, K.-H.; Wang, D.; Lu, X.; Zhang, X.; Xie, Z.; Liu, Y.; Su, B.-J.; Chen, J.-M.; Su, D.-S.; Qi, W.; et al. Highly selective hydrogen peroxide electrosynthesis on carbon: In situ interface engineering with surfactants. *Chem* **2020**, *6*, 1443–1458. [CrossRef]
31. Wang, L.; Liang, K.; Deng, L.; Liu, Y.-N. Protein hydrogel networks: A unique approach to heteroatom self-doped hierarchically porous carbon structures as an efficient ORR electrocatalyst in both basic and acidic conditions. *Appl. Catal. B Environ.* **2019**, *246*, 89–99. [CrossRef]
32. Chen, Y.P.; Yang, L.M.; Paul Chen, J.; Zheng, Y.M. Electrospun spongy zero-valent iron as excellent electro-Fenton catalyst for enhanced sulfathiazole removal by a combination of adsorption and electro-catalytic oxidation. *J. Hazard. Mater.* **2019**, *371*, 576–585. [CrossRef]
33. Deng, F.; Olvera-Vargas, H.; Garcia-Rodriguez, O.; Zhu, Y.; Jiang, J.; Qiu, S.; Yang, J. Waste-wood-derived biochar cathode and its application in electro-Fenton for sulfathiazole treatment at alkaline pH with pyrophosphate electrolyte. *J. Hazard. Mater.* **2019**, *377*, 249–258. [CrossRef]
34. Haider, M.R.; Jiang, W.-L.; Han, J.-L.; Sharif, H.M.A.; Ding, Y.-C.; Cheng, H.-Y.; Wang, A.-J. In-situ electrode fabrication from polyaniline derived N-doped carbon nanofibers for metal-free electro-Fenton degradation of organic contaminants. *Appl. Catal. B Environ.* **2019**, *256*, 117774. [CrossRef]
35. Song, X.; Zhang, H.; Bian, Z.; Wang, H. In situ electrogeneration and activation of H$_2$O$_2$ by atomic Fe catalysts for the efficient removal of chloramphenicol. *J. Hazard. Mater.* **2021**, *412*, 125162. [CrossRef]
36. Qin, X.; Zhao, X.; Quan, X.; Cao, P.; Chen, S.; Yu, H. Highly efficient metal-free electro-Fenton degradation of organic contaminants on a bifunctional catalyst. *J. Hazard. Mater.* **2021**, *416*, 125859. [CrossRef]
37. Chen, Y.; Wang, M.; Tian, M.; Zhu, Y.; Wei, X.; Jiang, T.; Gao, S. An innovative electro-Fenton degradation system self-powered by triboelectric nanogenerator using biomass-derived carbon materials as cathode catalyst. *Nano Energy* **2017**, *42*, 314–321. [CrossRef]
38. Jia, N.; Yang, T.; Shi, S.; Chen, X.; An, Z.; Chen, Y.; Yin, S.; Chen, P. N,F-codoped carbon nanocages: An efficient electrocatalyst for hydrogen peroxide electroproduction in alkaline and acidic solutions. *ACS Sustain. Chem. Eng.* **2020**, *8*, 2883–2891. [CrossRef]
39. Sheng, H.; Janes, A.N.; Ross, R.D.; Kaiman, D.; Huang, J.; Song, B.; Schmidt, J.R.; Jin, S. Stable and selective electrosynthesis of hydrogen peroxide and the electro-Fenton process on CoSe$_2$ polymorph catalysts. *Energy Environ. Sci.* **2020**, *13*, 4189–4203. [CrossRef]

40. Tang, H.; Zhu, Z.; Shang, Q.; Tang, Y.; Zhang, D.; Du, Y.; Liu, M.; Yin, K.; Liu, C. Highly efficient continuous-flow electro-Fenton treatment of antibiotic wastewater using a double-cathode system. *ACS Sustain. Chem. Eng.* **2021**, *9*, 1414–1422. [CrossRef]
41. Fan, W.; Zhang, B.; Wang, X.; Ma, W.; Li, D.; Wang, Z.; Dupuis, M.; Shi, J.; Liao, S.; Li, C. Efficient hydrogen peroxide synthesis by metal-free polyterthiophene via photoelectrocatalytic dioxygen reduction. *Energy Environ. Sci.* **2020**, *13*, 238–245. [CrossRef]
42. Zhai, L.-F.; Sun, Y.-M.; Guo, H.-Y.; Sun, M. Surface modification of graphite support as an effective strategy to enhance the electro-Fenton activity of Fe_3O_4/graphite composites in situ fabricated from acid mine drainage using an air-cathode fuel cell. *ACS Sustain. Chem. Eng.* **2019**, *7*, 8367–8374. [CrossRef]
43. Li, Z.; Shen, C.; Liu, Y.; Ma, C.; Li, F.; Yang, B.; Huang, M.; Wang, Z.; Dong, L.; Wolfgang, S. Carbon nanotube filter functionalized with iron oxychloride for flow-through electro-Fenton. *Appl. Catal. B Environ.* **2020**, *260*, 118204. [CrossRef]

Article

Biomass-Derived Porous Carbon with a Good Balance between High Specific Surface Area and Mesopore Volume for Supercapacitors

Yanbo Wang [1,2], Yiqing Chen [3], Hongwei Zhao [1], Lixiang Li [1], Dongying Ju [3], Cunjing Wang [2,*] and Baigang An [1,3,*]

1. Key Laboratory of Energy Materials and Electrochemistry Research Liaoning Province, School of Chemical Engineering, University of Science and Technology Liaoning, Anshan 114051, China
2. School of Chemistry and Materials Engineering, Xinxiang University, Xinxiang 453003, China
3. State Key Laboratory of Metal Material for Marine Equipment and Application, Anshan 114009, China
* Correspondence: wangcunjing@126.com (C.W.); bgan@ustl.edu.cn (B.A.)

Abstract: Porous carbon has been one desirable electrode material for supercapacitors, but it is still a challenge to balance the appropriate mesopore volume and a high specific surface area (SSA). Herein, a good balance between a high SSA and mesopore volume in biomass-derived porous carbon is realized by precarbonization of wheat husk under air atmosphere via a chloride salt sealing technique and successive KOH activation. Due to the role of molten salt generating mesopores in the precarbonized product, which can further serve as the active sites for the KOH activation to form micropores in the final carbon material, the mesopore–micropore structure of the porous carbon can be tuned by changing the precarbonization temperature. The appropriate amount of mesopores can provide more expressways for ion transfer to accelerate the transport kinetics of diffusion-controlled processes in the micropores. A high SSA can supply abundant sites for charge storage. Therefore, the porous carbon with a good balance between the SSA and mesopores exhibits a specific gravimetric capacitance of 402 F g^{-1} at 1.0 A g^{-1} in a three-electrode system. In a two-electrode symmetrical supercapacitor, the biomass-derived porous carbon also delivers a high specific gravimetric capacitance of 346 F g^{-1} at 1.0 A g^{-1} and a good cycling stability, retaining 98.59% of the initial capacitance after 30,000 cycles at 5.0 A^{-1}. This work has fundamental merits for enhancing the electrochemical performance of the biomass-derived porous carbon by optimizing the SSA and pore structures.

Keywords: biomass; porous carbon; supercapacitors; mesopore–micropore; salt template

Citation: Wang, Y.; Chen, Y.; Zhao, H.; Li, L.; Ju, D.; Wang, C.; An, B. Biomass-Derived Porous Carbon with a Good Balance between High Specific Surface Area and Mesopore Volume for Supercapacitors. *Nanomaterials* 2022, *12*, 3804. https://doi.org/10.3390/nano12213804

Academic Editors: Antonio Di Bartolomeo and Sergio Brutti

Received: 30 August 2022
Accepted: 25 October 2022
Published: 28 October 2022

Publisher's Note: MDPI stays neutral with regard to jurisdictional claims in published maps and institutional affiliations.

Copyright: © 2022 by the authors. Licensee MDPI, Basel, Switzerland. This article is an open access article distributed under the terms and conditions of the Creative Commons Attribution (CC BY) license (https://creativecommons.org/licenses/by/4.0/).

1. Introduction

Rechargeable battery technologies are urgently needed to tackle the current energy and environmental concerns of modern society [1–3]. As typical electrochemical energy storage and conversion devices, supercapacitors have shown extraordinary promise due to their high power density and outstanding durability [4–7]. Porous carbons with attractive advantages such as a high specific surface area (SSA), high conductivity, and outstanding chemical stability have been identified as ideal electrode materials for supercapacitors [8–13].

Porous carbons are generally obtained with coals, cokes, and pitches or biomass as precursors. In the past few decades, biomass has become a rapidly developed and widely used precursor to fabricate porous carbon electrode materials due to its advantages such as the abundance of raw materials, low cost, and renewability [14–19]. For carbon-based supercapacitors, the energy storage involves charge accumulation and separation at the interface of the electrolyte and electrode material during charge and discharge [20–23]. Based on this mechanism, the specific capacitance of a supercapacitor mainly depends on the SSA of the electrode materials. Therefore, enhancing the SSA of porous carbon materials is the most promising strategy to improve the capacitance performance of supercapacitors [24–26]. Although abundant micropores can increase the SSA of the electrode materials and micropores of a size similar to that of the ions can enhance the capacitance [27], the slow ion transport in the long narrow micropores may lead to an inferior rate performance of the supercapacitors [27]. To make a large SSA fully accessible, introducing mesopores is considered as an effective method since the mesopores can support more fast transport ways for ions [28–30].

Template methods, physical activation (activation with O_2, CO_2, or plasma treatment), and chemical activation (activation with KOH, $ZnCl_2$, etc.) have been successfully employed to fabricate porous carbons with mesopores, micropores, or hierarchical pores [28–31]. These porous carbon electrodes improved the electrochemical performance of the supercapacitors to a great extent. However, the relatively low energy density and inferior rate performance still significantly limit their future practical applications. Although a high SSA can provide abundant sites for ion storage and a large mesoporous volume is convenient for the ease of ion transportation, how to balance the high SSA and appropriate mesopore/micropore volume is still a challenge. Recent research has developed new ways to balance the high SSA and mesopore volume of porous carbons, which are useful for good performance in supercapacitors [32–34]. Chemical activation of carbonaceous matter adopting KOH as the activating agent is widely used, and the porosity of such carbons depends on the nature of the precursor and activation conditions [35]. Subjecting the precursor to an initial carbonization step before activation can increase the carbon content and reduce the O/C ratio in the material, meaning that fewer oxidizing gases are released upon activation, leading to more controlled activation and the improvement of the porosity of the carbons [35–37]. In this technique, biomass is a preferred choice for activated carbons due to being readily available, renewable, and essentially offering a "carbon neutral" route to porous carbons [36,37].

Producing value-added carbon electrode materials using wheat husk is a good way to reuse agricultural waste; however, precarbonizing wheat husk in N_2 atmosphere and further activating the precarbonized carbon with KOH often generate abundant micropores in the carbon material with slow ion transportation kinetics [31–33]. To overcome this critical problem, taking into account the template and etching role of molten salt and oxygen on porous carbon produced in air using the salt sealing technique [19–21], we developed a facile strategy to construct porous carbon with a good balance between a high SSA and mesopore volume by precarbonization of wheat husk in air and successive KOH activation. The precarbonization process employs a mixture of KCl and NaCl as the salt template to generate mesopores in the carbon framework; meanwhile, the mixed salt can act as a shielding agent to prevent the carbon framework from oxidation in air at a high temperature. The mesopores in the carbon framework can provide active sites for successive KOH activation to produce a large amount of micropores. Benefiting from the synergistic pore-generating effect of molten salt and KOH, the obtained porous carbon exhibits a hierarchical pore structure with a good balance between a high SSA and mesopore volume. The high SSA provides a sufficient electrode/electrolyte interface for charge accumulation, and the high mesopore volume supplies more expressways for fast ion transfer. The biomass-derived porous carbon exhibits a high specific gravimetric capacitance of 402 F g^{-1} at 1.0 A g^{-1} in a three-electrode system. In a two-electrode symmetrical supercapacitor, the biomass-derived porous carbon also delivers an excellent specific gravimetric capacitance of 346 F g^{-1} at 1.0 A g^{-1} and a good cycling stability with a retention of 98.59% after 30,000 cycles at 5.0 A^{-1}.

2. Material and Methods

2.1. Synthesis of Porous Carbon

Firstly, the mixture of KCl and NaCl in a mass ratio of 1:1 was ground. The wheat husks were washed with distilled water and dried at 60 °C in an oven. Then, 4.0 g of wheat husk was mixed with 16.0 g of a KCl and NaCl mixture in a 50 mL porcelain crucible, precarbonized at 700 °C, 800, and 900 °C for 1 h in a muffle furnace in air atmosphere. The obtained precarbonized product was washed with distilled water and dried at 80 °C, mixed with KOH at a weight ratio of 1:2 (carbon/KOH) in an agate mortar. The precarbonized product was further activated by KOH in a tubular furnace at 800 °C for 1 h under nitrogen atmosphere to obtain the porous carbon. The porous carbons as prepared were washed with 3.0 M HCl solution and distilled water until the pH of the washing effluent reached 6–7, then the filtered product was dried at 80 °C in an oven overnight. According to the carbonization temperature, the porous carbons derived from wheat husk were named HPC-700, HPC-800, and HPC-900, respectively. In addition, the precarbonized products at 700 °C, 800, and 900 °C in a muffle furnace in air atmosphere for 1 h were named MHPC-700, MHPC-800, and MHPC-900. Wheat husk was also precarbonized at 700 °C in nitrogen atmosphere without molten salt for 1 h, and the precarbonized product was named as NHPC-700. The carbon yield of the different synthesis processes is summarized in Table 1.

Table 1. Carbon yield of the different synthesis processes.

Samples	The Mass of Wheat Husk or Carbonized Product (g)	The Mass of Salt or KOH (g)	The Mass of Carbon Product (g)	Carbon Yield (wt%)
MHPC-700	4 (wheat husk)	16 (salt)	0.81	20.25
MHPC-800	4 (wheat husk)	16 (salt)	0.57	14.25
MHPC-900	4 (wheat husk)	16 (salt)	0.19	4.75
NHPC-700	2 (wheat husk)	0 (salt)	0.57	28.5
HPC-700	0.1 (carbon)	0.2 (KOH)	0.056	56
HPC-800	0.1 (carbon)	0.2 (KOH)	0.055	55
HPC-900	0.1 (carbon)	0.2 (KOH)	0.055	55

2.2. Material Characterizations

Field emission scanning electron microscopy (SEM, FEI Quanta FEG 250, Hillsboro, OR, USA), transmission electron microscopy (TEM, JEM-2100, JEOL USA, Peabody, MA, USA), X-ray diffraction (XRD; Bruker D8 Advance, Billerica, MA, USA) equipped with Cu Kα radiation (λ = 1.5418 Å), X-ray photoelectron spectroscopy (XPS) using a Perkin-Elmer PHI-5700 ESCA System (Waltham, MA, United States) multifunctional photoelectron spectrometer with Al Kα radiation (1486.6 eV), and a NOVA2200e physisorption analyzer were used to examine the morphology, microstructure, crystallographic structure, surface chemical species, and porous texture of the obtained porous carbons. In addition, the specific surface area and the pore size distribution of the carbon materials were calculated by the Brunauer–Emmett–Teller (BET) equation from the nitrogen adsorption data in the relative pressure (P/P_0) of 0.03~0.30 and the nonlocal density functional theory (NLDFT) equilibrium model for cylinder/slit pores from N_2 sorption data. The total pore volume (V_{total}) was determined at a relative pressure $p/p_0 = 0.990$ and the micropore volume (V_{micro}) using the t-plot method.

2.3. Electrochemical Measurements

Electrodes were prepared by painting a paste containing the porous carbon, carbon black, and polytetrafluoroethylene (PTFE) in a weight ratio of 80:15:5 onto a current collector of stainless steel mesh; the mass loading of the electrode materials was 3 mg cm^{-2}. The electrochemical tests of the individual electrode including cyclic voltammetry (CV), galvanostatic charge/discharge (GCD), and electrochemical impedance spectroscopy (EIS) were measured on a CHI660C electrochemical work station in a 1 mol L^{-1} H$_2$SO$_4$ solution in a three-electrode system. The porous carbon electrodes, a platinum foil, and a Ag/AgCl electrode served as the working, counter, and reference electrodes, respectively. The CV at scan rates from 10–100 mV s^{-1} and GCD tests at current densities ranging from 1.0 to 20.0 A g^{-1} were recorded between 0 and 1 V (vs. Ag/AgCl). EIS was obtained in the frequency range from 100 kHz to 0.1 Hz with a 5 mV AC voltage amplitude. The symmetric supercapacitors were assembled with a glassy fibrous separator and tested in a 1 mol L^{-1} H$_2$SO$_4$ solution with a LAND CT2001A instrument.

The gravimetric specific capacitance of a single electrode, C_G (F g^{-1}), was calculated from the discharge curve according to the equation

$$C_G = \frac{I}{m \times (\Delta U / \Delta t)} \tag{1}$$

where I is the constant charge/discharge current, Δt is the discharge time, ΔU is the potential window during the discharge process, and m is the mass of the active materials in a single electrode.

The gravimetric specific capacitance of a single electrode in a two-electrode system, C_{sG} (F g^{-1}), was calculated from the discharge curve based on the equation.

$$C_{sG} = 4 \times \frac{I}{m \times (\Delta U / \Delta t)} \tag{2}$$

where m is the total mass of active materials in the two electrodes. The gravimetric energy density of the device E_G (Wh kg^{-1}) was estimated by using the following equation:

$$E_G = \frac{1}{28.8} C_{sG} \Delta U^2 \tag{3}$$

and the gravimetric power density of the device P_G (W kg^{-1}) was calculated according to the equation

$$P_G = \frac{3600 E_G}{\Delta t} \tag{4}$$

3. Results and Discussion

Assuming that molten salt can penetrate into the carbon skeleton to serve as a "cutting" reagent or template [19–21], as illustrated in Scheme 1, the wheat husk was first precarbonized under air atmosphere by a simple salt sealing technique employing low-cost and non-toxic mixed salt of KCl and NaCl as a dual function agent to prevent the carbon structures from oxidation at 700~900 °C above the melting temperature of the salt mixture of NaCl and KCl (669 °C for the equal mass mixture [20]) to build mesopore structures during the precarbonization process. Then, the mesopores generated in the carbon materials further served as active sites for the successive KOH activation to produce a large amount of micropores. The morphologies and microstructure of the carbon samples were examined by SEM and TEM.

Scheme 1. Schematic synthesis process for the HPC.

Figure 1a,b display the SEM images of HPC-700, in which irregular carbon sheets are observed and the surface of the carbon sheets looks very smooth. The surface of the carbon sheets of HPC-800 (Figure 1c,d) looks rougher than HPC-700, and obvious macropores are observed in the carbon sheets of HPC-900 (Figure 1e,f). The changes of the sheet morphologies of HPC-800 and HPC-900 can be ascribed to the stronger etching effect of the molten salt at a higher temperature. For the mesoporous–microporous carbon electrode materials, the sheet structure can reduce the ion transport distance in the carbon electrodes and accelerate the ion transfer kinetics in the micropores limited by diffusion control [19–21]. The sheet carbon structure was further confirmed by low-resolution TEM images of the three samples shown in Figure 2a,c,e. Consistent with the SEM results, the surface of HPC-700 looks smoother and transparent. In addition, obvious mesopores in HPC-700 are observed in Figure 2a. The corresponding high-resolution TEM images of the samples in Figure 2b,d,f show that the carbon sheets consist of multilayer discontinuous graphite stripes with a low graphitization degree, consistent with the selected area electron diffraction (SAED) images with the diffuse rings in the insets. To investigate the formation process of the sheet morphologies of the carbon samples, Figure 3a–c show the SEM images of the carbon samples MHPC-700, MHPC-800, and MHPC-900 precarbonized in molten salt. By comparison with the SEM image of the carbon sample NHPC-700 precarbonized in nitrogen atmosphere without molten salt (Figure 3d), it can be observed that the obvious sheet morphology dominates in MHPC-700, while a thick carbon block is prominent in NHPC-700, demonstrating that a sheet structure of the carbon samples HPC-700, HPC-800, and HPC-900 was formed during the precarbonization process in molten salt. In this precarbonization process, wheat husk firstly experiences a steady transition from sp^3C-X (X: e.g., C, O, H) bonds to the aromatic sp^2 C-C bonds to form the carbon skeleton; when the temperature is above the melting temperature of the salt mixture of NaCl and KCl, the molten salt diffusing into the carbon skeleton functions as a "cutting" reagent to prohibit the stacking of the sp^2 coordinated carbon layers along the C-axis due to the van der Waals force [20]; meanwhile, the high energy Cl^- ions in the molten salt media continue to etch the carbon structures, leading to the final formation of carbon sheets with multilayer discontinuous graphite stripes [20].

Figure 1. SEM images of (**a**,**b**) HPC-700, (**c**,**d**) HPC-800, and (**e**,**f**) HPC-900.

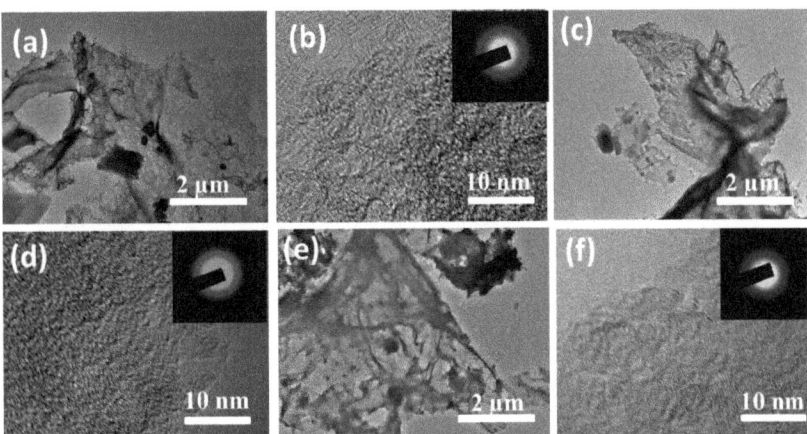

Figure 2. TEM images of (**a**,**b**) HPC-700, (**c**,**d**) HPC-800, and (**e**,**f**) HPC-900 (the inset is the corresponding SAED images).

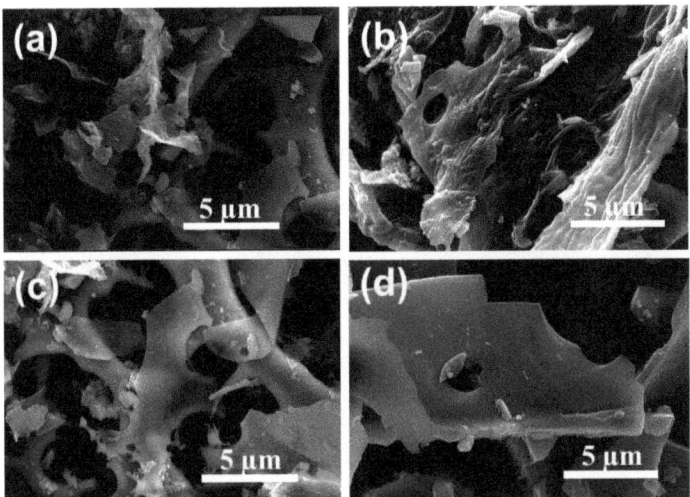

Figure 3. SEM images of (**a**) MHPC-700, (**b**) MHPC-800, (**c**) MHPC-900, and (**d**) NHPC-700.

The XRD patterns of the samples are shown in Figure 4a. The broad and weak peaks at about ~25° and ~44° attributed to the (002) and (100) planes of graphite further reveal the low graphitization degree of the three carbon samples [36,37]. To analyze the surface composition of the samples, the XP survey spectra of the carbon samples are shown in Figure 4b. In the XP spectra, two peaks at binding energy of 284.6 and 532.4 eV were assigned to C1s and O1s. The content of carbon and oxygen was evaluated to be HPC-700 (89.34 and 10.66 at%), HPC-800 (89.98 and 10.02 at%), and HPC-900 (90.56 and 9.44 at%). The high-resolution C1s and O1s spectra of HPC-700 (Figure 4c,d) show three peaks at 284.5, 285.8, and 287.8 eV, attributed to C=C, C–O, and C=O bonds, and two peaks at 532.8 and 531.6 eV, attributed to C–O and C=O bonds. These functional groups can introduce pseudocapacitance to the carbon electrodes during the charge/discharge process [19–21].

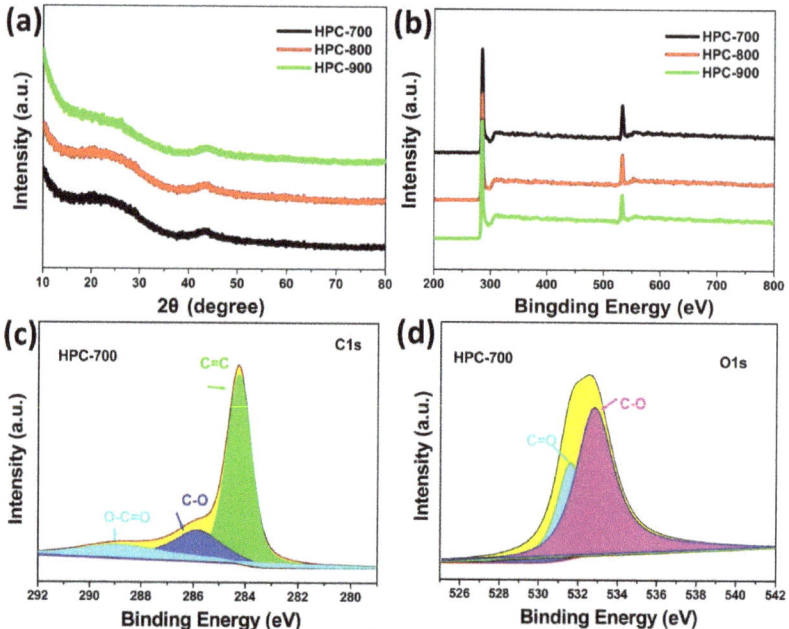

Figure 4. (a) XRD patterns and (b) XPS survey spectra of HPC-700, HPC-800, and HPC-900; (c) C1s and (d) O1s spectra of HPC-700.

The pore structures of the samples were further investigated by nitrogen adsorption-desorption measurements at 77 K; the results are shown in Figure 5a. The typical hysteresis loops at $p/p_0 > 0.4$ and high nitrogen uptake at low relative pressure indicate the mesoporous feature and a large amount of micropores. By comparison, HPC-700 has the highest total nitrogen uptake and the biggest hysteresis loop, indicating the largest SSA and mesopore volume. The detailed data in Table 2 suggest that HPC-700 has the highest SSA, V_{total}, V_{micro}, and V_{meso}, possessing a good balance between high SSA and V_{meso}, which is an advantage for improving the electrochemical performance of the supercapacitors. Figure 5b shows the pore size distribution (PSD) curves of samples based on a nonlocal density functional theory (NDFT) model. No vast differences in the pore size distribution were observed among these samples. These samples all show distributed pore sizes centered at 1.3~6 nm and contain many hierarchical mesopores–micropores.

As for the formation process of such porous carbon structures with a high SSA and V_{meso}, it is assumed that the synergistic pore-generating effect of the molten salt and KOH is crucial. As reported in the literature, molten salt plays important template and etching roles in generating mesopores in porous carbon, and traces of oxygen penetrated into the system can generate micropores [19–21]. By comparison, the obviously higher total nitrogen uptake of MHPC-700, MHPC-800, and MHPC-900 than that of NHPC-700 and the big hysteresis loops in their nitrogen adsorption–desorption isotherms (Figure 5c) confirm the role of molten salt and the trace of oxygen in generating mesopores and micropores. The hysteresis loops of NHPC-700 were ascribed to inorganic salts in the wheat husk. The detailed data in Table 2 further suggest that NHPC-900 has the lowest SSA and MHPC-700 has a higher SSA and mesopore volume than those of MHPC-800 and MHPC-900. Moreover, their PSDs in Figure 5d also reveal their prominent mesopore features. These data further suggest that, when the temperature of precarbonization is 700 °C, the mesopore amount generated in the carbon skeleton is just suitable for the successive KOH activation, thus leading to the highest SSA of HPC-700. With the temperature of precarbonization increased to 800 and

900 °C, a part of the mesopores collapses, resulting in the obvious decrease in the V_{meso} and SSA of MHPC-800, MHPC-900, HPC-800, and HPC-900.

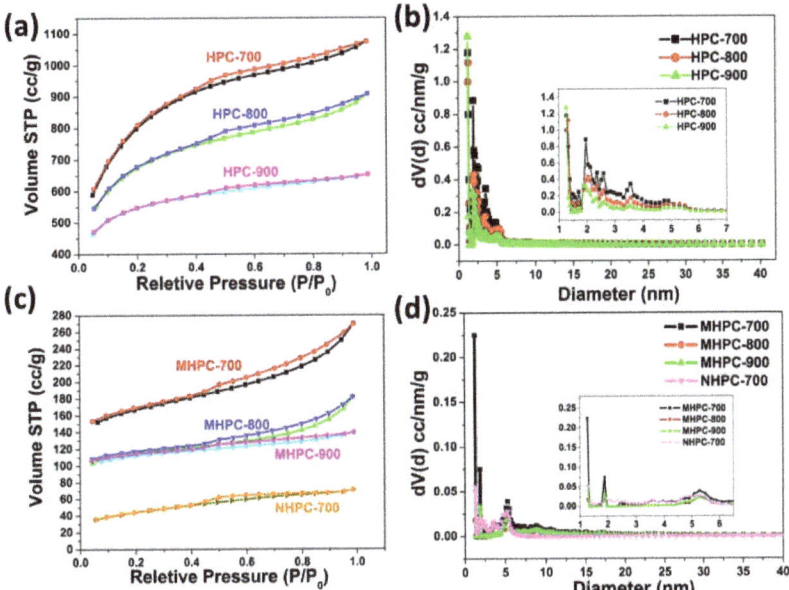

Figure 5. (**a**) N$_2$ adsorption/desorption isotherms and (**b**) PSDs of HPC-700, HPC-800, and HPC-900; (**c**) N$_2$ adsorption/desorption isotherms and (**d**) PSDs of MHPC-700, MHPC-800 MHPC-900, and NHPC-700.

Table 2. Pore textural properties of the as-obtained porous carbon materials.

Samples	S_{total} (m^2 g^{-1})	V_{total} (cm^3 g^{-1})	V_{micro} (cm^3 g^{-1})	V_{meso} (cm^3 g^{-1})
HPC-700	2721.	1.66	0.79	0.87
HPC-800	2202	1.41	0.75	0.65
HPC-900	1739	1.01	0.69	0.32
NHPC-700	149	0.11	0.03	0.08
MHPC-700	524	0.42	0.19	0.23
MHPC-800	361	0.28	0.15	0.14
MHPC-900	309	0.26	0.11	0.15

S_{total}: total BET specific surface area; V_{total}: total pore volume; V_{micro}: micropore volume; V_{meso}: mesopore volume.

Considering the high SSA and mesopore volume suitable for charge accumulation and rapid ion transfer, these carbon samples are expected to exhibit excellent electrochemical performance for supercapacitors. Figure 6a shows the CV at a scan rate of 10 mV s^{-1} with a quasi-rectangular shape and the broad redox peaks around 0.2~0.6 V, indicating an ideal electrical double-layer capacitance of the porous carbon electrode materials [19]. Compared with the other samples, HPC-700 displays the largest enclosed CV curve area, indicating the highest specific capacitance. The GCD curves of the samples at 1.0 A g^{-1} are further shown in Figure 6b. HPC-700 shows a longer discharge time, signifying a higher specific capacitance of 402 F g^{-1} than that of HPC-800 (210 F g^{-1}) and HPC-900 (199 F g^{-1}). The excellent specific capacitance of HPC-700 results from the good balance between its large accessible SSA, providing a more efficient electrode–electrolyte interface and the rational V_{meso} supplying more rapid ion transportation pathways [38]. Based on the advantage of this balance, the wheat-husk-derived carbon HPC-700 exhibits a remarkable capacitance compared with the other biomass-derived carbon electrodes, as listed in Table 3.

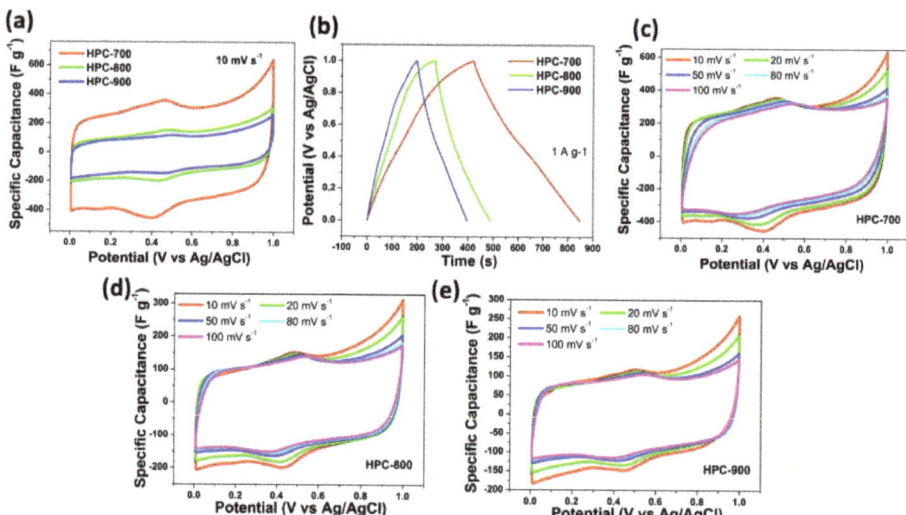

Figure 6. (a) CV curves at 10 mV s^{-1} and (b) GCD curves at 1.0 A g^{-1} of HPC-700, HPC-800, and HPC-900; CV curves of HPC-700 (c), HPC-800 (d), and HPC-900 (e) at different scanning rates.

Table 3. Specific capacitance of HPC-700 and some biomass-derived porous carbon reported in the literature.

Carbon Precursors	C_G (F g^{-1})	Electrolyte	Current Density (A g^{-1})	Refs.
Wheat husk	402	1 M H$_2$SO$_4$	1.0	This work
Sugarcane bagasse	371	1 M H$_2$SO$_4$	0.1	[38]
Sucrose	143	6 M KOH	1.0	[39]
Peach gum	426	1 M H$_2$SO$_4$	0.5	[40]
Flaxseed residue	369	6 M KOH	0.5	[41]
Rice husk	163	6 M KOH	0.2	[42]
Tobacco waste	197	1 M H$_2$SO$_4$	0.2	[43]
Grape marcs	446	1 M H$_2$SO$_4$	0.5	[44]
Mangifera indica peels	402	1 M H$_2$SO$_4$	1.0	[45]
Dead plant leaves	345	1 M H$_2$SO$_4$	0.5	[46]
Foxtail grass seeds	358	6 M KOH	0.5	[47]
Egg white	304	1 M H$_2$SO$_4$	1.0	[48]
Pomelo peel	314	6 M KOH	1.0	[49]

In addition, the high V_{meso} of the carbon samples can improve the rate performance limited by the slow diffusion-controlled reaction kinetics in the micropores of porous carbon. The CV curves of HPC-700 at scan rates from 10 to 100 mV s^{-1} are shown in Figure 6c. The maintained rectangular shapes combined with wide peaks around 0.2–0.6 V suggest the ideal capacitive behavior of HPC-700 at high scan rates. The specific capacitance of HPC-700 at a high scan rate is obviously higher than that of HPC-800 and HPC-900 according to the area of the CV curves (Figure 6d,e). As shown in Figure 7a–c, the GCD curves of the samples at the different current densities were measured to study their rate performance as the electrode materials of supercapacitors. Based on the GCD data, Figure 7d shows that HPC-700 supplies a capacitance of 346 F g^{-1} at 20.0 A g^{-1} with a capacitance retention of 86.1%, higher than the 134 F g^{-1} of HPC-800 with a capacitance retention of 63.6% and the 122 F g^{-1} of HPC-900 with a capacitance retention of 61.31% at 20 A g^{-1}. The excellent rate performance of HPC-700 confirms that the high V_{meso} can provide the rapid transport pathways to enhance the diffusion kinetics of ions in the micropores of the carbon electrode at high charge/discharge rates effectively.

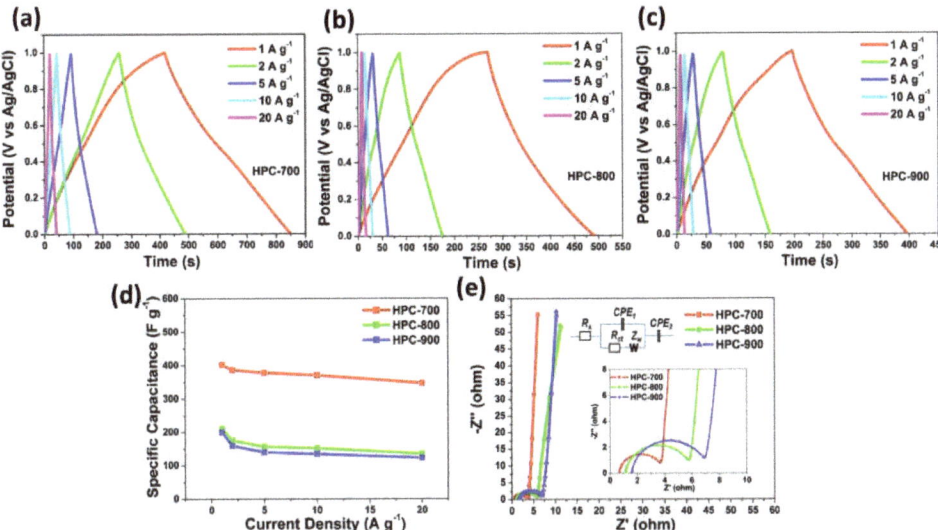

Figure 7. GCD curves of HPC-700 (**a**), HPC-800 (**b**), and HPC-900 (**c**) at different current densities; (**d**) the rate performance and (**e**) Nyquist plots of HPC-700, HPC-800, and HPC-900.

The ion transfer kinetics in carbon electrodes were further investigated by EIS. Figure 7e displays the Nyquist diagrams of the porous carbon electrodes; the inset is the corresponding equivalent circuit. The fast ion adsorption/desorption kinetics at the interfaces of the electrolyte/electrodes is proven by the steep lines almost perpendicular to the X-axis in the low-frequency region [48] owing to the very small diffusion polarization. The smaller diameter of the semicircle in the high-frequency region corresponds to the lower charge transfer resistance (R_{ct}, 3.1 Ω) of HPC-700 than HPC-800 (R_{ct}, 4.9 Ω) and HPC-900 (R_{ct}, 5.2 Ω) [48]. In addition, the fast electrolyte ion diffusion process in the HPC-700 electrode is also verified by the lower Warburg resistance (Z_w, 0.42 Ω) of HPC-700 than HPC-800 (Z_w, 0.53 Ω) and HPC-900 (Z_w, 0.58 Ω) in the middle-frequency region. The EIS results further indicate that the high V_{meso} of HPC-700 can reduce the ion transfer and diffusion resistance to accelerate the interface reaction kinetics in the carbon electrode. Therefore, a good balance between the high SSA and the rational V_{meso} enables the excellent capacitance and rate performance of HPC-700.

The HPC-700//HPC-700 symmetric supercapacitor was further assembled. The GCD curves (Figure 8a) of the supercapacitor show highly symmetric triangles at current densities from 1.0 to 20.0 A g^{-1}, indicating the highly reversible charge and discharge process. Figure 8b manifests that HPC-700 still shows a high specific capacitance of 346 F g^{-1} at 1.0 A g^{-1} and 262 F g^{-1} at 20.0 A g^{-1} in the supercapacitor, with a high Coulombic efficiency of 99.2%. The symmetric supercapacitor supplies an energy density of 12.02 Wh kg^{-1} at a power density of 250 W kg^{-1} (Figure 8c), which are higher than the results for bio-derived carbon-based supercapacitors reported in the literature [50–59]. Meanwhile, the supercapacitor also possesses a prolonged cycling life with 98.59% of the initial capacitance after 30,000 cycles at 5.0 A g^{-1} (Figure 8d). The excellent electrochemical performance of the supercapacitor using HPC-700 as the electrode material is due to the good balance of the high SSA and V_{meso} in HPC-700, which can make a large amount of the micropores accessible for charge storage through the expressways of the mesopores for ion transport.

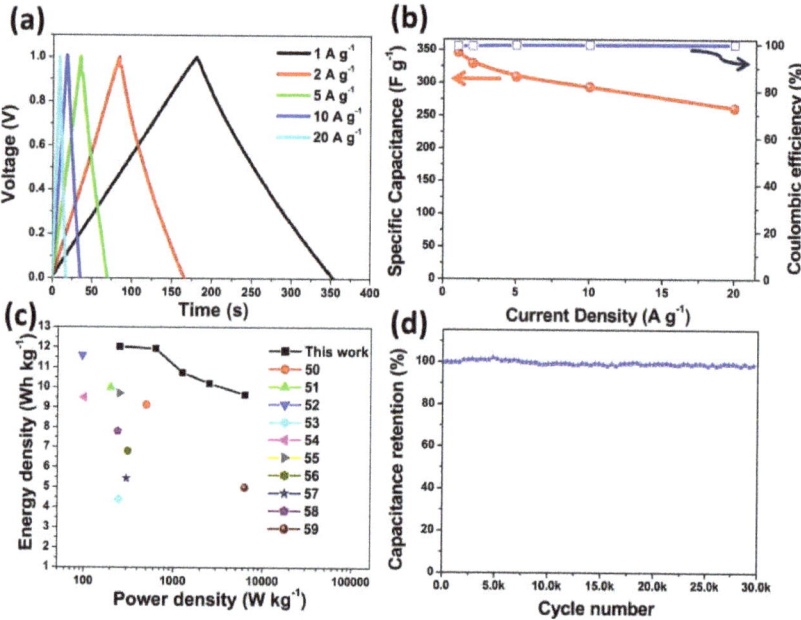

Figure 8. GCD curves (**a**) and specific capacitance, as well as Coulombic efficiency (**b**) of HPC700 at different current densities in HPC-700//HPC-700. (**c**) Ragone plot and (**d**) cycling stability of HPC700//HPC-700 at 5.0 A g^{-1} after 30,000 cycles.

4. Conclusions

In conclusion, a good balance between a high SSA and V_{meso} in the biomass-derived porous carbon was achieved by using molten salt as a mesopore-generating agent during the precarbonization process and KOH as a further micropore-forming agent. The mesopore/micropore structure of the porous carbon can be tuned by changing the precarbonization temperatures. The mesopores generated by the molten salt in the precarbonized product can further serve as the active sites for the KOH activation to produce micropores. The high SSA and V_{meso} of the porous carbon provide a sufficient electrode/electrolyte interface to facilitate the electrolyte ion penetration and ion transfer expressways to accelerate the transport kinetics by diffusion control in the micropores. Consequently, the obtained carbon electrode exhibits a high specific gravimetric capacitance of 402 F g^{-1} at 1.0 A g^{-1} in a three-electrode system. In a symmetric supercapacitor, the obtained carbon delivers an excellent specific gravimetric capacitance of 346 F g^{-1} at 1.0 A g^{-1}, as well as a 98.59% capacitance retention after 30,000 cycles at 5.0 A^{-1}. This work has fundamental merits for enhancing the electrochemical performance of the biomass-derived porous carbon by optimizing the pore structures and SSA.

Author Contributions: Investigation, methodology, writing—original draft, Y.W.; data curation, visualization, formal analysis, H.Z.; investigation, L.L.; methodology, D.J.; writing—original draft, funding acquisition, C.W.; conceptualization, writing—review and editing, funding acquisition, Y.C. and B.A. All authors have read and agreed to the published version of the manuscript.

Funding: This research was supported by National Natural Science Foundation of China (51672118, 51672117, 51972156, 51872131, and 51902278) and the Joint Fund projects of the University of Science and Technology Liaoning State-Key Laboratory of Metal Material (No. HGSKL-USTLN (2020)08).

Institutional Review Board Statement: Not applicable.

Informed Consent Statement: Not applicable.

Data Availability Statement: The data presented in this study are available upon request from the corresponding author. The data are not publicly available due to privacy. Data are contained within the article.

Acknowledgments: The authors acknowledge National Natural Science Foundation of China (51672118, 51672117, 51972156, 51872131, and 51902278) for the financial support and the Joint Fund projects of the University of Science and Technology Liaoning State-Key Laboratory of Metal Material for Marine Equipment and Application (No. HGSKL-USTLN (2020)08).

Conflicts of Interest: The authors declare that they have no known competing financial interest or personal relationships that could have appeared to influence the work reported in this paper.

References

1. Liang, C.; Chen, Y.; Wu, M.; Zheng, S.; Pan, H. Green synthesis of graphite from CO_2 without graphitization process of amorphous carbon. *Nat. Commun.* **2021**, *12*, 119–130. [CrossRef]
2. Cano, Z.P.; Banham, D.; Ye, S.; Hintennach, A.; Lu, J.; Fowler, M.; Chen, Z. Batteries and fuel cells for emerging electric vehicle markets. *Nat. Energy* **2018**, *3*, 279–289. [CrossRef]
3. Zhang, C.; Pan, H.; Sun, L.; Xu, F.; Ouyang, Y.; Rosei, F. Progress and perspectives of 2D materials as anodes for potassium-ion batteries. *Energy Storage Mater.* **2021**, *38*, 354–378. [CrossRef]
4. Zhao, Y.; Dong, C.; Sheng, L.; Xiao, Z.; Jiang, L.; LI, X.; Jiang, M.; Shi, J. Heteroatom-Doped Pillared Porous Carbon Architectures with Ultrafast Electron and Ion Transport Capabilities under High Mass Loadings for High-Rate Supercapacitors. *ACS Sustain. Chem. Eng.* **2020**, *8*, 8664–8674. [CrossRef]
5. Gopi, C.V.V.M.; Vinodh, R.; Sambasivam, S.; Obaidat, I.M.; Kim, H. Recent progress of advanced energy storage materials for flexible and wearable supercapacitor: From design and development to applications. *J. Energy Storage* **2020**, *27*, 101035.
6. Xu, Z.; Zhang, X.; Liang, Y.; Lin, H.; Zhang, S.; Liu, J.; Jin, C.; Choe, U.; Sheng, K. Green synthesis of nitrogen-doped porous carbon derived from rice straw for high-performance supercapacitor application. *Energy Fuels* **2020**, *34*, 8966–8976. [CrossRef]
7. Lim, T.; Kim, T.; Suk, J.W. Activated graphene deposited on porous Cu mesh for supercapacitors. *Nanomaterials* **2021**, *11*, 893. [CrossRef]
8. Chernysheva, D.; Pudova, L.; Popov, Y.; Smirnova, N.; Leontyev, I. Non-isothermal decomposition as efficient and simple synthesis method of NiO/C nanoparticles for asymmetric supercapacitors. *Nanomaterials* **2021**, *11*, 187. [CrossRef]
9. Liu, Q.; Ning, J.; Guo, H.; Xia, M.; Wang, B.; Feng, X.; Wang, D.; Zhang, J.; Hao, Y. Tungsten-modulated molybdenum selenide/graphene heterostructure as an advanced electrode for all-solid-state supercapacitors. *Nanomaterials* **2021**, *11*, 1477. [CrossRef]
10. Dos Reis, G.S.; Larsson, S.H.; de Oliveira, H.P.; Thyrel, M.; Lima, E.C. Sustainable biomass activated carbons as electrodes for battery and supercapacitors—A mini-review. *Nanomaterials* **2020**, *10*, 1398. [CrossRef]
11. Liu, C.; Chen, W.; Hong, S.; Pan, M.; Jiang, M.; Wu, Q.; Mei, C. Fast microwave synthesis of hierarchical porous carbons from waste palm boosted by activated carbons for supercapacitors. *Nanomaterials* **2019**, *9*, 405. [CrossRef]
12. Taer, E.; Apriwandi, A.; Dalimunthe, B.K.L.; Taslim, R. A rod-like mesoporous carbon derived from agro-industrial cassava petiole waste for supercapacitor application. *J. Chem. Technol. Biotechnol.* **2021**, *96*, 662–671. [CrossRef]
13. Ding, C.; Yan, X.; Ryu, S.; Yu, Y.; Yang, X. Camphor wood waste-derived microporous carbons as high-performance electrode materials for supercapacitors. *Carbon Lett.* **2019**, *29*, 213–218. [CrossRef]
14. Divya, M.; Natarajan, S.; Lee, Y.S.; Aravindan, V. Biomass-derived carbon: A value-added journey towards constructing high-energy supercapacitors in an asymmetric fashion. *ChemSusChem* **2019**, *12*, 4353–4382. [CrossRef]
15. Dubey, P.; Shrivastav, V.; Maheshwari, P.H.; Sundriyal, S. Recent advances in biomass derived activated carbon electrodes for hybrid electrochemical capacitor applications: Challenges and opportunities. *Carbon* **2020**, *170*, 1–29. [CrossRef]
16. Durairaj, A.; Sakthivel, T.; Ramanathan, S.; Obadiah, A.; Vasanthkumar, S. Conversion of laboratory paper waste into useful activated carbon: A potential supercapacitor material and a good adsorbent for organic pollutant and heavy metals. *Cellulose* **2019**, *26*, 3313–3324. [CrossRef]
17. Elaiyappillai, E.; Srinivasan, R.; Johnbosco, Y.; Devakumar, P.; Murugesan, K.; Kesavan, K.; Johnson, P.M. Low cost activated carbon derived from Cucumis melo fruit peel for electrochemical supercapacitor application. *Appl. Surf. Sci.* **2019**, *486*, 527–538. [CrossRef]
18. Elanthamilan, E.; Sriram, B.; Rajkumar, S.; Dhaneshwaran, C.; Nagaraj, N.; Merlin, J.P.; Vijayan, A.; Wang, S.F. Couroupita guianansis dead flower derived porous activated carbon as efficient supercapacitor electrode material. *Mater. Res. Bull.* **2019**, *112*, 390–398. [CrossRef]
19. Wang, C.J.; Wu, D.P.; Wang, H.J.; Gao, Z.Y.; Xu, F.; Jiang, K. Nitrogen-doped two-dimensional porous carbon sheets derived from clover biomass for high performance supercapacitors. *J. Power Sources* **2017**, *363*, 375–383. [CrossRef]
20. Wang, C.J.; Wu, D.P.; Wang, H.J.; Gao, Z.Y.; Xu, F.; Jiang, K. A green and scalable route to yield porous carbon sheets from biomass for supercapacitors with high capacity. *J. Mater. Chem. A* **2018**, *6*, 1244–1254. [CrossRef]
21. Wang, C.J.; Wu, D.P.; Wang, H.J.; Gao, Z.Y.; Xu, F.; Jiang, K. Biomass derived nitrogen-doped hierarchical porous carbon sheets for supercapacitors with high performance. *J. Colloid Interface Sci.* **2018**, *523*, 133–143. [CrossRef] [PubMed]

22. Abioye, A.M.; Ani, F.N. Recent development in the production of activated carbon electrodes from agricultural waste biomass for supercapacitors: A review. *Renew. Sustain. Energy Rev.* **2015**, *52*, 1282–1293. [CrossRef]
23. Ahmed, S.; Ahmed, A.; Rafat, M. Supercapacitor performance of activated carbon derived from rotten carrot in aqueous, organic and ionic liquid based electrolytes. *J. Saudi Chem. Soc.* **2018**, *22*, 993–1002. [CrossRef]
24. Gao, Y.; Li, L.; Jin, Y.; Wang, Y.; Yuan, C.; Wei, Y.; Chen, G.; Ge, J.; Lu, H. Porous carbon made from rice husk as electrode material for electrochemical double layer capacitor. *Appl. Energy* **2015**, *153*, 41–47. [CrossRef]
25. Ashraf, C.M.; Anilkumar, K.; Jinisha, B.; Manoj, M.; Pradeep, V.; Jayalekshmi, S. Acid washed, steam activated, coconut shell derived carbon for high power supercapacitor applications. *J. Electrochem. Soc.* **2018**, *165*, A900–A909. [CrossRef]
26. Deng, P.; Lei, S.; Wang, W.; Zhou, W.; Ou, X.; Chen, L.; Xiao, Y.; Cheng, B. Conversion of biomass waste to multi-heteroatom-doped carbon networks with high surface area and hierarchical porosity for advanced supercapacitors. *J. Mater. Sci.* **2018**, *53*, 14536–14547. [CrossRef]
27. Guo, N.; Li, M.; Wang, Y.; Sun, X.; Wang, F.; Yang, R. Soybean root-derived hierarchical porous carbon as electrode material for high-performance supercapacitors in ionic liquids. *ACS Appl. Mater. Interfaces* **2016**, *8*, 33626–33634. [CrossRef]
28. Jin, H.; Hu, J.; Wu, S.; Wang, X.; Zhang, H.; Xu, H.; Lian, K. Three-dimensional interconnected porous graphitic carbon derived from rice straw for high performance supercapacitors. *J. Power Sources* **2018**, *384*, 270–277. [CrossRef]
29. Li, Z.; Bai, Z.; Mi, H.; Ji, C.; Gao, S.; Pang, H. Biowaste-derived porous carbon with tuned microstructure for high-energy quasi-solid-state supercapacitors. *ACS Sustain. Chem. Eng.* **2019**, *7*, 13127–13135. [CrossRef]
30. Liu, X.; Zhang, S.; Wen, X.; Chen, X.; Wen, Y.; Shi, X.; Mijowska, E. High yield conversion of biowaste coffee grounds into hierarchical porous carbon for superiorcapacitive energy storage. *Sci. Rep.* **2020**, *10*, 3518. [CrossRef]
31. Lian, Y.M.; Ni, M.; Zhou, L.; Chen, R.J.; Yang, W. Synthesis of biomass-derived carbon induced by cellular respiration in yeast for supercapacitor applications. *Chem. Eur. J.* **2018**, *24*, 18068–18074. [CrossRef] [PubMed]
32. Wan, Z.; Sun, Y.; Tsang, D.C.W.; Khan, E.; Yip, A.C.K.; Ng, Y.H.; Rinklebe, J.; Ok, Y.S. Customised fabrication of nitrogen-doped biochar for environmental and energy applications. *Chem. Eng. J.* **2020**, *401*, 126136. [CrossRef]
33. Niu, L.; Shen, C.; Yan, L.; Zhang, J.; Lin, Y.; Gong, Y.; Li, C.; Sun, C.Q.; Xu, S. Waste bones derived nitrogen-doped carbon with high micropore ratio towards supercapacitor applications. *J. Colloid Interface Sci.* **2019**, *547*, 92–101. [CrossRef] [PubMed]
34. Shen, F.; Su, J.; Zhu, L.; Qi, X.; Zhang, X. Comprehensive utilization of dairy manure to produce glucose and hierarchical porous carbon for supercapacitors. *Cellulose* **2017**, *24*, 2571–2579. [CrossRef]
35. Blankenship, L.S.; Mokaya, R. Modulating the porosity of carbons for improved adsorption of hydrogen, carbon dioxide, and methane: A review. *Mater. Adv.* **2022**, *3*, 1905–1930. [CrossRef]
36. Altwala, A.; Mokaya, R. Predictable and targeted activation of biomass to carbons with high surface area density and enhanced methane storage capacity. *Energy Environ. Sci.* **2020**, *13*, 2967–2978. [CrossRef]
37. Altwala, A.; Mokaya, R. Modulating the porosity of activated carbons via pre-mixed precursors for simultaneously enhanced gravimetric and volumetric methane uptake. *J. Mater. Chem. A* **2022**, *10*, 13744–13757. [CrossRef]
38. Wang, B.; Wang, Y.; Peng, Y.; Wang, Y.; Wang, N.; Wang, J.; Zhao, J. Nitrogen-doped biomass-based hierarchical porous carbon with large mesoporous volume for application in energy storage. *Chem. Eng. J.* **2018**, *348*, 850–859. [CrossRef]
39. Skorupska, M.; Kamedulski, P.; Lukaszewicz, J.P.; Ilnicka, A. The Improvement of Energy Storage Performance by Sucrose-Derived Carbon Foams via Incorporating Nitrogen Atoms. *Nanomaterials* **2021**, *11*, 760. [CrossRef]
40. Lin, Y.; Chen, Z.; Ying, C.; Zhong, W. Facile synthesis of high nitrogen-doped content, mesopore-dominate biomass-derived hierarchical porous graphitic carbon for high-performance supercapacitors. *Electrochim. Acta* **2020**, *334*, 135615. [CrossRef]
41. Li, Y.; Zhang, D.; Zhang, Y.; He, J.; Wang, Y.; Wang, K.; Xu, Y.; Li, H.; Wang, Y. Biomass-derived microporous carbon with large micropore size for high-performance supercapacitors. *J. Power Sources* **2020**, *448*, 227396. [CrossRef]
42. Liu, Y.; Tan, H.; Tan, Z.; Cheng, X. Rice husk derived capacitive carbon prepared by one-step molten salt carbonization for supercapacitors. *J. Energy Storage* **2022**, *55*, 105437. [CrossRef]
43. Liu, Y.; Cheng, X.; Zhang, S. Hierarchically porous carbon derived from tobacco waste by one-step molten salt carbonization for supercapacitor. *Carbon Lett* **2022**, *32*, 251–263. [CrossRef]
44. Zhang, J.; Chen, H.; Bai, J.; Xu, M.; Luo, C.; Yang, L.; Bai, L.; Wei, D.; Wang, W.; Yang, H. N-doped hierarchically porous carbon derived from grape marcs for high-performance supercapacitors. *J. Alloys Compd.* **2021**, *854*, 157207. [CrossRef]
45. Jalalah, M.; Sivasubramaniam, S.S.; Aljafari, B.; Irfan, M.; Almasabi, S.S.; Alsuwian, T.; Khazi, M.I.; Nayak, A.K.; Harraz, F.A. Biowaste assisted preparation of self-nitrogen-doped nanoflakes carbon framework for highly efficient solid-state supercapacitor application. *J. Energy Storage* **2022**, *54*, 105210. [CrossRef]
46. Divya, P.; Rajalakshmi, R. Renewable, low cost green functional mesoporous electrodes from Solanum lycopersicum leaves for supercapacitors. *J. Energy Storage* **2020**, *27*, 101149. [CrossRef]
47. Liang, X.; Liu, R.; Wu, X. Biomass waste-derived functionalized hierarchical porous carbon with high gravimetric and volumetric capacitances for supercapacitors. *Microporous Mesoporous Mater.* **2021**, *310*, 110659. [CrossRef]
48. Wang, C.; Yuan, X.; Guo, G.; Liang, H.; Ma, Z.; Li, P. Salt template tuning morphology and porosity of biomass-derived N-doped porous carbon with high redox-activation for efficient energy storage. *Colloid Surf. A* **2022**, *650*, 129552. [CrossRef]
49. Li, G.; Li, Y.; Chen, X.; Hou, X.; Lin, H.; Jia, L. One step synthesis of N, P co-doped hierarchical porous carbon nanosheets derived from pomelo peel for high performance supercapacitors. *J. Colloid Interface Sci.* **2022**, *605*, 71–81. [CrossRef]

50. Mitravinda, T.; Nanaji, K.; Anandan, S.; Jyothirmayi, A.; Chakravadhanula, V.S.K.; Sharma, C.S.; Rao, T.N. Facile synthesis of corn silk derived nanoporous carbon for an improved supercapacitor performance. *J. Electrochem. Soc.* **2018**, *165*, A3369–A3379. [CrossRef]
51. Raj, C.J.; Rajesh, M.; Manikandan, R.; Yu, K.H.; Anusha, J.R.; Ahn, J.H.; Kim, D.-W.; Park, S.Y.; Kim, B.C. High electrochemical capacitor performance of oxygen and nitrogen enriched activated carbon derived from the pyrolysis and activation of squid gladius chitin. *J. Power Sources* **2018**, *386*, 66–76. [CrossRef]
52. Raj, F.R.M.S.; Jaya, N.V.; Boopathi, G.; Kalpana, D.; Pandurangan, A. S-doped activated mesoporous carbon derived from the Borassus flabellifer flower as active electrodes for supercapacitors. *Mater. Chem. Phys.* **2020**, *240*, 122151–122161.
53. Rani, M.U.; Nanaji, K.; Rao, T.N.; Deshpande, A.S. Corn husk derived activated carbon with enhanced electrochemical performance for high-voltage supercapacitors. *J. Power Sources* **2020**, *471*, 228387. [CrossRef]
54. Okonkwo, C.A.; Lv, T.; Hong, W.; Li, G.; Huang, J.; Deng, J.; Jia, L.; Wu, M.; Liu, H.; Guo, M. The synthesis of micro mesoporous carbon derived from nitrogen-rich spirulina extract impregnated castor shell based on biomass self-doping for highly efficient supercapacitor electrodes. *J. Alloys Compd.* **2020**, *825*, 154009. [CrossRef]
55. Zhao, N.; Zhang, P.; Luo, D.; Xiao, W.; Deng, L.; Qiao, F. Direct production of porous carbon nanosheets/particle composites from wasted litchi shell for supercapacitors. *J. Alloy. Compd.* **2019**, *788*, 677–684. [CrossRef]
56. Lu, C.; Qian, X.Z.; Zhu, H.Y.; Hu, Y.X.; Zhang, Y.S.; Zhang, B.M.; Kong, L.B.; Liu, M.C. 3D hierarchical porous carbon derived from direct carbonization and in-situ chemical activation of potatoes toward high-performance supercapacitors. *Mater. Res. Express* **2019**, *6*, 115615–115625. [CrossRef]
57. Yang, Z.; Xiang, M.; Wu, Z.; Hui, J.; Huang, Q.; Zhang, J.; Qin, H. A three-dimensional carbon electrode derived from bean sprout for supercapacitors. *Ionics* **2020**, *26*, 5705–5714. [CrossRef]
58. Wang, Y.; Shao, C.; Qiu, S.; Zhu, Y.; Qin, M.; Meng, Y.; Wang, Y.; Chu, H.; Zou, Y.; Xiang, C. Nitrogen-doped porous carbon derived from ginkgo leaves with remarkable supercapacitance performance. *Diam. Relat. Mater.* **2019**, *98*, 107475. [CrossRef]
59. Zhang, Y.H.; Wu, C.; Dai, S.; Liu, L.F.; Zhang, H.; Shen, W.; Sun, W.; Li, C.M. Rationally tuning ratio of micro- to meso-pores of biomass-derived ultrathin carbon sheets toward supercapacitors with high energy and high power density. *J. Colloid Interface Sci.* **2022**, *606*, 817–825. [CrossRef]

Article

Sustainable Carbon Derived from Sulfur-Free Lignins for Functional Electrical and Electrochemical Devices

Bony Thomas [1], Mohini Sain [1,2] and Kristiina Oksman [1,2,3,*]

[1] Division of Materials Science, Department of Engineering Sciences and Mathematics, Luleå University of Technology, SE-97187 Luleå, Sweden
[2] Mechanical & Industrial Engineering (MIE), University of Toronto, Toronto, ON M5S 3G8, Canada
[3] Wallenberg Wood Science Center (WWSC), Luleå University of Technology, SE-97187 Luleå, Sweden
* Correspondence: kristiina.oksman@ltu.se; Tel.: +46-(0)920-493371

Abstract: Technical lignins, kraft, soda, lignoboost, and hydrolysis lignins were used for the production of carbon particles at different carbonization temperatures, 1000 °C and 1400 °C. The results showed that the lignin source and carbonization temperature significantly influenced the carbon quality and microstructure of the carbon particles. Soda lignin carbonized up to 1400 °C showed higher degree of graphitization and exhibited the highest electrical conductivity of 335 S·m^{-1}, which makes it suitable for applications, such as electromagnetic interference shielding and conductive composite based structural energy storage devices. The obtained carbon particles also showed high surface area and hierarchical pore structure. Kraft lignin carbonized up to 1400 °C gives the highest BET surface area of 646 m^2 g^{-1}, which makes it a good candidate for electrode materials in energy storage applications. The energy storage application has been validated in a three-electrode set up device, and a specific capacitance of 97.2 F g^{-1} was obtained at a current density of 0.1 A g^{-1} while an energy density of 1.1 Wh kg^{-1} was observed at a power density of 50 W kg^{-1}. These unique characteristics demonstrated the potential of kraft lignin-based carbon particles for electrochemical energy storage applications.

Keywords: lignin; carbonization; microstructure; energy storage; electrical conductivity

Citation: Thomas, B.; Sain, M.; Oksman, K. Sustainable Carbon Derived from Sulfur-Free Lignins for Functional Electrical and Electrochemical Devices. *Nanomaterials* **2022**, *12*, 3630. https://doi.org/10.3390/nano12203630

Academic Editor: Dapeng Wu

Received: 12 September 2022
Accepted: 10 October 2022
Published: 16 October 2022

Publisher's Note: MDPI stays neutral with regard to jurisdictional claims in published maps and institutional affiliations.

Copyright: © 2022 by the authors. Licensee MDPI, Basel, Switzerland. This article is an open access article distributed under the terms and conditions of the Creative Commons Attribution (CC BY) license (https://creativecommons.org/licenses/by/4.0/).

1. Introduction

Over the past decade, the development of biomass-based carbon materials and their conversion into different kind of carbon nanomaterials has gained immense attention in the research community and different biomass feedstocks have been converted into value added end products by researchers around the globe [1]. Out of different biomass feedstocks, one of the most promising feedstocks is lignin, which is an abundant biopolymer, waste product from paper and pulp industries with over 50 million metric tons isolated every year [2]. Lignin is pyrolyzed to get a carbon rich biochar which is used as soil amendment [2]. Recent studies have revealed that the lignin can be utilized to prepare high value-added carbon materials that can be used for different applications like conductive fillers for composites or electrodes for energy storage applications [1,2]. In addition to the compositional complexity of the feedstock, type of isolation technique plays a vital role in defining the final chemical structure of the technical lignins [3]. Kraft pulping, soda pulping, sulfite pulping, and organosolv pulping, etc. are the most common type of lignin isolation techniques [4]. Kraft and sulfite pulping are classified as sulfur bearing processes while organosolv and soda pulping are coming under sulphur free process. Kraft pulping uses a mixture of sodium hydroxide and sodium sulphide known as white liquor while soda pulping uses only sodium hydroxide [4]. One of the recent advancements in the kraft pulping process is the lignoboost process, which provides more efficient extraction with high purity lignin, higher yield, low amounts of ash and lower carbohydrate content [5].

Hydrolysis lignin is another type of lignin obtained as a byproduct from the bio-fuel industries when the lignin containing biomass is treated with hydrolytic enzymes [6]. Different carbon materials, including carbon nanofibers [7,8], carbon aerogels [5,9], and carbon particulates [10], have been produced from extracted lignins using different isolation processes. To make carbon nanofibers, researchers are following different processing techniques like melt spinning, solution spinning and electrospinning [7]. Recently, carbon aerogels were prepared using ice templating followed by freeze drying and carbonization [5,9]. Direct carbonization has been performed to make lignin-based carbon particles as a replacement for carbon black [10]. Different activation strategies [11] have been performed to increase the properties of lignin derived carbon materials which are not environment friendly and cost effective [12]. Metal catalyzed hydrothermal carbonization has been performed for the preparation of graphitic biocarbon from lignin [13]. Even though most these studies have shown that a specific type of lignin can be used for preparing carbon materials with specific properties, there is a need for a comprehensive study where the properties of carbons from most common technical lignins can be compared at different carbonization temperatures (without any chemical or physical activation and in the absence of any catalysts).

In the current study, four different commonly available technical lignins, namely kraft lignin (KL), soda lignin (SL), lignoboost lignin (LB), and hydrolysis lignin (HL), were carbonized at two different carbonization temperatures 1000 °C and 1400 °C. A simple, green, and low-cost strategy such as a direct carbonization of the lignin, was adopted. Neither activation steps nor any special templating agents, techniques, additional additives, or extra processing steps were used to enhance the porosity or microstructure of the carbon particles. Resulting carbon particles were systematically analyzed for their suitability for high value applications, such as supercapacitor electrodes and conductive graphitic carbon additives. The carbon particles with the highest specific surface area were used for making supercapacitor electrodes and their electrochemical performances were analyzed. Carbon particles were also evaluated for their electrical conductivities to determine technical lignin and temperature of carbonization for achieving highest degree of graphitization for their effective use as a conductive filler.

Remarkable electrochemical and electrical properties were achieved, and the resulting graphitic carbon obtained in this study is another important step forward towards achieving functional graphitic carbon from biomass. Hence, this study presents the importance of types of lignin and their carbonization process, and how they impact on the final properties of carbon materials. This work gives new insights towards the effective utilization and value addition of an abundant and unique natural resource, reducing the carbon footprint, economic growth potential, and improving ecosystem sustainability.

2. Experimental

2.1. Materials

Four different lignins were used; (1) a low sulfonate content kraft lignin (KL) supplied from Sigma-Aldrich (St. Louis, MI, USA); (2) soda lignin (SL) Protobind 2000, from Greenvalue Enterprises LLC, (Media, PA, USA); (3) Lignoboost lignin, Biochoice^{TM} lignin (LB), based on softwood was kindly supplied by Domtar Plymouth pulp mill (Plymouth, NC, USA) and (4) hydrolysis lignin (HL), which was kindly supplied by a Finnish energy company, St1 biofuels (Kajaani, Finland). Sulphuric acid (H_2SO_4, 95–98%) and ethanol was purchased from VWR International AB (Stockholm, Sweden). Polytetrafluoroethylene (PTFE) (60 wt.% aqueous suspension) was purchased from Sigma-Aldrich (Stockholm, Sweden).

2.2. Preparation of Carbon Particles

Before the carbonization, all the lignin powders were powdered and sieved using an 80 μm sieve size to the similar particle size (<80 μm). A horizontal tube furnace Nabertherm RHTC-230/15, Nabertherm GmbH (Lilienthal, Germany) was used for the carbonization in nitrogen (N_2) atmosphere. The setup used was as follows: first one hour at 100 °C, then one hour at 400 °C, and then two hours at the final temperatures 1000 °C or 1400 °C. All the heating steps during the carbonization process were performed at a rate of 5 °C/min. After carbonization, the samples were cooled down to room temperature and taken out carefully for further characterizations. Table 1 shows the different carbons and their samples codes.

Table 1. Table showing the type of lignin, carbonization temperature and the sample code for the prepared carbons.

	Carbonization Temperature (°C)	Sample Codes
Kraft lignin (KL)	1000	KL1000
	1400	KL1400
Soda lignin (SL)	1000	SL1000
	1400	SL1400
Lignoboost lignin (LB)	1000	LB1000
	1400	LB1400
Hydrolysis lignin (HL)	1000	HL1000
	1400	HL1400

2.3. Electrode Preparation

Working electrode for three electrode measurement was prepared by mixing prepared carbon particles, polytetrafluoroethylene (PTFE), and carbon black in ethanol, in the weight ratio 75:15:10, where PTFE acted as polymeric binder and carbon black as a conductive filler. This slurry was then left for 6 h in oven at 100 °C to evaporate ethanol (which was used to disperse the carbon particles and PTFE in the slurry) and water (from the aqueous suspension of PTFE). After drying, the working electrode was weighed (Approximately 20 mg) and used for the electrochemical measurements.

2.4. Characterization

TGA-Q500 (TA Instruments, New Castle, DE, USA) was used for the thermo gravimetric analysis (TGA) in nitrogen (N_2) atmosphere, from room temperature to 1000 °C, maintaining a heating rate of 10 °C/min.

Equation (1) was used for the calculation of carbon yield,

$$\text{Carbon yield} = \frac{\text{mass of carbon obtained}}{\text{mass of lignin initally taken}} \times 100 \qquad (1)$$

where the mass of the lignin before and mass of carbon after carbonization was measured for the determination of the yield.

The microstructure of the carbon particles was studied using scanning electron microscopy (SEM) JSM-IT300, (JEOL, Tokyo, Japan) with an acceleration voltage of 20 kV under a high vacuum. The samples preparation was performed by diluting aqueous dispersions of the carbon particles to approximately 0.01 wt.%, the dispersion was drop casted onto a freshly cleaved mica sheet attached to the sample holder, and the dispersion was dried before sputter coating with platinum to a thickness of 15 nm.

The elemental analysis of the carbon materials was carried out using energy dispersive X-ray spectroscopy (EDX) on the same SEM instrument equipped with a silicon drift detector Oxford X-MaxN 50 mm^2, (Oxford Instruments, Abingdon, UK).

The particle sizes were determined from SEM images taken at random locations of the samples. For each sample, the sizes of at least 200 particles were measured by the software ImageJ (University of Wisconsin, Madison, WI, USA).

A Bruker Senterra Raman microscope, (Bruker Corporation, Billerica, MA, USA), using green laser (λ = 532 nm) with a power of 5 mW, was used to study carbon structure of the prepared materials. The measurements were performed at a magnification of 20×, and the spectra were recorded from 50 to 3500 cm^{-1} with an acquisition time of 2 s.

X-Ray diffraction (XRD) was performed with a PANalytical EMPYREAN diffractometer (Malvern Instruments, Malvern, UK) using Cu K alpha radiation. Intensity data were recorded in intervals of 0.03° between the scattering angles $2\theta = 5°$ and $50°$. Small angle X-ray scattering was carried out with a PixCel3D detector and a graphite monochromator. Based on the full width at half maximum (FWHM), the XRD crystallite size was calculated. Thus, the radial expansion of the carbon crystal and the stacking height of graphene layers were determined using the Scherrer equation, as shown in Equation (2) [14].

$$L_i = \frac{K_i \lambda}{\beta_i \cos \theta_i}, \quad i = a, c \tag{2}$$

where L_i (nm) can take two indices: L_a which is the radial expansion of the carbon crystal contributed by the (001) peak and L_c which is the stacking height of the graphene layers can be determined from (002) peak. K_i is a structural constant and values used were $K_a = 1.84$ and $K_c = 0.89$ taken from the literature [15]. λ is the wavelength of Cu K_α radiation, $\lambda = 0.1542$ nm [15]. Finally, β_i (rad) is the FWHM for each peak and θ_i (rad) is the diffraction angle for each peak.

The distance (d) between two graphene layers was calculated using Bragg's law for $n = 1$,

$$2d \sin \theta_c = n\lambda \tag{3}$$

The number of graphene layers per stack unit was calculated by dividing the stacking height (L_c) with interplanar distance (d).

Specific surface area (SSA), average pore size, and pore volume of the carbon particles were measured using Micromeritics Gemini VII 2390a Brunauer–Emmett–Teller (BET) analyzer at 77 K (Micromeritics Instrument Corporation, Norcross, GA, USA). The samples were degassed at 300 °C for 3 h prior the BET measurement in a Micromeritics FlowPrep060 to remove entrapped moisture and increase the accessibility to the pores. Micropore area was calculated using the standard t-plot method and total pore volume was determined from the amount of N_2 adsorbed at a relative pressure $P/P_0 = 0.99$.

Electrochemical properties of KL1400 electrode were measured using a Princeton Applied Research VerstaSTAT 3 Potentiostat/Galvanostat (AMETEK Scientific Instruments, Wokingham, UK) connected with a three-electrode cell kit (Pine Research Instrumentation, Durham, NC, USA). The working electrode was prepared as described in the previous section. Platinum was used as the counter electrode, Ag/AgCl as reference electrode and the electrolyte used was 1M sulphuric acid (H_2SO_4) solution. To analyze the rate capabilities of the electrodes, cyclic voltammetry measurements (CV) were carried out at different scan rates between 2 and 100 mV s^{-1} in the potential range from 0 to 1 V. Specific capacitances obtained from CV tests can be calculated using Equation (4),

$$C = \frac{1}{mv(V_2 - V_1)} \int_{V_1}^{V_2} I dV \tag{4}$$

where C (F g^{-1}) is the specific capacitance; m (mg) is the mass of active materials loaded in the working electrode; v (V s^{-1}) is the scan rate; I (A) is the discharge current; V_2 and V_1 (V) are high and low potential limits of the CV tests. Further galvanostatic charge discharge method (GCD) was conducted using at different current densities in the range from 0.1 to 1 Ag^{-1} to investigate the electrochemical performance. Specific capacitance from GCD measurements can be calculated using Equation (5),

$$C = \frac{I \Delta t}{m \Delta V} \tag{5}$$

where C (F g^{-1}) is the specific capacitance; I (A) is the discharge current; Δt (s) is the discharge time; ΔV (V) is the potential window; and m (mg) is the total mass of electrode material.

To calculate the specific energy density (E) and specific power density (P), a practical supercapacitor was assembled using the prepared electrodes and two electrode method was used for the measurements and the details of which is given in supporting information. From the galvanostatic tests, E and P can be calculated using Equations (6) and (7),

$$E = \frac{C\Delta V^2}{2*3.6*4} \quad (6)$$

$$P = \frac{E*3600}{\Delta t} \quad (7)$$

where E (Wh kg^{-1}) is the average energy density; C (F g^{-1}) is the specific capacitance; ΔV (ΔV) is the potential window; P (W kg^{-1}) is the average power density and Δt (s) is the discharge time. Electrochemical impedance spectroscopy (EIS) measurements were carried out in the frequency range of 10^{-2} to 10^5 Hz to determine the resistances offered by the electrode.

The electrical conductivity of the carbon particles was measured at room temperature using a Hioki IM 3536 LCR meter (Hioki E.E. Corporation, Nagano, Japan). The conductivity values were recorded for frequencies ranging from 1 kHz to 5 MHz.

3. Results and Discussions

Figure 1a shows the schematic of the preparation process of carbon particles from different technical lignin. Kraft (KL), soda (SL), lignoboost (LB), and hydrolysis lignin (HL) powders were carbonized in a horizontal tubular furnace maintained in N$_2$ atmosphere to 1000 and 1400 °C to obtain the different carbon particles as listed in Table 1.

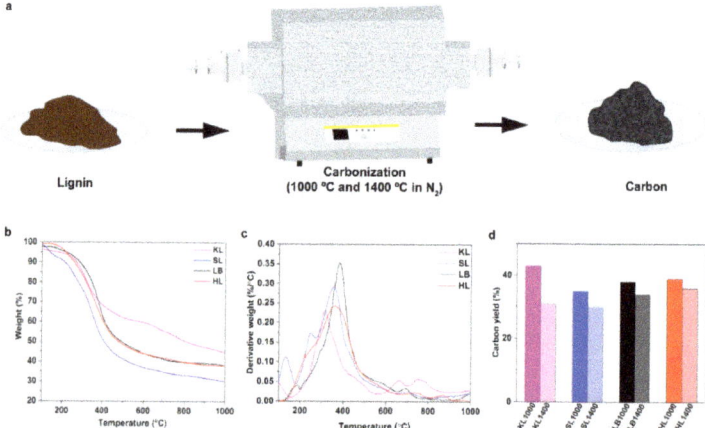

Figure 1. (a) Schematic showing the preparation of lignin-based carbon particles using the horizontal carbonization furnace. (b) Thermal behavior of and residues obtained for lignins at 1000 °C after the TGA analysis in N$_2$ atmosphere. (c) Derivative thermogravimetric (DTG) curve for lignins. (d) Carbon yield obtained after carbonization at 1000 °C and 1400 °C for different lignins used in the study.

Figure 1b shows the results obtained from thermogravimetric analysis (TGA) for different lignins. KL resulted in highest amount of char residue, 45% at 1000 °C. LB and HL lignins showed approximately same amount of char residue, 38% while SL had lowest percentage of char residue of 31%. To analyze in more depth the thermal decomposition behaviors, the derivative thermogravimetric (DTG) curve for all lignins has been plotted and shown in Figure 1c. All the lignins had their major decomposition peaks (DTG$_{max}$) in the temperature range between to 300 °C to 400 °C, as shown in Table 2, which corresponds to the degradation of lignin. SL and LB showed slight decomposition below 200 °C which corresponds to the early release of volatiles and SL released much more volatiles than LB, which could be the reason for lower amount of residue for SL in comparison with all other lignins. KL and HL had no decomposition peaks and hence no release of volatiles for temperatures less than 200 °C. Smaller decomposition peaks observed between 200 °C and 250 °C for SL, LB, and HL (SL—250 °C, LB—219 °C, HL—221 °C) could be due to the presence hemicellulose. Above 600 °C, lignins produce aromatics and CO_2 as major decomposition products during TGA and KL showed relatively stronger peaks between 600 °C and 700 °C, indicating the release of more CO and CO_2 [9] which could result in carbon particles with high specific surface area (SSA) compared to other lignins. In the TG-IR studies conducted by Zhang et al., it has been observed that KL produced relatively smaller amounts of volatiles, and no carbonyl groups were released during the temperature range below 1000 °C, leading to approximately 46% char residue [16].

Table 2. Table showing the thermal properties of used materials, DTG_{max} and percentual amount of the char residue, obtained from TGA.

	DTG Max (°C)	Char Residue (%)
KL	315	45
SL	355	38
LB	383	38
HL	360	31

Experimental carbon yield obtained after carbonization of lignin particles are represented in Figure 1d. It was observed that the carbon yields were higher at 1000 °C compared to that at 1400 °C and the values for 1000 °C were in accordance with the amount of residue obtained from TGA analysis. At 1400 °C, KL and SL had almost similar carbon yield (KL—31% and SL—30%). KL showed a significant reduction in carbon yield (from 43% at 1000 °C to 31% at 1400 °C) with increase in temperature, indicating the release of more carbon as volatiles, which could lead to the formation of numerous micropores (<2 nm) which results in higher SSA and hierarchical pore structure. LB and HL did not show remarkable differences for carbon yields between 1000 °C and 1400 °C. It has been reported by Liu et. al. [17] that the first step of the depolymerization of lignin chain is the breaking of the β-O-4 linkage. Collision between these free radicals potentially leads to the chemical bond formation leading to formation of stable compounds, and the random bonding of the radicals results in polyaromatic biochar for temperatures higher than 350 °C [17]. This indicates all lignins formed stable carbons already during the isothermal heating period at 1000 °C. The heat energy supplied during the isothermal processing at 1400 °C for 2 h could be attributed to the reorganization of the carbon structure to develop graphitic domains. The later proposed mechanism yet to be confirmed. The carbon content and molecular weight (MW) of the lignins can affect the final carbon yield. In one of our previous studies, Wei et al. [18] reported that HL and SL had considerably higher overall ratios of aliphatic to aromatic carbons than KL and LB. KL and HL have similar MW and showed similar carbon yield [18], which is in conformity to the results discussed here.

The morphology of carbon particles was analyzed using scanning electron microscopy (SEM) and the results are shown in Figure 2. KL1000 and KL1400 showed rough surface texture compared to all other lignins which showed flat and smooth surfaces. Comparing KL1000 and KL1400 the surface morphology did not change noticeably while SL1400 clearly showed a smoother and layered structure compared to SL1000. LB1000 had smooth surface while well-defined layered structure was visible for LB1400. HL showed similar surface for HL1000 and HL1400, but it was observed that many smaller carbon particles were attached to the bigger sized particles. A direct comparison of particle size was difficult in the case of SL, LB, and HL because the lignin powder melted and fused together to form bigger agglomerates which were broken down to uniform size using a mortar and pestle. KL retained particle morphology even after carbonization, which is confirmed using the particle size measurements, performed qualitatively using ImageJ. KL particles (average diameter was 5.33 µm) experienced a volume shrinkage during the carbonization process (average diameter for KL1000 was 4.3 µm, and for KL1400 2.2 µm). Thus, it can be concluded that KL1000 and KL1400 show relatively rougher surfaces with nanostructured wrinkles and surface cavities. The latter can provide higher SSA compared to all other lignins at the same time as SL and LB and is expected to have higher degrees of graphitization compared to other lignins, which is yet to be confirmed.

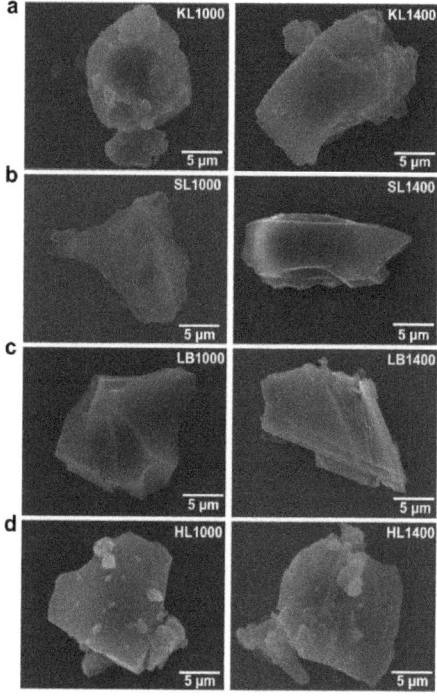

Figure 2. Microstructure of lignin-based carbon particles at 1000 °C and 1400 °C respectively (**a**) KL1000 and KL1400, (**b**) SL1000 and SL1400, (**c**) LB1000 and LB1400, (**d**) HL1000 and HL1400.

Elemental analysis of carbon particles was performed using energy dispersive X-ray spectroscopy (EDX) and the results obtained are listed in Table 3. The atomic percentage of carbon was between 86% and 97% for all carbon particles and the percentage of carbon was increased with increase in the carbonization temperature. The amount of oxygen was varied between 3% to 12% between the carbon particles. The atomic percentage of oxygen was decreased with the increase in carbonization temperature indicating the breakage of more oxygen containing bonds utilizing the higher supplied energy, and more carbonaceous gases, such as CO and CO_2, were released. Minor amounts of sodium (Na) and sulfur (s) were present in KL and SL based carbon particles, which were derived from the lignin isolation process. Minor amounts of potassium (K) and silicon (Si) were present in SL based carbon particles. These were due to the impurities from the wheat straw arising from the starting materials for isolating soda lignin. The presence of ash also can contribute to the presence of silicon in the carbonized lignins. A High purity of lignoboost lignin [5] could be the reason for THE presence of very small amounts of inorganic elements in LB1000 and LB1400. HL lignin carbon particles were also free from inorganic elements which might have been released as volatiles during the carbonization. It was observed that at higher carbonization temperatures, there was no considerable difference in elemental percentages detected by different elemental analysis techniques, such as SEM-EDX and XPS. [19]. EDX provides bulk elemental compositions because of the higher X-ray generation depth of around 1 μm, while the results from the XPS are more at the surface, within the depth around 10 nm. Since the bulk elemental composition is more important for electrochemical properties, SEM-EDX analysis is expected to provide reliable information about the bulk composition of carbon particles [19].

Table 3. Elemental composition in at.% of different lignins carbonized at 1000 and 1400 °C obtained from SEM-EDX analysis.

Sample	C	O	Na	S	Si	K
KL1000	86.0	12.1	0.30	0.90	0.70	-
KL1400	88.0	9.90	0.40	0.20	1.50	-
SL1000	96.3	3.10	0.10	0.10	0.30	0.10
SL1400	94.7	4.40	0.40	0.40	0.10	-
LB1000	92.3	7.00	0.10	0.30	0.20	0.10
LB1400	96.5	3.50	-	-	-	-
HL1000	90.4	9.60	-	-	-	-
HL1400	96.9	3.10	-	-	-	-

Raman spectra of carbon particles are represented in Figure 3. All carbon particles showed characteristic bands for carbon materials at 1582 cm^{-1} (G-band) and at 1330 cm^{-1} (D-band). Vibrations of sp^2 bonded carbon atoms in the 2D hexagonal lattice is the reason for G-band and in-plane terminated disordered tangling bonds in graphite, as represented by the D-band [15,20]. Raman spectra are also evaluated using the intensity ratio between D and G bands (I_D/I_G ratio) [20–22].

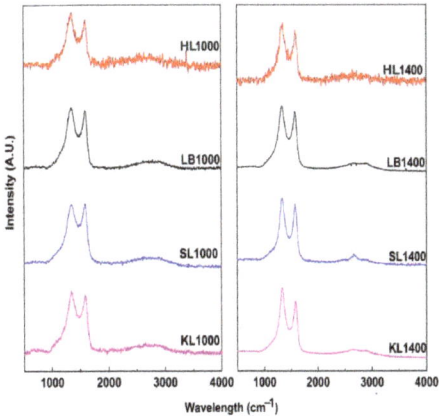

Figure 3. Raman spectra of carbonized HL, LB, SL and KL at 1000 °C and 1400 °C respectively.

Figure 4 shows the I_D/I_G ratio and full width at half maximum (FWHM) of D-band of the carbon particles carbonized at 1000 °C and 1400 °C calculated from the Raman spectra. In Figure 4a, it is seen that I_D/I_G ratio increased significantly when the temperature was increased from 1000 °C to 1400 °C. This clearly indicates the growth of aromatic clusters and graphitic layers in the carbon structure [15,21,23]. In addition to the increase, the shape of Raman spectra changed significantly with the increase in temperature. To analyze the changes in full width at half maximum (FWHM) and band positions with respect to carbonization temperature, simple curve fitting involving only D and G bands [24] was used (Figure 4b). During curve fitting, D bands were fitted using the Gaussian function and G bands were fitted with the Lorentzian function, which is in accordance with literature [24]. For all carbon particles the D-band became narrower as can be seen in Figure 4b, the values are given in Table 4. It is also seen that the valley intensity between D and G bands was also decreased (Figure 3). The FWHM for G bands (Table 4) was increased when the carbonization temperature was increased except for HL. This could be due to the conversion of aromatic rings to small graphite crystallite when the carbonization temperature was increased [24]. These results indicate the increase of ordering in the carbon structure during the high temperature carbonization as the non-crystalline carbon was evolved as volatiles [24].

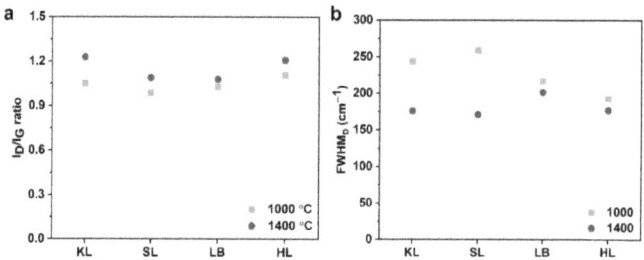

Figure 4. (a) I_D/I_G ratio of carbon particles carbonized at 1000 °C and 1400 °C and calculated from the Raman spectra. (b) Full width at half maximum (FWHM) of D-band for KL, SL, LB and HL carbon particles calculated from the Raman spectra shown in Figure 3.

Table 4. Full width at half maximum (FWHM) of D and G Raman bands for carbon particles with respect to carbonization temperatures.

Sample	D_{FWHM} (cm^{-1})	G_{FWHM} (cm^{-1})	Sample	D_{FWHM} (cm^{-1})	G_{FWHM} (cm^{-1})
KL1000	244.3 ± 1.6	84.9 ± 0.9	KL1400	176.1 ± 1.1	92.4 ± 0.8
SL1000	259.7 ± 2.0	85.7 ± 1.0	SL1400	171.4 ± 1.1	87.7 ± 0.7
LB1000	217.8 ± 1.0	85.9 ± 0.6	LB1400	202.3 ± 1.2	90.1 ± 0.7
HL1000	193.6 ± 1.6	90.3 ± 1.1	HL1400	177.4 ± 1.3	89.1 ± 1.0

The positions of D and G bands are given in Table 5. The D and G bands showed a red shift with the increase in carbonization temperature. The shift in D band peak position was prominent in case of KL, SL and LB based carbon particles while the position didn't change for HL based carbon particles. The red shift in the peak position of D band also demonstrates the increase in graphitization due to the formation of larger aromatic clusters which agree with the biomass-based chars reported in the literature [24–26]. The red shift observed in the G band peak position could be due to the introduction of more defective aromatic rings into the already generated graphite layers as the carbonization temperature was increased [22]. A similar observation of red shift of G band peak has been observed in the nanocrystalline carbon and the presence of defects and impurities in the microstructure led to anharmonic contribution to the lattice potential, which resulted in the red shift [27]. All the carbon particles except HL1400 showed more defined 2D peaks than the broad 2D peak observed at 1000 °C. It is important to note that the parameters, such as peak positions of D and G bands and FWHM of D and G bands, can differ significantly with respect to the biomass feedstock [24]. The chemical structure of the biomass such as combination of aromatic rings and the connections between different chemical groups vary significantly for each lignin which can affect the final structure of the carbon [24].

Table 5. Temperature dependence of peak positions for D and G bands obtained from Raman spectroscopy analysis.

Sample	D Band Position (cm^{-1})	G Band Position (cm^{-1})	Sample	D Band Position (cm^{-1})	G Band Position (cm^{-1})
KL1000	1350	1582	KL1400	1331	1575
SL1000	1351	1582	SL1400	1337	1576
LB1000	1347	1584	LB1400	1339	1579
HL1000	1347	1584	HL1400	1347	1583

XRD patterns of carbon particles were recorded and represented in Figure 5. For all carbon particles produced at 1000 °C the first diffraction peak was observed between $2\theta = 22$–$23°$ and the second peak was located between $2\theta = 43$–$44°$. The first diffraction peak ($2\theta = 22$–$23°$) corresponds to the (002) plane which represents the reflections from stacked graphene layers. The second maximum ($2\theta = 43$–$44°$) corresponds to the (100) plane, which originates from the aromatic ring structures present within the graphene layers [15]. With the increase in the carbonization temperature, the peak intensities of both the planes were increased and the peaks became more prominent at 1400 °C. This indicated the growth of graphitic layers with the carbonization temperature. Among all carbon particles at 1400 °C, SL1400 was found to have more prominent (002) and (100) peaks, showing the higher degree of graphitization, as indicated by the lowest FWHM of the Raman D band as shown in Figure 4b.

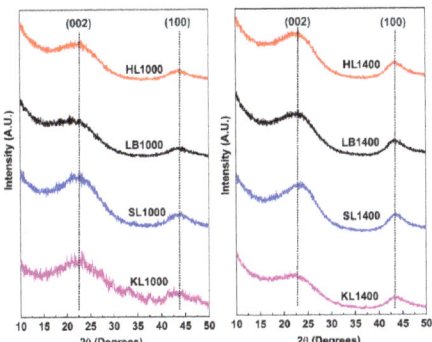

Figure 5. X-ray diffraction patters for KL, SL, LB, HL carbonized at 1000 °C and 1400 °C.

To quantitatively analyze the (002) and (100) peaks, Bragg's Law ($n = 1$) and Scherrer-Equation were used and mean distance between graphene layers d, stacking height L_c and radial expansion L_a were calculated as shown in Table 6. It has been observed that the radial expansion L_a increased with increase in the carbonization temperature and the highest expansion ($L_a = 5.65$ nm and $L_c = 1.25$ nm) was observed for SL1400, which indicated the higher level of graphitization achieved, as discussed earlier. The distance between the stacked layers were calculated and the value is ranging between 0.377 nm–0.395 nm (Table 6). This is in accordance with the results obtained in the graphitization analysis undertaken by Schneider et al. [15], where the biochar derived from the high temperature pyrolysis of beech wood showed similar inter layer distance of 0.387 nm at 1600 °C. The lowest interlayer distance was found for SL1400 (0.377 nm), which is close to the pure graphite which has 0.354 nm as the interlayer distance between the layers and which can be achieved for temperatures above 2100 °C [15,28]. The number of layers in the stack was calculated by dividing L_c value by the interlayer distance d [15]. It was observed that the stacks consist of approximately 2–3 layers in all the carbon particles. A similar observation of constant stack height and stack order has been reported in the literature and the carbonization temperature needs to be increased to make a noticeable difference in these parameters [15,28,29]. Overall, XRD analysis confirmed the graphitization occurred during the high temperature carbonization of lignin particles.

Table 6. Results of quantitative analysis of XRD spectrum. Values for radial expansion (L_a), stacking height (L_c) and distance between graphene layers (d) obtained for all carbon particles at 1000 °C and 1400 °C.

Sample	L_a (nm)	L_c (nm)	d (nm)	Sample	L_a (nm)	L_c (nm)	d (nm)
KL1000	4.10	1.05	0.387	KL1400	5.46	1.13	0.387
SL1000	4.75	1.06	0.387	SL1400	5.65	1.25	0.377
LB1000	3.04	0.99	0.395	LB1400	4.33	1.12	0.382
HL1000	3.32	1.05	0.385	HL1400	4.18	1.08	0.384

The specific surface area (SSA), average pore diameter (d_a), and pore volume (V_p) of the carbon particles were measured using Brunauer–Emmett–Teller (BET) analysis. Nitrogen (N_2) adsorption isotherms for all the samples carbonized at 1000 °C and 1400 °C are represented in Figure 6a,b. Detailed information about the microstructure of carbon particles is shown in Table 7. Except for HL, all the carbon particles produced at 1400 °C exhibited higher SSA than those produced at 1000 °C, as shown in Figure 6c. This can be attributed to the generation of more micropores since more and more carbon was eliminated as volatiles during the high temperature carbonization. KL1000 showed highest SSA of 266 m^2 g^{-1} among the samples produced at 1000 °C while KL1400 showed 646 m^2 g^{-1} among the samples carbonized at 1400 °C.

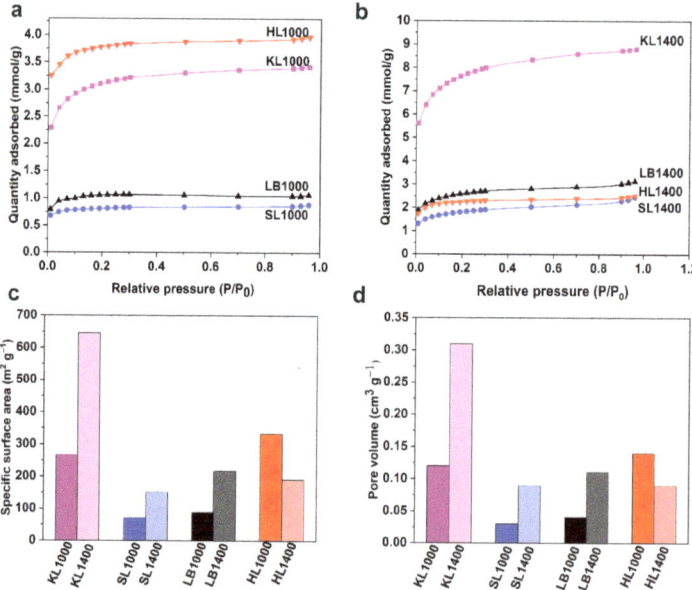

Figure 6. Results obtained from BET analysis (**a**,**b**) Nitrogen adsorption isotherms for carbon particles produced at 1000 °C and 1400 °C respectively, (**c**) BET specific surface area for all carbon particles, (**d**) Comparison of pore volumes obtained for all carbons.

Table 7. SSA provided by micropores, contribution towards the SSA by meso and macropores, average pore diameter and pore volume obtained from BET surface area analysis.

Sample	SSA ($m^2 g^{-1}$)	S_{micro} ($m^2 g^{-1}$)	$S_{meso+macro}$ ($m^2 g^{-1}$)	d_a (nm)	V_p ($cm^3 g^{-1}$)
KL1000	266	186	80	1.79	0.12
KL1400	646	414	232	1.89	0.31
SL1000	71	58	13	1.74	0.03
SL1400	151	88	63	2.25	0.09
LB1000	89	70	19	1.65	0.04
LB1400	216	136	80	2.01	0.11
HL1000	333	290	43	1.65	0.14
HL1400	191	157	34	1.78	0.09

(SSA—specific surface area, S_{micro}—micropore area, d_a—average pore diameter, V_p—pore volume).

The pore volume of carbon particles is graphically represented in Figure 6d. Pore volume followed a similar tendency to SSA with KL1400 having the highest pore volume of 0.31 cm^{-3}. This indicates that KL is generating more volatiles during both carbonization temperatures, leading to the generation of a more porous microstructure with average pore size of 1.89 nm, which could be suitable for electrochemical applications.

LB (71 $m^2 g^{-1}$ at 1000 °C and 151 $m^2 g^{-1}$ at 1400 °C) and SL (81 $m^2 g^{-1}$ at 1000 °C and 216 $m^2 g^{-1}$ at 1400 °C) showed quite similar adsorption behavior during the BET analysis, but SL showed the lowest adsorption and hence lowest SSA among all the samples for both carbonization temperatures. This could be due to the higher graphitization and the growth of crystallites [30] in comparison with other carbon particles, as discussed in the Raman and XRD analysis. HL1400 showed lower SSA than HL1000 and it was observed that the SSA contributed by micropores was reduced in HL1400 (157 $m^2 g^{-1}$) compared to HL1000 (290 $m^2 g^{-1}$) while the contribution from meso and macropores remained almost the same for both HL1000 and HL1400. To gain a better understanding of the SSA, pore size distributions of all carbon particles are shown in Figure S1a,b in Supplementary Materials. In addition to the higher contribution of SSA from micropores (<2 nm), KL1400, which has highest surface area among all carbon particles, showed a higher contribution of SSA from pore sizes ranging from 2 to 6 nm (Figure S1a,b) compared to others.

The electrochemical properties of the KL1400 based electrode were studied using cyclic voltammetry (CV) and galvanostatic charge discharge (GCD) measurements using a three-electrode system with 1M sulphuric acid (H_2SO_4) solution as the electrolyte. CV measurements were carried out at different scan rates and the obtained results are shown in Figure 7a. Near rectangular shaped CVs were observed at lower scan rates and slight deviation from rectangular shapes were observed at high scan rates, indicating the good rate capability of the KL1400 electrode.

At 2 mV s^{-1}, the specific capacitance observed was 151 F g^{-1} while the specific capacitance was 16 F g^{-1} at the scan rate of 100 mV s^{-1} which is proportional to the area under the cyclic voltammograms shown in Figure 7a. Galvanostatic charge discharge measurements were carried out for different current densities and the results are shown in Figure 7b. The highest specific capacitance obtained was 97.2 F g^{-1} at a current density of 0.1 A g^{-1}. Even at 1 A g^{-1}, electrode retained a specific capacitance of 45 F g^{-1}, which showed the capability of electrode material to store energy even at higher current densities. Figure 7c shows the specific capacitance values obtained from the GCD measurements.

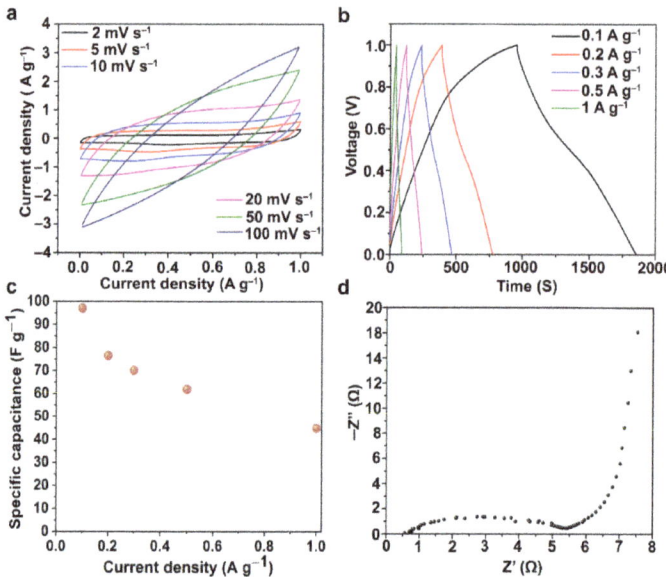

Figure 7. Results from the electrochemical analysis performed for KL1400 electrode. (**a**) Cyclic voltammograms (CVs) at different scan rates; (**b**) Galvano-static charge discharge (GCD) diagrams at different current densities; (**c**) specific capacitance values obtained at different current densities and (**d**) Nyquist plot obtained from electrochemical impedance analysis.

Figure S2a (in Supplementary Materials) shows the GCD curves for the practical supercapacitor measured using 2 electrode method and Figure S2b shows the obtained specific capacitance values. At 0.05 A g^{-1}, the supercapacitor made up of KL1400 electrodes showed a specific capacitance of 50 F g^{-1}. The highest energy density obtained was 1.7 Wh kg^{-1} at a power density of 25 W kg^{-1}, while at a higher power density of 50 W kg^{-1} the energy density was 1.1 Wh kg^{-1}. Electrochemical impedance spectroscopy (EIS) has been carried out in the frequency range between 10^{-2} and 10^5 Hz and the Nyquist plot for KL1400 is shown in Figure 7d. KL1400 exhibited negligible equivalent series resistance (<1 Ω) and low charge transfer resistance (5.4 Ω). The lower charge transfer resistance indicates the lesser resistance experienced by the electrolytes in the pores of the electrodes and better contact between electrode and current collector [31]. The nearly vertical portion in the Nyquist plot (Figure 7d) at the low frequency region indicates the proper infiltration of pores with the electrolyte ions. Thus, the KL1400 electrode exhibited ideal capacitative behavior at the low frequencies. The Ragone plot for KL1400 is shown in Figure 8a. All these results indicate the suitability of KL1400 to be used as electrodes in supercapacitors. Electrical conductivity is the primary material property of concern when using the carbon particles as conductive reinforcements in composites for energy storage or electromagnetic interference shielding applications [32–34].

Figure 8b shows the electrical conductivity values obtained for all the carbon particles at a frequency of 1 kHz. It has been observed that with the increase in the carbonization temperature the electrical conductivity values were increased. This is in accordance with the increase in the graphitization observed during the high temperature carbonization, which was also proved using Raman spectroscopy and XRD analysis. SL1400, which showed highest degree of graphitization, showed the highest electrical conductivity of 335 S m^{-1} while KL1000 showed the lowest electrical conductivity of the order of 10^{-6} S m^{-1}. The direct comparison of electrical conductivity values between different studies is difficult due to different factors, e.g., the difference in lignin resource, carbonization procedures, and conductivity measurement conditions, such as the equipment and conditions of the measurement.

However, the conductivity values obtained for SL1400 is much higher than the conductivity values reported by Snowden et al. [10] for the ball milled carbonized lignin (0.9 S m^{-1}) and that of lignin-based carbon fillers (142 S m^{-1}) reported by Gindl-Altmutter et al. [35]. Table 8 illustrates the superior electrical conductivity of SL1400 compared to other carbon materials produced by direct carbonization of lignin. In conclusion, SL based carbon particles showed a higher degree of electrical conductivities at both carbonization temperatures, indicating the suitability of SL for making conductive carbon particles for the preparation of conductive additives for composites.

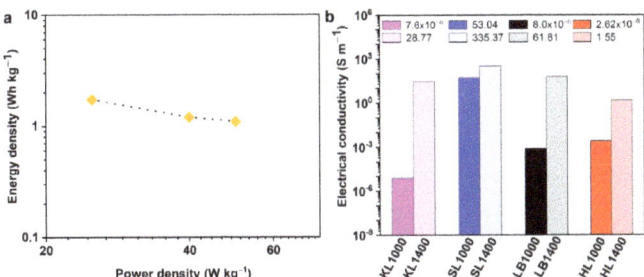

Figure 8. (a) Ragone plot for KL1400 electrode showing the energy density at different power densities and (b) Electrical conductivity values obtained for all carbon particles.

Table 8. Comparison of electrical conductivity values of lignin-based carbon particles produced by direct carbonization.

Sample	Carbonization Temperature	Electrical Conductivity	Reference Number
SL1400	1400 °C	335 S m^{-1}	This study
Carbonized lignin particles	900 °C	0.9 S m^{-1}	[10]
Carbon fillers	2000 °C	142 S m^{-1}	[35]
L-900	900 °C	33.3 S m^{-1}	[36]

4. Conclusions

In the current study, carbon particles were prepared from four different types of technical lignins. Two different final carbonization temperatures were chosen to study the impact of carbonization temperature on the microstructural evolution. It has been concluded that, except for hydrolysis lignin, the specific surface area was increased with the increase in the carbonization temperature. This is attributed to the evolution of more volatiles at higher temperatures. Kraft lignin carbonized at 1400 °C showed the highest surface area and was used as electrode material together with PTFE as binder to show promising electrochemical properties and hence suitability as electrodes in supercapacitor. It was also observed that the extent of graphitization was influenced by the type of lignin and carbonization temperature. Soda lignin at 1400 °C showed the highest electrical conductivity and could be utilized as functional carbon additives in composites, e.g., in EMI shielding applications. Hence, the current study presents new insights regarding the importance of the proper selection of technical lignins and their carbonization conditions for achieving the requirements of specific applications.

Supplementary Materials: The following supporting information can be downloaded at: https://www.mdpi.com/article/10.3390/nano12203630/s1. Figure S1: Pore size distribution of carbon particles obtained from DFT pore size analysis. (a) carbon particles at 1400 °C and (b) at 1000 °C, Two electrode measurement, Figure S2: Results from two electrode measurement of supercapacitor (a) galvanostatic charge discharge diagrams at different current densities (b) specific capacitance obtained at different current densities.

Author Contributions: Conceptualization and methodology: B.T. and K.O. Validation: B.T. Formal analysis and investigation: B.T. Writing—original draft preparation: B.T. Visualization: B.T. Writing—review and editing: B.T., M.S. and K.O. Supervision: K.O. and M.S. Funding acquisition: K.O. Project administration: K.O. All authors have read and agreed to the published version of the manuscript.

Funding: This research was funded by Bio4Energy, a Swedish strategic program and Swedish Research Council (Carbon Lignin 2017-04240).

Data Availability Statement: The data presented in this study will be available on reasonable request from the corresponding author.

Acknowledgments: The authors acknowledge Bio4Energy, a Swedish strategic program and Swedish Research Council (Carbon Lignin 2017-04240), for financial support this research. Authors also acknowledge Anthony Schreyeck and Shiyu Geng for their initial contributions in the study.

Conflicts of Interest: The authors declare no conflict of interest.

References

1. Tiwari, S.K.; Bystrzejewski, M.; de Adhikari, A.; Huczko, A.; Wang, N. Methods for the Conversion of Biomass Waste into Value-Added Carbon Nanomaterials: Recent Progress and Applications. *Prog. Energy Combust. Sci.* **2022**, *92*, 101023. [CrossRef]
2. Kane, S.; Ulrich, R.; Harrington, A.; Stadie, N.P.; Ryan, C. Physical and Chemical Mechanisms That Influence the Electrical Conductivity of Lignin-Derived Biochar. *Carbon Trends* **2021**, *5*, 100088. [CrossRef]
3. Yuan, T.-Q.; Sun, S.-N.; Xu, F.; Sun, R.-C. Characterization of Lignin Structures and Lignin–Carbohydrate Complex (LCC) Linkages by Quantitative 13C and 2D HSQC NMR Spectroscopy. *J. Agric. Food Chem.* **2011**, *59*, 10604–10614. [CrossRef] [PubMed]
4. Mandlekar, N.; Cayla, A.; Rault, F.; Giraud, S.; Salaün, F.; Malucelli, G.; Guan, J.P. An Overview on the Use of Lignin and Its Derivatives in Fire Retardant Polymer Systems. In *Lignin Trends and Applications*; Cayla, A., Ed.; IntechOpen: Rijeka, Croatia, 2018. [CrossRef]
5. Thomas, B.; Geng, S.; Wei, J.; Lycksam, H.; Sain, M.; Oksman, K. Ice-Templating of Lignin and Cellulose Nanofiber-Based Carbon Aerogels: Implications for Energy Storage Applications. *ACS Appl. Nano Mater.* **2022**, *5*, 7954–7966. [CrossRef]
6. Raud, M.; Tutt, M.; Olt, J.; Kikas, T. Dependence of the Hydrolysis Efficiency on the Lignin Content in Lignocellulosic Material. *Int. J. Hydrogen Energy* **2016**, *41*, 16338–16343. [CrossRef]
7. Wang, S.; Bai, J.; Innocent, M.T.; Wang, Q.; Xiang, H.; Tang, J.; Zhu, M. Lignin-Based Carbon Fibers: Formation, Modification and Potential Applications. *Green Energy Environ.* **2022**, *7*, 578–605. [CrossRef]
8. Zhu, M.; Liu, H.; Cao, Q.; Zheng, H.; Xu, D.; Guo, H.; Wang, S.; Li, Y.; Zhou, J. Electrospun Lignin-Based Carbon Nanofibers as Supercapacitor Electrodes. *ACS Sustain. Chem. Eng.* **2020**, *8*, 12831–12841. [CrossRef]
9. Thomas, B.; Geng, S.; Sain, M.; Oksman, K. Hetero-Porous, High-Surface Area Green Carbon Aerogels for the next-Generation Energy Storage Applications. *Nanomaterials* **2021**, *11*, 653. [CrossRef] [PubMed]
10. Snowdon, M.R.; Mohanty, A.K.; Misra, M. A Study of Carbonized Lignin as an Alternative to Carbon Black. *ACS Sustain. Chem. Eng.* **2014**, *2*, 1257–1263. [CrossRef]
11. Li, W.; Zhang, Y.; Das, L.; Wang, Y.; Li, M.; Wanninayake, N.; Pu, Y.; Kim, D.Y.; Cheng, Y.-T.; Ragauskas, A.J.; et al. Linking Lignin Source with Structural and Electrochemical Properties of Lignin-Derived Carbon Materials. *RSC Adv.* **2018**, *8*, 38721–38732. [CrossRef] [PubMed]
12. Abioye, A.M.; Ani, F.N. Recent Development in the Production of Activated Carbon Electrodes from Agricultural Waste Biomass for Supercapacitors: A Review. *Renew. Sustain. Energy Rev.* **2015**, *52*, 1282–1293. [CrossRef]
13. Demir, M.; Kahveci, Z.; Aksoy, B.; Palapati, N.K.R.; Subramanian, A.; Cullinan, H.T.; El-Kaderi, H.M.; Harris, C.T.; Gupta, R.B. Graphitic Biocarbon from Metal-Catalyzed Hydrothermal Carbonization of Lignin. *Ind. Eng. Chem. Res.* **2015**, *54*, 10731–10739. [CrossRef]
14. Hajizadeh, Z.; Taheri-Ledari, R.; Asl, F.R. 3-Identification and analytical methods. In *Micro and Nano Technologies, Heterogeneous Micro and Nanoscale Composites for the Catalysis of Organic Reactions*; Maleki, A., Ed.; Elsevier: Amsterdam, The Netherlands, 2022; pp. 33–51, ISBN 9780128245279. [CrossRef]
15. Schneider, C.; Walker, S.; Phounglamcheik, A.; Umeki, K.; Kolb, T. Effect of Calcium Dispersion and Graphitization during High-Temperature Pyrolysis of Beech Wood Char on the Gasification Rate with CO_2. *Fuel* **2021**, *283*, 118826. [CrossRef]
16. Zhang, M.; Resende, F.L.P.; Moutsoglou, A.; Raynie, D.E. Pyrolysis of Lignin Extracted from Prairie Cordgrass, Aspen, and Kraft Lignin by Py-GC/MS and TGA/FTIR. *J. Anal. Appl. Pyrolysis* **2012**, *98*, 65–71. [CrossRef]
17. Liu, W.J.; Jiang, H.; Yu, H.Q. Thermochemical Conversion of Lignin to Functional Materials: A Review and Future Directions. *Green Chem.* **2015**, *17*, 4888–4907. [CrossRef]
18. Wei, J.; Shah, F.U.; Johansson Carne, L.; Geng, S.; Antzutkin, O.N.; Sain, M.; Oksman, K. Oriented Carbon Fiber Networks by Design from Renewables for Electrochemical Applications. *ACS Sustain. Chem. Eng.* **2021**, *9*, 12142–12154. [CrossRef]
19. Wei, J.; Geng, S.; Kumar, M.; Pitkänen, O.; Hietala, M.; Oksman, K. Investigation of Structure and Chemical Composition of Carbon Nanofibers Developed from Renewable Precursor. *Front. Mater.* **2019**, *6*, 334. [CrossRef]

20. Ferrari, A.C.; Basko, D.M. Raman Spectroscopy as a Versatile Tool for Studying the Properties of Graphene. *Nat. Nano* **2013**, *8*, 235–246. [CrossRef]
21. Enengl, C.; Lumetzberger, A.; Duchoslav, J.; Mardare, C.C.; Ploszczanski, L.; Rennhofer, H.; Unterweger, C.; Stifter, D.; Fürst, C. Influence of the Carbonization Temperature on the Properties of Carbon Fibers Based on Technical Softwood Kraft Lignin Blends. *Carbon Trends* **2021**, *5*, 100094. [CrossRef]
22. Ferrari, A.C.; Robertson, J. Interpretation of Raman Spectra of Disordered and Amorphous Carbon. *Phys. Rev. B* **2000**, *61*, 14095–14107. [CrossRef]
23. Ferrari, A.C.; Meyer, J.C.; Scardaci, V.; Casiraghi, C.; Lazzeri, M.; Mauri, F.; Piscanec, S.; Jiang, D.; Novoselov, K.S.; Roth, S.; et al. Raman Spectrum of Graphene and Graphene Layers. *Phys. Rev. Lett.* **2006**, *97*, 187401. [CrossRef] [PubMed]
24. Xu, J.; Liu, J.; Ling, P.; Zhang, X.; Xu, K.; He, L.; Wang, Y.; Su, S.; Hu, S.; Xiang, J. Raman Spectroscopy of Biochar from the Pyrolysis of Three Typical Chinese Biomasses: A Novel Method for Rapidly Evaluating the Biochar Property. *Energy* **2020**, *202*, 117644. [CrossRef]
25. Smith, M.W.; Dallmeyer, I.; Johnson, T.J.; Brauer, C.S.; McEwen, J.S.; Espinal, J.F.; Garcia-Perez, M. Structural Analysis of Char by Raman Spectroscopy: Improving Band Assignments through Computational Calculations from First Principles. *Carbon N. Y.* **2016**, *100*, 678–692. [CrossRef]
26. Guizani, C.; Jeguirim, M.; Valin, S.; Limousy, L.; Salvador, S. Biomass Chars: The Effects of Pyrolysis Conditions on Their Morphology, Structure, Chemical Properties and Reactivity. *Energies* **2017**, *10*, 796. [CrossRef]
27. Rosenburg, F.; Ionescu, E.; Nicoloso, N.; Riedel, R. High-Temperature Raman Spectroscopy of Nano-Crystalline Carbon in Silicon Oxycarbide. *Materials* **2018**, *11*, 93. [CrossRef]
28. Franklin, R.E. The Structure of Graphitic Carbons. *Acta Crystallogr.* **1951**, *4*, 253–261. [CrossRef]
29. Russell, N.v.; Gibbins, J.R.; Williamson, J. Structural Ordering in High Temperature Coal Chars and the Effect on Reactivity. *Fuel* **1999**, *78*, 803–807. [CrossRef]
30. Chung, D.D.L. Carbon Fibers, Nanofibers, and Nanotubes. In *Carbon Composites*, 2nd ed.; Chung, D.D.L., Ed.; Butterworth-Heinemann: Oxford, UK, 2017; pp. 1–87, ISBN 978-0-12-804459-9.
31. Hu, Y.; Tong, X.; Zhuo, H.; Zhong, L.; Peng, X.; Wang, S.; Sun, R. 3D Hierarchical Porous N-Doped Carbon Aerogel from Renewable Cellulose: An Attractive Carbon for High-Performance Supercapacitor Electrodes and CO_2 Adsorption. *RSC Adv.* **2016**, *6*, 15788–15795. [CrossRef]
32. Han, Z.; Li, H.; Xiao, J.; Song, H.; Li, B.; Cai, S.; Chen, Y.; Ma, Y.; Feng, X. Ultralow-Cost, Highly Sensitive, and Flexible Pressure Sensors Based on Carbon Black and Airlaid Paper for Wearable Electronics. *ACS Appl. Mater. Interfaces* **2019**, *11*, 33370–33379. [CrossRef]
33. Zhang, W.; Dehghani-Sanij, A.A.; Blackburn, R.S. Carbon Based Conductive Polymer Composites. *J. Mater. Sci.* **2007**, *42*, 3408–3418. [CrossRef]
34. Pantea, D.; Darmstadt, H.; Kaliaguine, S.; Sümmchen, L.; Roy, C. Electrical Conductivity of Thermal Carbon Blacks: Influence of Surface Chemistry. *Carbon N. Y.* **2001**, *39*, 1147–1158. [CrossRef]
35. Gindl-Altmutter, W.; Fürst, C.; Mahendran, A.R.; Obersriebnig, M.; Emsenhuber, G.; Kluge, M.; Veigel, S.; Keckes, J.; Liebner, F. Electrically Conductive Kraft Lignin-Based Carbon Filler for Polymers. *Carbon N. Y.* **2015**, *89*, 161–168. [CrossRef]
36. Liu, F.; Wang, Z.; Zhang, H.; Jin, L.; Chu, X.; Gu, B.; Huang, H.; Yang, W. Nitrogen, Oxygen and Sulfur Co-Doped Hierarchical Porous Carbons toward High-Performance Supercapacitors by Direct Pyrolysis of Kraft Lignin. *Carbon N. Y.* **2019**, *149*, 105–116. [CrossRef]

Review

Green Production of Biomass-Derived Carbon Materials for High-Performance Lithium–Sulfur Batteries

Chao Ma [1], Mengmeng Zhang [2], Yi Ding [2], Yan Xue [1,*], Hongju Wang [3], Pengfei Li [2] and Dapeng Wu [2,*]

1. College of Mechanical and Electrical Engineering, School of 3D Printing, Xinxiang University, Xinxiang 453003, China
2. School of Business, Henan Normal University, Xinxiang 453007, China; lipengfei@htu.edu.cn (P.L.)
3. Key Laboratory for Yellow River and Huai River Water Environmental and Pollution Control, School of Environment, Ministry of Education, Collaborative Innovation Center of Henan Province for Green Manufacturing of Fine Chemicals, Henan Normal University, Xinxiang 453007, China
* Correspondence: xueyan6066@163.com (Y.X.); dapengwu@htu.edu.cn (D.W.)

Abstract: Lithium–sulfur batteries (LSBs) with a high energy density have been regarded as a promising energy storage device to harness unstable but clean energy from wind, tide, solar cells, and so on. However, LSBs still suffer from the disadvantages of the notorious shuttle effect of polysulfides and low sulfur utilization, which greatly hider their final commercialization. Biomasses represent green, abundant and renewable resources for the production of carbon materials to address the aforementioned issues by taking advantages of their intrinsic hierarchical porous structures and heteroatom-doping sites, which could attribute to the strong physical and chemical adsorptions as well as excellent catalytic performances of LSBs. Therefore, many efforts have been devoted to improving the performances of biomass-derived carbons from the aspects of exploring new biomass resources, optimizing the pyrolysis method, developing effective modification strategies, or achieving further understanding about their working principles in LSBs. This review firstly introduces the structures and working principles of LSBs and then summarizes recent developments in research on carbon materials employed in LSBs. Particularly, this review focuses on recent progresses in the design, preparation and application of biomass-derived carbons as host or interlayer materials in LSBs. Moreover, outlooks on the future research of LSBs based on biomass-derived carbons are discussed.

Keywords: carbon materials; biomass; working mechanism; lithium–sulfur batteries; cathodes; interlayer

1. Introduction

Nowadays, human society is facing more and more critical social problems as it is confronted with the ever-growing energy demands and serious environmental crises. In order to efficiently store clean and renewable energy, such as solar, wind, tide, geothermal and other energy sources for sustainable development, research studies on advanced energy storage systems have attracted intense attention worldwide [1–6]. Among them, secondary batteries with a high energy density represent a cutting-edge energy storage technology. Traditional lithium batteries, which adopt graphite as the anode material and lithium metal oxide ($LiCoO_2$, $LiNi_xCo_yMn_{1-x-y}O_2$) or lithium phosphate ($LiFePO_4$) as the cathode material, demonstrate a theoretical energy density of 400 Wh kg^{-1}, which has been widely used in portable electronic devices. After years of research and development, the electrochemical properties of the electrodes of lithium-ion batteries have been reaching their theoretical values, but this still cannot fully meet the needs for energy storage devices to power electric vehicles or to store the huge volume of electricity generated from clean-energy harnessing facilities.

Therefore, researchers have turned their attention to other cathode materials with a high theoretical capacity. Among various potential cathode materials, lithium–sulfur batteries (LSBs) have attracted much attention as a potential low-cost and efficient energy storage system due to the advantages of high theoretical capacity (1675 mAhg^{-1}), high energy density (2600 Whkg^{-1}), wide sources and low cost of elemental sulfur [7,8]. LSBs consist of elemental sulfur as the cathode, an electrolyte, a separator and lithium metal as the anode. Through the multi-electron electrochemical conversion between sulfur and lithium, LSBs could give rise to a much higher specific capacity than that of traditional lithium batteries.

Carbon-based materials represent a group of important raw materials for different industries. Duo to their high electroconductivity and high thermal and chemical stability, as well as bio-compatibility, many types of carbon materials, such as graphene or reduced graphene oxides [4–15], carbon fibers [16–18], carbon dots [19], active carbons [20–22] and biomass-derived carbons [23–32], have been deliberately prepared for different application aims. Biomass has many definitions, such as biodegradable products, wastes and residues from agriculture, forestry and related industries, including fisheries and aquaculture, as well as biodegradable parts of industrial and municipal wastes. In addition, biomass is also regarded as a class of organic macromolecular materials derived from organisms [33]. During carbon preparation, biomass precursors undergo thermochemical transformation under high-temperature conditions, which could be classified into two distinctive processes. The first one is solid-state carbonization, which represents the decomposition of biomass precursors under a high temperature and an inert atmosphere to generate carbon-rich and thermal stable products [34]. The second one is hydrothermal carbonization, which could convert wet biomass into a carbon-rich product (hydrochar). In these processes, the organic components in the biomass experience drastic chemical changes, such as carbon skeleton recombination and functional group decomposition, which finally yield a carbon atom network with high-conductivity sp^2 domains and rich surface functional groups [34]. Generally, activation processes are introduced to further treat the as-obtained carbon materials to increase the surface area and to regulate the pore structures. These activation processes are usually carried out under high-temperature conditions and with the assistance of physical (water vapor or carbon dioxide) or chemical activators (KOH, H$_3$PO$_4$, ZnCl$_2$, NaOH, etc.) [35,36]. In addition, the templating method is also one of the commonly used strategies to optimize the surface area and pore structures of as-obtained carbon materials. Generally, templates are firstly introduced into a biomass precursor, which is then carbonized under a high temperature with inert atmosphere. Finally, the templates are removed by immersion in a NaOH or HF solution to generate carbon materials with ordered pore distribution [37].

Activated carbon production from biomass has long been regarded as important research in materials science. At the very beginning, biomass-activated carbon was only used for adsorption. Thanks to their rich pore structures, huge specific surface area, and good physical and chemical stability, biomass-derived carbon materials have gradually developed into a wide range of adsorbents for application in purification, deodorization, decolorization and separation [38]. China is a traditional agricultural country, and there are abundant biomasses produced from agriculture, forestry, animal husbandry and aquaculture annually. In addition, illegally discarded or wrongly treated biomass has become an important source that causes environmental pollution. Therefore, the production of functional carbon materials from biomass could not only lead to economic benefits but also alleviate such environmental problems, which has attracted more and more attention.

At present, biomass-derived carbon materials are employed in energy storage applications, such as supercapacitors [39], lithium-ion batteries [40,41], lithium–sulfur batteries (LSBs) [42,43], and so on. As for the application of biomass-derived carbons in LSBs, many review works have been devoted to highlight the significance of these materials from different aspects [44–52]. This review work will focus on discussing recent advancements in the design, preparation and application of biomass-derived carbons as host or interlayer materials in LSBs. Moreover, outlooks on rational preparation based on biomass-derived carbons and potential future research on LSBs are also discussed.

2. Fundamentals

2.1. Working Principles of LSBs

In the discharge curve shown below, two typical discharge platforms at 2.3 and 2.1 V could be clearly observed, corresponding to the solid (S_8), liquid (Li_2S_n) and solid (Li_2S_2/Li_2S) processes, respectively [53] (Figure 1a). S_8 is firstly converted into long-chain Li_2S_n, and then the long-chain Li_2S_n is converted to Li_2S_2/Li_2S. For the higher-voltage discharge platform located at 2.3 V, S_8 is reduced to long-chain polysulfide, corresponding to the theoretical capacity of 418 mAh g^{-1}. For the lower-voltage discharge platform at 2.1 V, polysulfide is further reduced to insoluble Li_2S_2/Li_2S, corresponding to the theoretical capacity of 1254 mAh g^{-1}. During the charging process, 2.3–2.4 V could be regarded as a platform, corresponding to the transformation process from Li_2S_2/Li_2S to S_8. The intermediate polysulfide dissolves and diffuses into the electrolyte due to the concentration gradients and electric field forces (Figure 1b). In addition, a small polarization voltage can be observed during the initial phase of charging, which is due to the final product of the discharge phase (Li_2S_2/Li_2S) being an insulator, thus leading to a large inter-phase transition barrier.

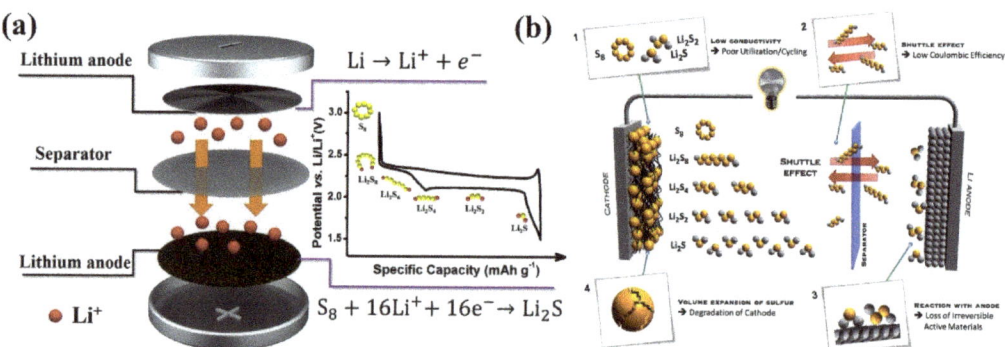

Figure 1. (a) Structure of an LSB and the typical charge–discharge process and intermediate product conversion of the battery [53]. Reprinted with permission from Ref. [53]. Copyright 2018 Wiley-VCH. (b) Conversion of soluble polysulfide and insoluble Li_2S_2/Li_2S during charge and discharge in a secondary lithium–sulfur battery consisting of a sulfur positive electrode and lithium negative electrode [8]. Reprinted with permission from Ref. [8]. Copyright 2017 Wiley-VCH.

2.2. Obstacles for LSBs

Due to their high theoretical capacity and energy density, LSBs have attracted extensive attention from both domestic and overseas researchers. However, LSBs still suffer from various inherent technical bottlenecks as listed below, which greatly restricts their application. The first bottleneck is the insulation of the charge and discharge products. Under room temperature, elemental sulfur (S_8) is an insulator for Li$^+$ and electrons (5 × 10^{-28} S cm^{-1}). Likewise, Li_2S is also an insulator for both Li$^+$ and electrons (<10^{-14} S cm^{-1}) [54]. In the working process, the low Li$^+$ and electron conductivity of the charge–discharge product hinders the charge–discharge process of the battery. When sulfur is used alone as the electrode material, such property leads to two serious problems which restrict the development of lithium–sulfur batteries: the low utilization rate of active substances and the slow reaction kinetics. The intrinsic insulation of S_8/Li_2S severely limits the utilization rate of S active substances, resulting in low capacity and low-rate capacity. Although the solid–liquid reaction with a discharge platform of about 2.3 V has a faster reaction kinetics, it is still difficult to achieve the theoretical capacity. Additionally, previous reports have demonstrated that after multiple cycles, there is still S_8 that is not involved in the reaction. In addition, due to the limited Li$^+$ and electro-kinetics, in the charge–discharge curve of lithium–sulfur batteries, a large polarization phenomenon is often observed on the lower discharge platform, which further leads to the attenuation of the actual energy density.

The second bottleneck is the shuttle effect of intermediate products. Due to the transport of Li^+ between the two electrodes, the electrode reaction of lithium–sulfur batteries involves a series of reversible lithiation and delithiation processes. In this series of transformation, the production of soluble Li_2S_n ($n \geq 4$) is inevitable, which leads to a series of negative effects [55]. Due to the concentration gradient, the high-valence polysulfide Li_2S_n ($n \geq 4$) will migrate to the anode, react with lithium metal at the anode, and then generate Li_2S_2/Li_2S. Therefore, during the charge–discharge process, polysulfide cannot completely migrate back to the cathode, which results in an irreversible loss of active materials. Afterward, the continuous deposition of Li_2S_2/Li_2S on the surface of the lithium negative electrode leads to the continuous generation of solid electrolyte interface film (SEI film) on the anode surface, which hinders the release of Li^+ and the effective contact between the electrolyte and the negative electrode, resulting in a continuous increase in battery impedance with an increase in the number of cycles. As the dissolution and deposition of active substances redistribute after longtime recycles, a passivation layer forms on the surface of the positive electrode material, which greatly increases the surface resistance of the electrode. The production of soluble Li_2S_n ($n \geq 4$) increases the viscosity of the electrolyte and reduces the conduction rate of Li^+ and electrons in the electrolyte, which poses negative impact on the rate capability of the battery. Therefore, due to the severe shuttle effect, a battery can only maintain a limited number of cycles and experiences poor cycle stability, low coulomb efficiency and self-discharge. Because the density difference between S_8 and Li_2S is great, active species of the cathode exhibits a volume expansion of nearly 80% during the discharge process. Such a massive volume expansion leads to two invisible problems. The first one is that the severe volume expansion triggers pulverization and shedding of active materials from the current collector. Meanwhile, the volume expansion also causes a series of safety problems, which seriously limits the practical application of LSBs.

3. Recent Progresses in Carbon Materials for LSBs

As we have stated before, the cathode material of LSBs plays an important role in these storage devices. In order to solve the aforementioned technical bottleneck and realize the commercial production of lithium–sulfur batteries, a large number of scholars have dedicated efforts to optimizing the LSB system from the perspective of cathode materials, hoping to make up for the intrinsic defects of lithium–sulfur batteries based on the design of cathode materials. According to recent research on carbonous cathodes for LSBs, we summarize recent progresses into two categories: structural regulation and heteroatom doping of carbonous materials.

3.1. Structural Regulation of Carbons

Carbon materials exist widely in the natural environment and have stable physical and chemical properties. The introduction of carbon materials can significantly improve the electronic conductivity and ion transport and buffer the volume expansion of active materials, which could avoid the pulverization and shedding of the positive electrode structure in the process of charge and discharge. Therefore, as shown in Figure 2, the exploration of carbon/sulfur composite cathode materials has triggered great research attention, and carbon materials, including porous carbon, hollow carbon structure, graphene, and so on, are commonly used for the cathode in lithium–sulfur batteries [53].

Porous carbon. Porous carbon is a functional carbonaceous material with porous structure. According to the pore size distribution, porous carbons could be classified into microporous carbon materials (pore size less than 2 nm), mesoporous carbon materials (pore size between 2 and 50 nm), macroporous carbon materials (pore size greater than 50 nm), and hierarchical porous carbon materials (with a variety of pore structures). High porosity and high specific surface area are conducive to the storage and uniform distribution of sulfur. It is also proposed that a porous structure could lead to better inhibition of the dissolution and diffusion of polysulfides, which effectively reduces the shuttle effect and, thus, improves the electrochemical performances of LSBs [47,48].

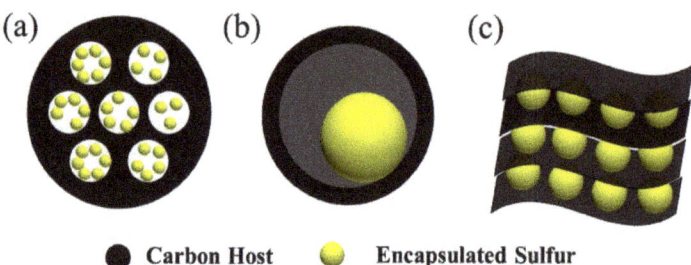

Figure 2. Carbon-based sulfur carriers with different nanostructures: (**a**) porous structure; (**b**) hollow structure.; and (**c**) lamellar graphene structure [53]. Reprinted with permission from Ref. [53]. Copyright 2018 Wiley-VCH.

As shown in Figure 3a, sucrose is used as a carbon source to obtain uniform microporous carbon spheres with a specific surface area of 843.5 $m^2\ g^{-1}$ and a pore size distribution of mainly 0.7 nm [56]. When the sulfur content is 42 wt%, the initial discharge capacity could amount to 1183.5 mAh g^{-1}, and after a long period of cycling, it still has a reversible capacity of 650 mAh g^{-1}. The electrode only shows a discharge platform at 1.8 V, which is not a typical lithium–sulfur battery with two discharge platforms, and the high-voltage discharge platform is absent. The unique phenomena could be interpreted as that due to their small size, sulfur molecules are the main species stored in the micropores of the carbon matrix, which avoids the generation of soluble polysulfide and results in a typical charge–discharge curve. As demonstrated by the theoretical simulation (Figure 3b) [7], when the size of microporous carbon is small enough (0.5 nm), S_8 molecules could be split into small sulfur molecules with a short chain length and stored in the micropores, which could inhibit the transformation from S_8 to S_4^{2-}, alleviate the shuttle effect of polysulfides, and, finally, result in the high capacity retention and coulombic efficiency during the charge–discharge process. Highly ordered mesoporous structures with a pore size of 3–4 nm are also synthesized as a conductive matrix to accommodate the S loading. During the charge and discharge process, this structure promotes the reaction between Li^+ and sulfur and inhibits the diffusion of soluble polysulfide by trapping polysulfide within the carbon framework [57].

Carbons with hierarchical porosities. Microporous carbon has a high specific surface area, which can ensure the dispersion and contact of elemental sulfur in the conductive matrix. Moreover, the strong physical adsorption capacity of micropores can effectively inhibit the shuttle effect. However, micropores can only provide a limited pore volume, which makes it difficult to accommodate higher active substances. When sulfur loading exceeds the critical amount, the extra sulfur will harm the electrical contact with the conductive matrix, which will reduce the utilization rate of active substances and limit the overall energy density of the battery. Large pores and mesoporous pores can house more sulfur than that of the micropores, which could substantially increase the sulfur loading amount of the electrode and effectively alleviate the volume expansion during the charge and discharge process. However, physical adsorption is not capable enough in inhibiting the dissolution and diffusion of polysulfides, thus resulting in an irreversible loss of active substances. Therefore, the design of hierarchical porous carbon materials with various pore size distributions could well balance the advantages of macroporous, mesoporous and microporous carbon materials and, thus, lead to improved electrochemical performances.

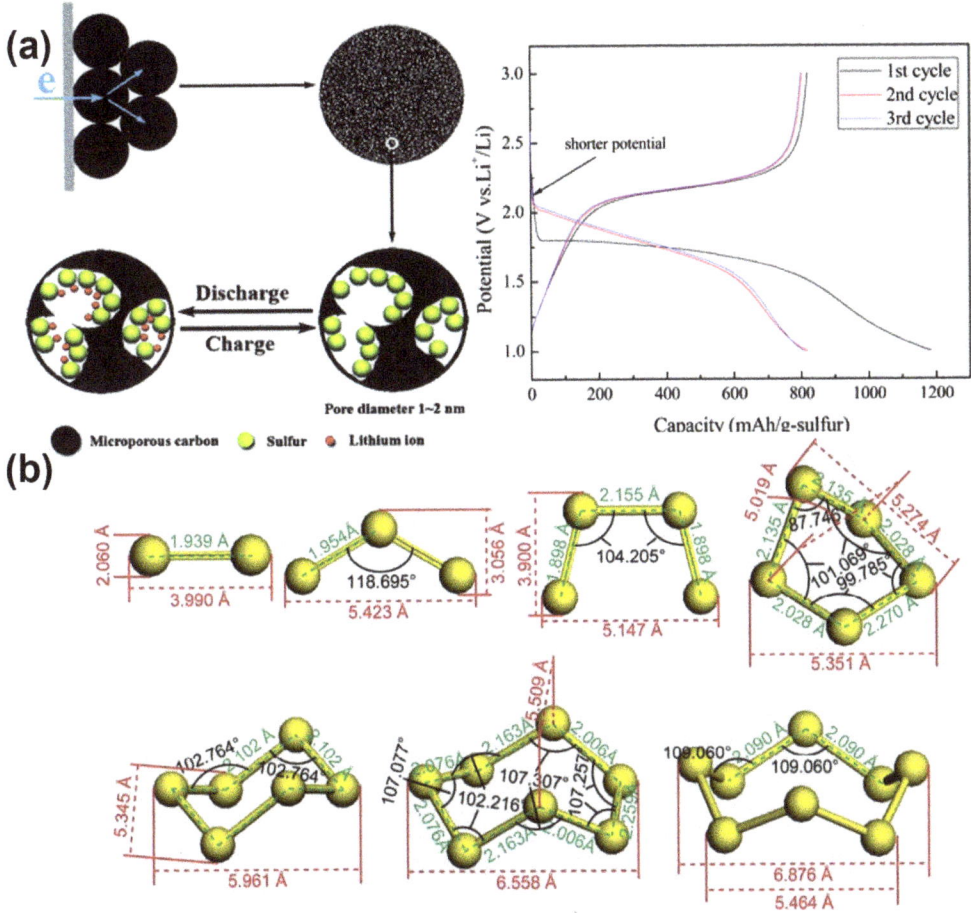

Figure 3. (a) The structure diagram and the discharge curve of microporous carbon spheres [56]. Reprinted with permission from Ref. [56]. Copyright 2008 Royal Society of Chemistry. (b) Models of various polysulfides formed during the charge and discharge process [7]. Reprinted with permission from Ref. [7]. Copyright 2012 American Chemical Society.

Liang et al. used ordered mesoporous carbon as the precursor and KOH as the activator to obtain hierarchical porous carbon with a dual-pore distribution (3 nm and 7.3 nm) (Figure 4a) [58]. Sulfur is found evenly distributed in the conductive carbon matrix, in which the micropores serve as the storage space of active sulfur to ensure the contact between sulfur and the conductive matrix. Mesoporous pores not only could effectively accommodate polysulfides dissolved in the electrolyte but also provide a fast transport channel for lithium ions. In addition, a layer of microporous carbon is coated on the surface of the highly ordered mesoporous structure to obtain a hierarchical porous carbon material with a core–shell structure (Figure 4b) [59]. The ordered mesoporous could greatly improve the sulfur loading amount and make full utilization of the carbon matrix. Meanwhile, the microporous carbon shell could function as a polysulfide barrier to reduce the capacity decay of the battery. Based on this, a rapid spray drying method was adopted to obtain carbon spheres with a hierarchical porous structure (Figure 4c) [60]. Mesoporous and microporous structures could be introduced inside the carbon spheres, which helps to achieve high sulfur accommodation. The outer microporous shell could be used as a physical site to anchor polysulfides to avoid the irreversible loss of active substances induced by the diffusion of polysulfides.

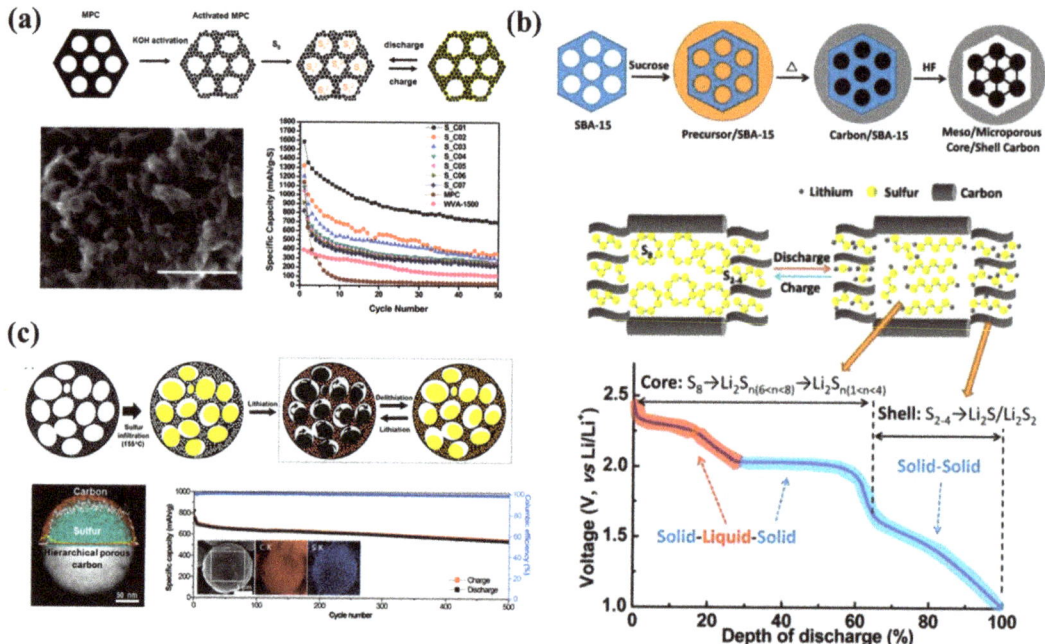

Figure 4. (**a**) The synthesis route, SEM images and corresponding cycle performance curves of carbon materials with different pore size distributions [58]. Reprinted with permission from Ref. [58]. Copyright 2009 American Chemical Society (the scale bar represents 50 nm). (**b**) Structure of core–shell hierarchical porous carbon materials and corresponding electrode reactions of different discharge platforms [60]. Reprinted with permission from Ref. [60]. Copyright 2009 American Chemical Society. (**c**) Structure of hierarchical porous carbon to obtain high sulfur accommodation and the cyclic performance curve at high current density [59]. Reprinted with permission from Ref. [59]. Copyright 2014 American Chemical Society.

Core–shell structure and hollow structure: Core–shell nanocomposites can be employed as physical prevention to inhibit the diffusion of polysulfide. On the other hand, polysulfide can be effectively bonded to core–shell structures, which could inhibit the shuttle effect of polysulfide. Meanwhile, the hollow structure provides a large inner cavity, and the shell usually acts as a barrier layer, which could greatly enhance the loading amount of active materials and effectively inhibit the dissolution and diffusion of polysulfides. However, the sulfur species trapped in the inner cavity of the hollow structure are not easily accessible. Therefore, many researchers have tried to optimize the internal cavity to realize the anchoring of polysulfides while ensuring the rapid transfer of the electrolyte, electrons and Li$^+$ ions toward the active species.

Due to its insulating properties and the shuttle effect of polysulfides, Li$_2$S exhibits poor utilization of active substances and a short cycle life. Based on these, core–shell Li$_2$S@C nanocomposites were prepared using plasma sparking and subsequent vulcanization process, through which Li$_2$S particles could be uniformly coated with a carbon layer of 0.8 nm [61]. It was demonstrated that the carbon coating effectively increases the conductivity of the composite material and effectively reduces the shuttle effect, resulting in superior electrochemical performances (Figure 5a). In addition, a solution evaporation method collaborated with a CVD process was adopted to prepare core–shell Li$_2$S@C composites (Figure 5b) [62]. The homogeneous amorphous carbon layer with a thickness of about 20 nm was evenly coated on Li$_2$S, which effectively prevents Li$_2$S from agglomeration. Compared to the pure Li$_2$S electrode, the core–shell Li$_2$S@C composite cathode shows high utilization rate of active species and excellent electrochemical performances. In addition, core–shell carbon spheres were prepared via the templating method, which compose of a mesoporous shell, a hollow cavity and a fixed carbon core (Figure 5c) [63]. When used as a carbonaceous sulfur carrier, the hollow cavity can improve the sulfur loading amount and reduce the volume expansion during the charge–discharge process. The mesoporous shell can provide a Li$^+$ transport channel and inhibit the shuttle effect of polysulfide. In general, this core–shell structure could function as both a physical buffer and a conductive matrix to maximize the potential capacity of active substances and alleviate the intrinsic deficiencies of LSBs.

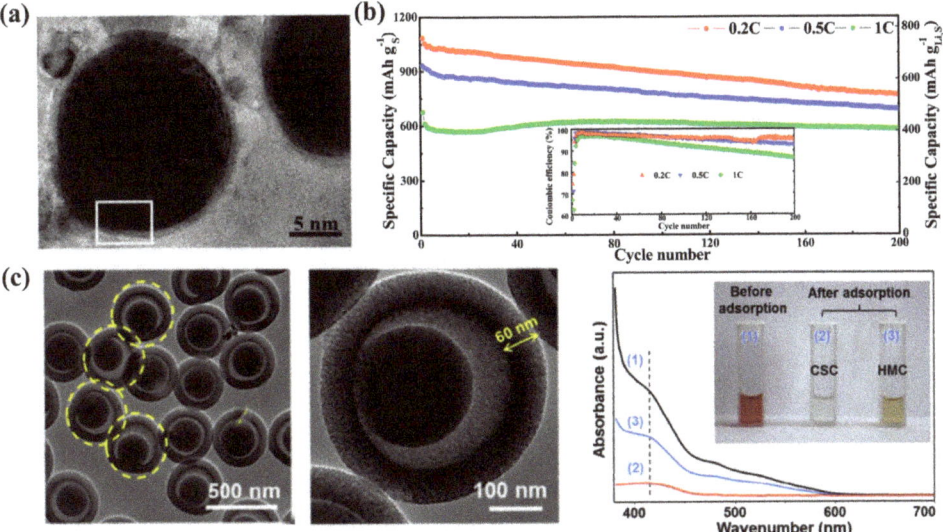

Figure 5. (a) TEM images of the core–shell structure Li$_2$S@C nanocomposites [61]. Reprinted with permission from Ref. [61]. Copyright 2012 Royal Society of Chemistry (the white square highlights the carbon layer on the Li$_2$S particle). (b) The cyclic performances of the nanocore–shell Li$_2$S@C composites [62]. Reprinted with permission from Ref. [62]. Copyright 2012 Royal Society of Chemistry. (c) TEM images and anchoring ability for polysulfides of hollow yolk-and-shell carbon spheres [63]. Reprinted with permission from Ref. [63]. Copyright 2017 American Chemical Society.

Graphene: Graphene represents a new kind of two-dimensional material with a hexagonal lattice structure composing of sp^2 carbon domains. The π-π conjugated bonding among carbon atoms endows graphene with excellent electrical conductivity. In addition, graphene with a high surface area, high conductivity and stable physical and chemical properties could serve as a potential candidate for LSBs. For example, a simple one-step method was developed to anchor sulfur nanocrystals to 3D cross-linked fibrous graphene (Figure 6a) [64]. When used as the cathode material for Li–sulfur batteries, the porous three-dimensional conductive network and uniformly distributed sulfur nanocrystals achieve rapid electron transport and shorten Li$^+$ diffusion distance. In addition, the presence of oxygen-containing functional groups enhances the anchoring ability of polysulfide and prevents the dissolution of polysulfide in the electrolyte, which amounts to the high capacity, high-rate performance and long cycle life. Moreover, graphene foam electrodes were proposed and prepared through an effective strategy to obtain flexible Li–sulfur batteries with high energy and power densities as well as long cycle life (Figure 6b) [65].

This research found that graphene foams can provide highly conductive networks, strong mechanical support and enough room for a high sulfur loading amount. In order to further enhance the electrochemical performances, hollow nanographene spheres (GSs) supported by carbon nanotubes (CNTs) were prepared using a room-temperature solubility-processable method (Figure 6c) [66]. Within the unique flexible electrode, the conductive carbon nanotubes could serve as flexible scaffolds and the hollow GSs provide a closed space to accommodate the sulfur species, which could adapt to the volume expansion and inhibit the shuttle and dissolution of polysulfides, leading to rapid electron and ion transport.

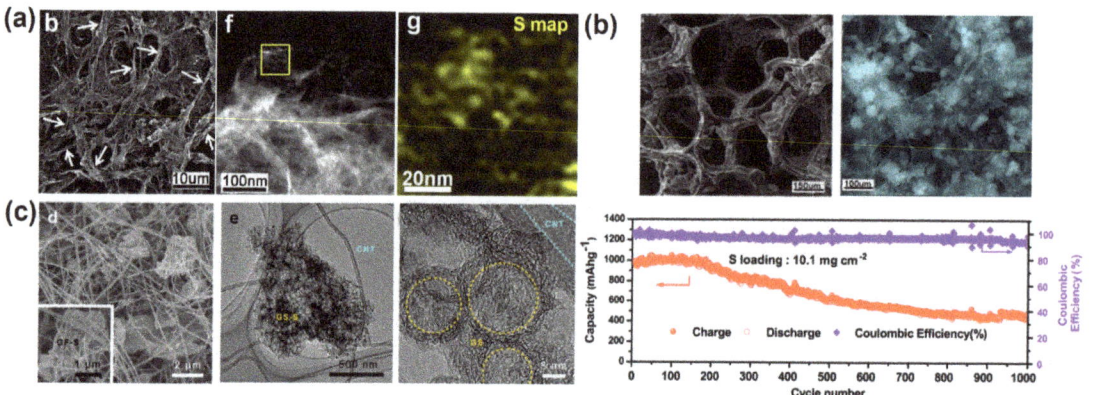

Figure 6. (a) SEM, TEM images and the corresponding S element map of the fibrous graphene/sulfur composites [64]. Reprinted with permission from Ref. [64] (the white arrows represent the fibrous graphene, and the yellow square indicates the sulfur map region of the samples). Copyright 2013 American Chemical Society. (b) The SEM and element map images of the self-supported graphene foam electrode (top), as well as their cycling performances of the corresponding LSB with high S loading density. [65]. Reprinted with permission from Ref. [65]. Copyright 2014 Elsevier Ltd. (c) SEM and TEM images of the carbon nanotube-supported hollow graphene spheres [66]. Reprinted with permission from Ref. [66]. Copyright 2014 Elsevier Ltd (the yellow circles represent the hollow graphene spheres).

3.2. Heteroatom-Doped Carbons for Cathode Materials

Nonmetallic heteroatoms (N, O, S, P, etc.) can be used as anchoring sites for polysulfides. Therefore, in addition to structural design, the introduction of heteroatoms into the conductive carbon matrix is also an effective way to improve the performance of batteries. Due to the electronegativity difference, electron-rich heteroatoms can lead to the surface polarization of the non-polar conductive carbon matrix, which helps to improve the chemical anchoring ability of the carbon matrix to polysulfide. At the same time, heteroatom impurity can introduce defect sites to enhance the catalytic performances of the electrodes so as to enhance the conversion of polysulfides. As shown in Figure 7a, nitrogen-doped mesoporous carbon microspheres permeated by carbon nanotubes as sulfur hosts were prepared through a self-assembly method, which greatly improves the performance of the batteries [67]. Due to the interaction between Li$^+$ and N atoms, soluble polysulfide in the electrolyte can be anchored by forming stable Li$_2$S$_x$-N chemical bonds, and polysulfide can be confined to the cathode materials, which significantly inhibits the shuttle effect and improves the electrochemical performance of the battery. Unlike conventional insulating adsorbents, nitrogen doping can inhibit the shuttle effect through the chemical interaction between the doping site and polysulfide. At the same time, the high conductivity of the nitrogen-doped carbon materials can directly trigger the conversion of polysulfides at the electrodes. For example, boron is a typical electron-deficient element, which can interact with electron-rich sulfur and polysulfide to form stable chemical bonds and, thus, can be used to absorb polysulfide and inhibit the shuttle effect. Han et al. prepared a boron-doped porous carbon material as the sulfur carrier [68] (Figure 7b). Compared to the undoped porous carbon, the boron-doped material showed high initial capacity (1300 mA h g^{-1}, 0.25 C), good cyclic stability and superior rate performance. Based on the mechanism study, boron, with a lower electronegativity than carbon, provides positively charged active sites, which could effectively adsorb negatively charged polysulfide.

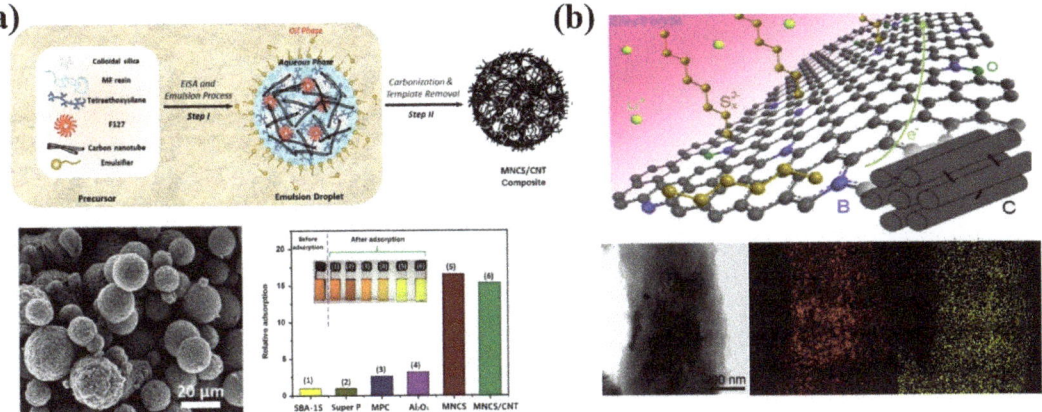

Figure 7. (a) Preparation routes, SEM images and polysulfide anchoring ability of carbon nanotube-penetrated nitrogen-doped mesoporous carbon microspheres [67]. Reprinted with permission from Ref. [67]. Copyright 2015 Wiley-VCH. (b) Mechanism diagram, STEM image and element distribution diagram of boron-doped ordered porous carbon/sulfur electrode material [68]. Reprinted with permission from Ref. [68]. Copyright 2014 American Chemical Society.

Compared to the single doping effect, the co-doping of heteroatoms can effectively combine the advantages of different heteroatoms, which can greatly improve the conductivity of carbon matrix and the adsorption of polysulfide. Diketoxime (DMG) and nickel chloride tetrahydrate (NiCl$_2$·4H$_2$O) were used as precursors for the preparation of nitrogen and oxygen co-doped porous carbon microrods with a large specific surface area and high porosity (Figure 8a) [69]. Based on the density functional theory (DFT) calculation, it is confirmed that the introduction of nitrogen and oxygen heteroatoms can effectively improve the adsorption capacity of the carbon matrix toward polysulfides and greatly inhibit the shuttle effect, which significantly improves the electrochemical performance of LSBs. Graphene oxide nanoribbons were modified by adding boric acid and urea to obtain nitrogen and boron co-doped curved graphene nanoribbons [70]. It was found that the reaction of the boric acid/urea precursor could promote the co-doping process of nitrogen and boron. In addition, the rich N-B motifs could significantly improve the electron conductivity, sulfur dispersion and polysulfide adsorption capacity of the electrodes. Wang et al. developed an effective strategy to improve the electrochemical performance of sulfur electrodes via the preparation of nitrogen/sulfur co-doped graphene matrix for the cathode material of LSBs [71]. In addition to the chemical anchoring of polysulfide by the nitrogen and sulfur defect sites, nitrogen and sulfur atoms with high electronegativity lead to the polarization of adjacent carbon atoms and oxygen-containing functional groups, which could increase the adsorption activity of sulfur and polysulfide. At the same time, the highly developed defects and edges and the porous structures obtained via the chemical activation of graphene not only achieve good dispersion of sulfur, but also act as a polysulfide reservoir to mitigate the shuttle effect. In addition, lithium iron phosphate nanoparticles were adopted as a hard template to prepare nitrogen, oxygen and phosphorus co-doped hollow carbon nanocapsules/graphene composites as the sulfur cathode [72]. The shuttle effect of polysulfide could be greatly inhibited by the physical and chemical adsorption of the abundant surface polar groups on the composite material.

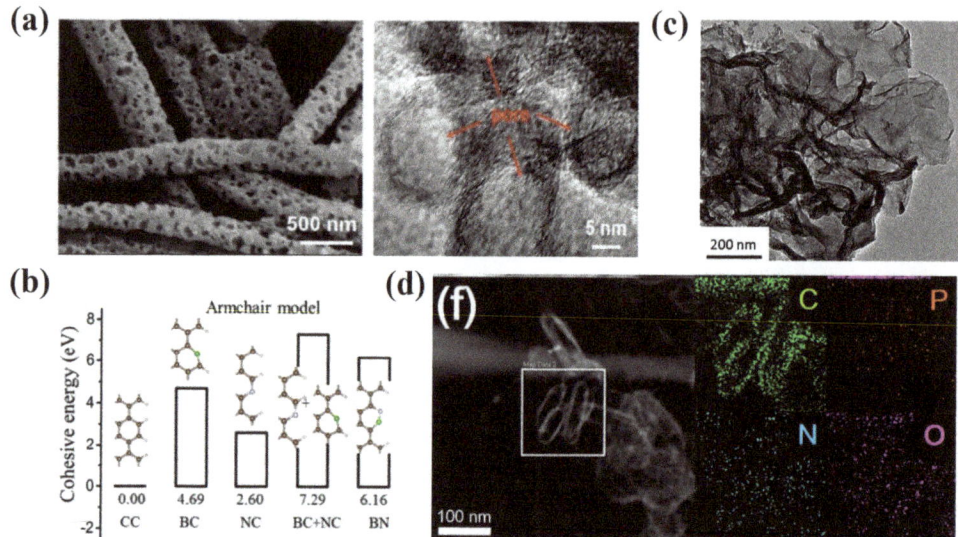

Figure 8. (a) SEM and TEM images of nitrogen–oxygen co-doped porous carbons [69]. Reprinted with permission from Ref. [69]. Copyright 2018 American Chemical Society. (b), The binding energy of polysulfide with different element doping patterns of the N-B co-doped curved graphene nanoribbon [70]. Reprinted with permission from Ref. [70]. Copyright 2012 Royal Society of Chemistry. (c) TEM image of nitrogen–sulfur co-doped porous graphene [71]. Reprinted with permission from Ref. [71]. Copyright 2012 Royal Society of Chemistry. (d) TEM image and element distribution of nitrogen, oxygen and phosphorus co-doped hollow carbon nanocapsules/graphene composites [72]. Reprinted with permission from Ref. [72]. Copyright 2018 American Chemical Society.

4. Biomass-Derived Carbon Materials for LSBs

4.1. Advantages of Biomass-Derived Carbon Materials

Compared to other non-renewable carbon sources, biomass can be used as an abundant and sustainable carbon source to prepare porous carbon materials for energy storage and conversion [73]. So far as we know, a large variety of biomasses have been used as biomass carbon sources to prepare carbon materials as sulfur carriers or functional separators in lithium–sulfur batteries [74]. Compared to other carbon sources, biomass precursors have the following advantages. First of all, after thousands of years of evolution, biomass usually possesses a unique structure and morphology. The inherent hierarchical channel structure obtained after carbonization is conducive to sulfur accommodation and adsorption, which could potentially alleviate the volume expansion of cathodes. Secondly, biomass precursors have diversified compositions. In the process of carbonization, a biomass's inherent heteroatoms could be doped into a carbon matrix, which could enhance the conductivity and the adsorption of polysulfides through the strong chemical interactions. Finally, due to the massive production of biomasses annually from different industries, biomass-derived carbons with a low price could substantially reduce the commercial production cost of LSBs.

4.2. Biomass-Derived Carbons for the Cathode of LSBs

Biomasses, such as agriculture wastes [75–79], forest wastes [80–83], weeds [84–86], food residues [87–100], and so on, have been developed as precursors to prepare high-performance carbons as cathode hosts for LSBs. As shown in Figure 9a, N, O co-doped carbon with a hierarchical porous structure was derived from bagasse, which could serve as a novel sulfur host for stable LSBs. It was found that the interconnected hierarchical porous structure facilitates the charge transport and alleviates the volume expansion of sulfur during the lithiation process, which finally amounts to highly stable LSBs [75]. In addition, a novel biomass waste, garlic peel, was used as a precursor to prepare carbons through two methods, pre-carbonization and hydrothermal treatment (Figure 9b). Due to the high surface area of carbon, the cathode exhibits high initial specific capacity and cycle retention. These structure advantages could also lead to the intimate contact between sulfur and the conductive carbon matrix, which could physically confine lithium polysulfide intermediates and reduce the shuttle effect [79]. Eucommia leaf residue was employed to prepare carbons with a hierarchical porous structure via the co-auxiliary activation of KCl and $CaCl_2$ with a low dosage of KOH (Figure 9c). The optimized pore distribution, high specific surface area and nitrogen-containing functional groups could enhance the utilization of sulfur and provide a chemical anchor for polysulfides, which gives rise to the excellent electrochemical performances [82]. Moreover, carbon materials with rational tailored morphology and structures could be obtained from balsa waste (Figure 9d). It was found that the mesopores in such carbon materials exhibit more merits than micro/macropores in improving sulfur utilization and restraining Li_2S_x, which could alleviate the notorious shuttle effect. In addition, the mechanism studies show that the conversion from long-chain polysulfide into solid S_8 and Li_2S could be accelerated by oxygen groups, which finally leads to improved sulfur immobilization and stable energy-storage capacity [83].

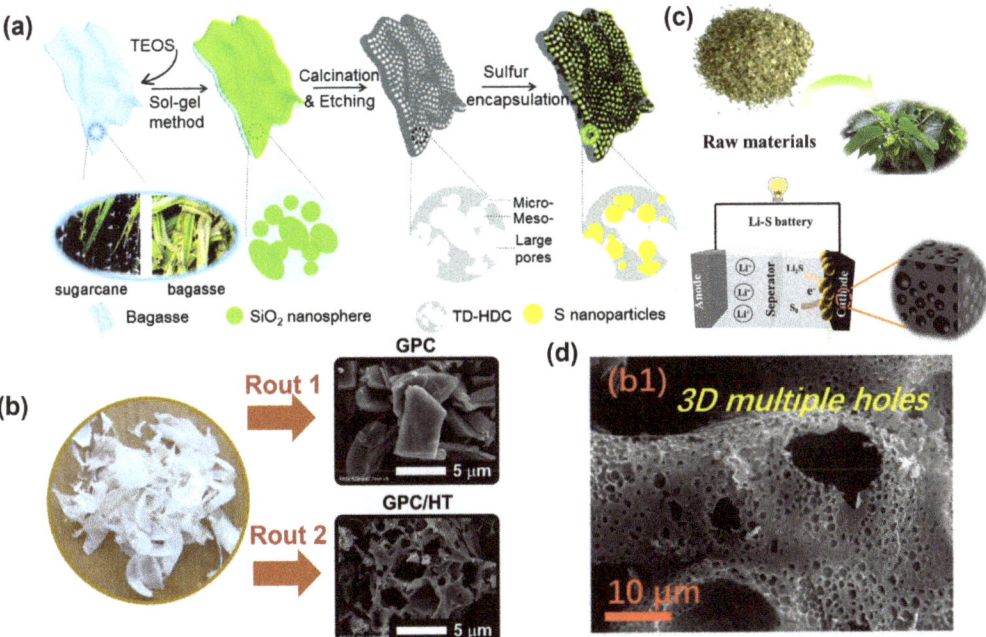

Figure 9. (a) Schematic illustration of the formation process of bagasse-derived carbon/sulfur composite [75]. Reprinted with permission from Ref. [75]. Copyright 2021 Royal Society of Chemistry. (b) Preparation of porous carbons using garlic peels and the SEM images of the carbons derived from different preparation routs [79]. Reprinted with permission from Ref. [79]. Copyright 2021 Elsevier Ltd. (c) The schematic illustration for the synthesis of Eucommia leaf residue-derived hierarchical porous carbon and their application in LSBs [82]. Reprinted with permission from Ref. [82]. Copyright 2023 Elsevier Ltd. (d) The SEM image of oxygen-doped carbon derived from balsa waste [83]. Reprinted with permission from Ref. [83]. Copyright 2021 Elsevier Ltd.

In addition to these biomasses from agriculture, forest wastes and food residues have also been employed as raw materials for the production of carbon as a host for LSBs. As depicted in Figure 10a, nitrogen/sulfur co-doped porous carbons were manufactured from cattail biomass through a one-step hydrothermal method. The stable foam-like porous structure, high specific surface area and N/S atom-doping sites could greatly inhibit the volume expansion of sulfur and the shuttle effect due to the physical confinement and chemical adsorption during the electrochemical process of LSBs [84]. In addition, nitrogen-doped porous carbon was prepared through the carbonization of pomelo peels to serve as a sulfur host material for LSBs (Figure 10b). The N-doping sites and the hierarchical porous architecture render the carbonous material with excellent sulfur confinement property due to the combination of physical and chemical adsorptions. Therefore, the sulfur composite cathodes exhibit ultrahigh initial capacity, high coulombic efficiency and stable sulfur electrochemistry [88]. Moreover, starch was adopted as a precursor to prepare porous carbon materials without using additional physical and chemical activators [97]. In order to avoid damage to the material structure caused by rapid generation and aggregation of water vapor during pyrolysis, the heating rate and airflow velocity during pyrolysis were carefully controlled (Figure 11a). The as-synthesized carbon material possesses the inherent porous structure of starch, which effectively increases the pore volume of the as-obtained carbon material. It was found that the narrow and long microporous channel not only avoids the direct contact between the electrolyte and active substances, but it also immobilizes polysulfide in the porous shell through physical adsorption, which effectively inhibits the shuttle effect of soluble polysulfide. To further optimize the pore structure of carbon products, mesoscale silica spheres with a uniform size were used as hard templates, which endows the as-obtained carbon with ordered structure and uniform pore size (Figure 11b). The abundant mesoporous structure provides enough space for the storage of active substances and alleviates the volume expansion during lithiation, which greatly overcomes the electronic insulation of sulfur and effectively inhibits the migration of polysulfides [98]. Nitrogen and oxygen co-doped porous carbon materials could also be obtained from soybeans [100]. The protein contained in this precursor is converted into different nitrogen-containing components (pyrrole nitrogen, pyridine nitrogen and graphite nitrogen) during pyrolysis, which leads to a higher nitrogen content in these electron donors to improve the overall electron density of the carbon materials and enhances the conductivity of the carbon materials. In addition, nitrogen-doping sites can be employed as electron-donating sites to bind electron-deficient polysulfides with strong chemical bonds, thus enhancing the anchoring ability of the carbon framework to polysulfides (Figure 11c). Based on the traditional Chinese expansion method, rice was employed as a precursor to prepare porous carbons composed of nickel-doped hybrid nanoflakes [99] (Figure 11d). After expansion, the dense starch structure becomes loose due to steam evaporation. It was found that the final carbon with a sheet-like structure provides a stable three-dimensional porous structure, which avoids the structural collapse caused by volume change during the charge and discharge process. The three-dimensional porous structure, as the space for storing active substances, ensures the electrical contact between the active substances and the conductive network. In addition, the stable chemical bond between Ni/NiO and polysulfide effectively alleviates the shuttle effect and ensures a high-capacity retention rate. Meanwhile, the embedded nickel nanoparticles not only significantly increase the electronic conductivity but also provide a shortened ion diffusion channel to accelerate the diffusion process. In order to further highlight the recent developments in biomass-derived carbons adopted as cathodes in LSBs, a systematic comparison table is provided in terms of the preparation method and the electrochemical performances of LSBs (Table S1).

Figure 10. (a) Photograph of the cattail biomass, SEM and TEM images of the cattail derived carbons prepared based on different processes. [84] Reprinted with permission from Ref. [84]. Copyright 2022 John Wiley and Sons. (b) Illustration of the pomelo peels, and the SEM images of the as-synthesized carbons before (left) and after (right) sulfur loading [88]. Reprinted with permission from Ref. [88]. Copyright 2020 Elsevier Ltd.

Figure 11. (a) Morphology of the hollow microporous carbon spheres prepared from starch via a multi-step pyrolysis method and their cycling performance at 0.1 C [97]. Reprinted with permission from Ref. [97]. Copyright 2017 American Chemical Society. (b) The TEM image of nitrogen–sulfur co-doped carbon materials prepared from soluble starch via a template method [98]. Reprinted with permission from Ref. [98]. Copyright 2012 Royal Society of Chemistry (The scale bar represents 20 nm). (c) Photograph of the crude soybean biomass and the TEM image of the nitrogen–oxygen co-doped porous carbon materials prepared by using soybean as a precursor [100]. Reprinted with permission from Ref. [100]. Copyright 2016 Royal Society of Chemistry. (d) The photograph of the rice before and after expansion treatment, as well as the photograph of the nickel-doped hybrid carbon prepared based on the expansion technology and calcination [99]. Reprinted with permission from Ref. [99]. Copyright 2017 WILEY-VCH.

4.3. Biomass-Derived Carbons for the Interlayer of LSBs

In order to enhance the performance of LSBs, free-standing or deposited interlayers, which could effectively prohibit the shuttle effect of polysulfides, have been designed and introduced between the electrodes without affecting the integrity of LSBs. So far as we know, many types of biomass-derived carbons have been adopted to modify the pristine separator or construct a free-standing interlayer between the electrodes of LSBs [101–104]. As depicted in Figure 12a, N, O co-doped carbons with a hierarchical porous structure, a high specific surface area, and good electrical conductivity were synthesized from chlorella biomass through a chemical activation process. As they are deposited on the polypropylene separator as an interlayer in LIBs, these carbons can improve the electrolyte wettability and Li$^+$ diffusion. In addition, N, O heteroatoms and the porous structure exhibit strong chemical adsorption and provide physical barriers confining lithium polysulfides, which result in enhanced cycling stability and rate performance. As shown in Figure 12b, Ginkgo Folium was employed as a biomass to synthesize carbon materials to decorate the separator of LSBs. Thanks to the three-dimensional interconnected porous structures and the defective graphite structure, the interlayer effectively inhibits the shuttle effect and boosts the electrochemical properties of the LSBs. In addition, rotten egg albumen-derived carbon was also fabricated via freeze drying and carbonization (Figure 12c). As modified on the separator, the layered carbon materials with high conductivity and the rich nitrogen-doped sites could enhance both lithium ion and electrical conductivity, and improve polysulfide adsorption, which substantially inhibits the shuttle effect of polysulfides and enhances the electrochemical performance of LSBs.

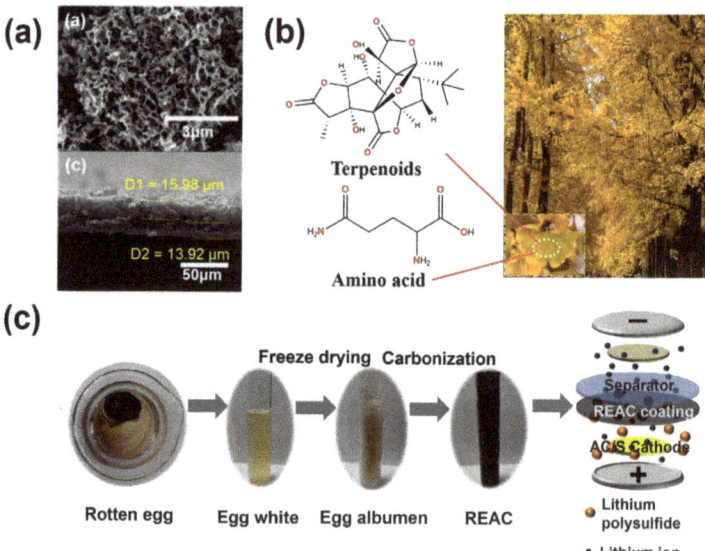

Figure 12. (a) SEM images of the top and cross-section morphology of chlorella-derived carbon-coated separator [101]. Reprinted with permission from Ref. [101]. Copyright 2020 Elsevier Inc. (b) Chemicals contained in the Ginkgo Folium biomass [103]. Reprinted with permission from Ref. [103]. Copyright 2022 Elsevier Inc. (c) Synthesis process of rotten egg-derived carbon-modified separator [104]. Reprinted with permission from Ref. [104]. Copyright 2021 Elsevier Inc.

As shown in Figure 13a, discarded crab shells were employed as a precursor to yield functional carbon materials through potassium hydroxide-assisted pyrolysis [105]. Due to the intrinsic rich keratin content contained in the crab shells, nitrogen doping could be easily introduced during pyrolysis. As deposited on the membrane surface of the separator, the surface wettability is greatly improved and strong chemisorption of polysulfides is formed, which could effectively enhance the electrolyte permeability, the transportation of lithium ions and the adsorption of the modified separator. As depicted in Figure 13b, bamboo chopsticks were used to prepare carbon fibers as an interlayer for LIBs with excellent electrochemical performances [106]. When used as an interlayer for LSBs, soluble polysulfides can be effectively adsorbed in the bamboo carbon fiber sandwich membrane, and the shuttle effect is substantially suppressed. At the same time, the cross-linked three-dimensional conductive framework constructed by the fiber structure also provides an interworking conductive network, which promotes both the electronic transport and the diffusion rate of lithium ions. In addition, sulfur-doped microporous carbon materials could be prepared from loofah pulp biomass. The as-prepared self-supporting porous carbons could effectively prevent the diffusion of polysulfides to improve the electrochemical performance of LSBs [107]. The mechanism studies indicated that the rich microporous structure and sulfur-doping sites could be used as both physical and chemical adsorption sites, which could effectively adsorb polysulfides in the conductive framework and avoid the irreversible loss of active substances, leading to high sulfur utilization and high cycle stability. Recent research progresses regarding biomass-derived carbons adopted as an interlay in LSBs are summarized in Table S2.

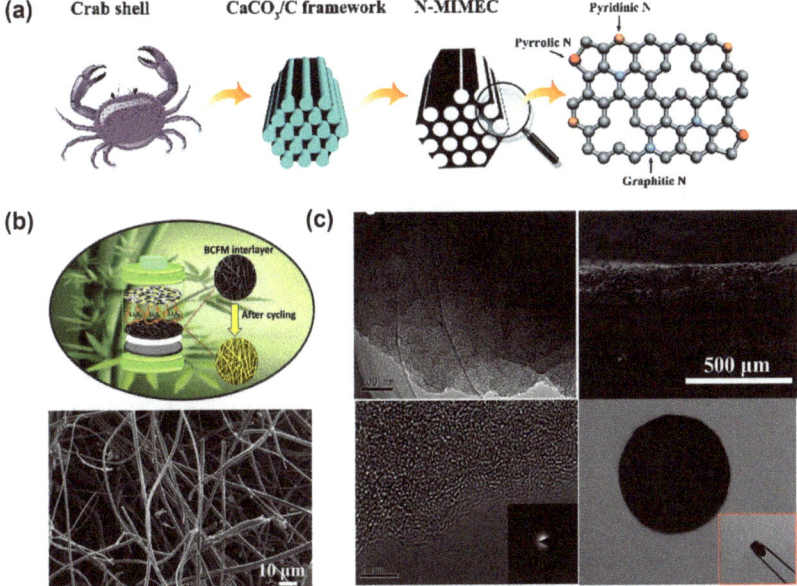

Figure 13. (a) Preparation route and the N doping sites of the hierarchical porous carbon materials prepared from crab shells [105]. Reprinted with permission from Ref. [105]. Copyright 2018 Royal Society of Chemistry. (b) Fabrication scheme and SEM image of hollow carbon fiber networks prepared from discarded bamboo chopsticks [106]. Reprinted with permission from Ref. [106]. Copyright 2012 Royal Society of Chemistry. (c) TEM, SEM images and the photograph of the self-supporting sulfur-doped porous carbon sandwich membrane prepared from loofah pulp [107]. Reprinted with permission from Ref. [107]. Copyright 2012 Royal Society of Chemistry.

4.4. Modified Biomass-Derived Carbons for LSBs

The main advantages of biomasses rely on their intrinsic heteroatoms and porous structure, which could endow the as-obtained carbonous materials with a high surface area, a hierarchical porous structure and heteroatom-doping sites. However, these merits highly depend on the inherent qualities of the biomass used, which is not favorable for further property enhancement of carbon materials. Therefore, many schemes have been developed to modify the biomass used prior to or after the carbonization process [108]. As shown in Figure 14a, in order to introduce P doping sites on the carbon materials, H_3PO_4 was added in cotton stalk biowaste, which could both enhance the surface area and introduce hierarchical pores. The P doping sites in the carbon networks not only provide more active sites but also improve the electrical conductivity, which finally results in excellent electrochemical performances [109]. In addition, CoO nanoparticles were decorated on hierarchical porous carbon by mixing a natural nori and Co precursor prior to the carbonization process. It was found that the carbon substrate could well accommodate CoO nanoparticles, which could enhance the adsorption immobilization of lithium polysulfides and facilitate their redox conversion. Therefore, the composite cathode possesses a high discharge capacity, excellent rate performance and cycling stability [110–116]. As shown in Figure 14b, rice straw was developed to fabricate conductive biochar and then decorated with highly dispersed CoO nanoparticles via a microwave-assisted method. The high LSB performances are derived from the excellent conductive framework of the biochar and the excellent adsorption capability of CoO nanoparticles, which greatly alleviate the shuttle effect as well as the conversion kinetics between the polysulfides [115].

Figure 14. (a) Schematic illustration of the preparation procedure of P-doped carbon materials from cotton stalk biowaste [108]. Reprinted with permission from Ref. [108]. Copyright 2022 American Chemical Society. (b) Highly dispersed CoO nanoparticles are formed in the inner channels of biochar via microwave-assisted treatment. [115] Reprinted with permission from Ref. [115]. Copyright 2023 Elsevier Inc.

Based on a similar concept, activated cotton textile was adopted as a scaffold to load Fe/Fe$_3$C-encapsulated multiwalled carbon nanotubes via a strategy combining vapor–liquid–solid and solid–liquid–solid processes (Figure 15a). The as-prepared composite was employed as a free-standing interlayer for LIBs, which leads to high cycling stability, ultralow capacity decay rate and remarkable specific capacity due to the enhanced electrode stability and suppressed shuttle effect of polysulfides [117]. In addition, phytoremediation residue-derived carbons were used as a host for LSBs. After the phosphorous acid-assisted pyrolysis of oilseed rape stems from phytoremediation, the as-yield carbon materials with a porous structure and abundant N, P, O doping sites could effectively enhance the sulfur loading amount and polysulfide adsorption, which lead to excellent electrochemical performances. This research not only proposed a promising approach for the safe disposal of phytoremediation residues but also high-performance cathode materials for LSBs [118]. As depicted in Figure 15b, carbon materials were prepared from brewing waste without any activation process. These carbons demonstrate high surface area and interconnected micro- and mesoporous distributions, which raises the capacity values and cyclability of

LSBs. This work demonstrated a promising and sustainable way to yield porous carbons while adopting a simple process without activation process.

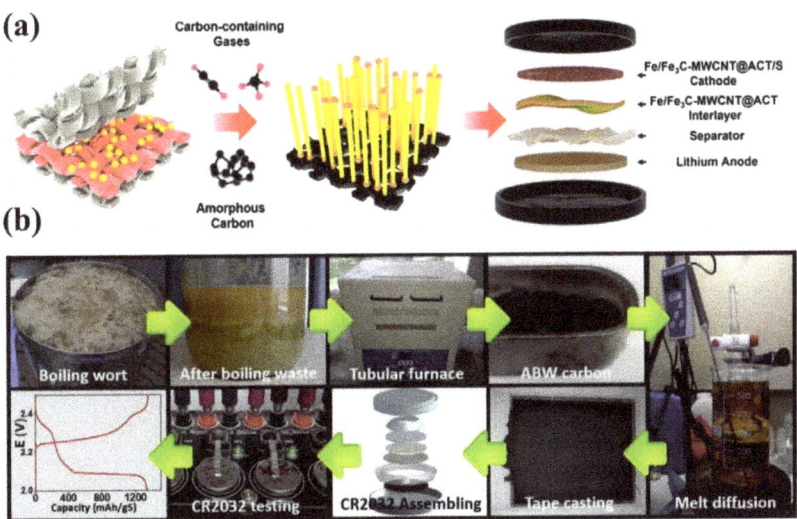

Figure 15. (**a**) Schematic of cotton-derived Fe/Fe₃C-encapsulated multiwalled carbon nanotubes' interlayer and their application in LSBs [117]. Reprinted with permission from Ref. [117]. Copyright 2022 American Chemical Society. (**b**) Scheme of cathode fabrication from beer production waste for LSBs [119]. Reprinted with permission from Ref. [119]. Copyright 2020 Wiley-VCH.

5. Conclusions and Outlook

Due to their low cost, abundance in resources and environmentally friendly qualities, biomasses generated from different scenarios have been employed to prepare carbon matrices as host or interlayer materials for LSBs. Many case works have proven that a hierarchical porous structure and heteroatom-doping sites inherited from the tissue of the biomass used are the main advantages to enhance sulfur utilization, prohibit the shuttle effect of polysulfides, and accelerate the sluggish redox conversion of soluble intermediates. Although remarkable research works have been devoted to the rational design and preparation of biomass-derived carbons, several bottlenecks still remain and require further investigation. The first challenge is that the activation of carbons usually needs corrosive or even poisonous chemical regents, such as NaOH, KOH, and $ZnCl_2$, which hinders the commercialization of biomass derived carbons. Although previous works have been devoted to preparing biomass-derived carbons without further activation or purification [119,120], more efforts should be devoted to developing green and low-toxic pore regulation strategies. Secondly, many of the reported sources of biomasses are rather limited in quantity or unstable in quality, which usually exhibit unexpected variation in their structure and composition, resulting in adverse impacts on the standard production of such carbon materials. The third obstacle is the inadequate test equipment used to characterize the structure evolution of biomass-derived carbons during their charge and discharge processes, which makes the underlying work mechanism of these carbons in LSBs difficult to be revealed. The final problem is the relative low loading amount of active species on the cathode of LSBs. It has been reported that the S loading amount could be enhanced to above 2 mg cm^{-2} [121,122] or even 5 mg cm^{-2} [123–125]. However, there is still a distance to meet the requirement for real application scenarios.

Based on these aforementioned challenges, future efforts in the design and production of biomass-derived carbons should be focused on the following aspects. First of all, low-cost, low-energy-consumption and ecofriendly strategies should be further explored for the massive production of biomass-derived carbons. For example, mild activation methods should be developed to reduce hazardous emission, potential safety concerns and corrosion on the equipment used. In addition, more fundamental understanding should be realized regarding the activation processes, such as the doping process of heteroatoms, the surface modification of different functional groups, and the regulation of porous structures and their conductivities. Further exploration of the formation mechanism of biomass-derived carbons could provide more specific instruction on the potential mass production of these high-performance carbons.

Secondly, in order to achieve the commercialization of biomass-derived carbons for LSBs, a standard production process should be established based on biomasses with high yield and abundant resources. For example, the composition and structure of biomasses vary dramatically as they are harvested in different seasons or at different locations. Even with the same biomass, different preparation methods usually lead to different structural qualities of carbons. Therefore, in order to provide stable and high-performance carbons for LSBs, many standards should not only be established for the collection and pretreatment of biomasses but also for the pyrolysis process and posttreatment.

Thirdly, advanced equipment composed of microscopy, spectroscopy and X-ray absorption should be designed and assembled to observe in situ the structure, composition and morphology evolution of biomass-derived carbons during the operation of LSBs, which could help to accurately select the appropriate carbons to attain a high-performance cathode or interlayer for LSBs. In particular, the characterization of the electrolyte's decomposition and conversion on the surface of the carbon electrode could provide straightforward evidence to enhance the coulombic efficiency, rate performance and cycling stability of LSBs.

Finally, the production of carbons from biomasses not only reduce the environmental burden from biomasses or biowastes generated in different industries but also pave an ecofriendly way for backing up the utilization of renewable energy with high-performance energy storage devices. However, an unneglectable fact that should not be omitted is that the activation, pyrolysis and purification processes during the production generate a large volume of liquid and gaseous effluents, and require high energy and chemical reagent consumptions, which will again lead to a huge environmental burden. Therefore, systematic environmental and economic sustainability studies should be devoted to the green and standardized mass production of biomass-derived carbons, which could provide more opportunities for the final commercialization of biomass carbon-derived LSBs.

Supplementary Materials: The following supporting information can be downloaded at https://www.mdpi.com/article/10.3390/nano13111768/s1. Table S1: Biomass-derived carbons for the cathode of LSBs; Table S2: Biomass-derived carbons for the interlayer of LSBs.

Author Contributions: C.M.: Data curation; Formal analysis; Writing-original draft; M.Z.: Conceptualization; Data curation; Validation; Writing-original draft. Y.D.: Data curation; Formal analysis. Y.X.: Formal analysis; Supervision; Writing-review & editing. H.W.: Formal analysis; Validation. P.L.: Data curation; Formal analysis. D.W.: Conceptualization; Funding acquisition; Supervision; Writing-review & editing. All authors have read and agreed to the published version of the manuscript.

Funding: This study was funded by the Key Science and Technology Project of Henan Province (21A430022), the Soft Science Research Program of Henan Province (212400410310), the Outstanding Youth Science Foundation of HTU (2021JQ03), and the Natural Science Foundation of Henan Province (212300410009).

Informed Consent Statement: Informed consent was obtained from all subjects involved in the study.

Data Availability Statement: The data is available upon request.

Conflicts of Interest: The authors declare no conflict of interest.

References

1. Zhang, W.C.; Lu, J.; Guo, Z.P. Approaching high-performance potassium-ion batteries via advanced design strategies and engineering. *Sci. Adv.* **2019**, *5*, eaav7412. [CrossRef] [PubMed]
2. Peng, H.J.; Huang, J.Q.; Cheng, X.B.; Zhang, Q. Review on High-Loading and High-Energy Lithium-Sulfur Batteries. *Adv. Energy Mater.* **2017**, *24*, 1700260. [CrossRef]
3. Zhang, W.C.; Lu, J.; Guo, Z.P. Challenges and future perspectives on sodium and potassium ion batteries for grid-scale energy storage. *Mater. Today* **2021**, *50*, 400–417. [CrossRef]
4. Zhang, F.L.; Zhang, W.C.; David, W.; Guo, Z.P. Recent Progress and Future Advances on Aqueous Monovalent-Ion Batteries towards Safe and High-Power Energy Storage. *Adv. Mater.* **2022**, *34*, 2107965. [CrossRef] [PubMed]
5. Chen, L.F.; Feng, Y.; Liang, H.W.; Wu, Z.Y.; Yu, S.H. Macroscopic-Scale Three-Dimensional Carbon Nanofiber Architectures for Electrochemical Energy Storage Devices. *Adv. Energy Mater.* **2017**, *7*, 1700826. [CrossRef]
6. Li, Y.; Wu, F.; Li, Y.; Liu, M.Q.; Feng, X.; Bai, Y.; Wu, C. Ether-based electrolytes for sodium ion batteries. *Chem. Soc. Rev.* **2022**, *51*, 4484–4536. [CrossRef]
7. Sen, X.; Lin, G.; Zhao, N.H. Smaller Sulfur Molecules Promise Better Lithium—Sulfur Batteries. *J. Am. Chem. Soc.* **2012**, *134*, 18510–18513. [CrossRef]
8. Zhang, J.; Huang, H.; Bae, J.; Chung, S.H.; Zhang, W.K.; Manthiram, A.; Yu, G.H. Nanostructured Host Materials for Trapping Sulfur in Rechargeable Li-S Batteries: Structure Design and Interfacial Chemistry. *Small Methods* **2018**, *2*, 1700279. [CrossRef]
9. Yang, D.X.; Ren, H.Y.; Wu, D.P.; Zhang, W.C.; Lou, X.D.; Wang, D.Q.; Cao, K.; Gao, Z.Y.; Xu, F.; Jiang, K. Bi-functional nitrogen-doped carbon protective layer on three-dimensional RGO/SnO$_2$ composites with enhanced electron transport and structural stability for high-performance lithium-ion batteries. *J. Colloid Interface Sci.* **2019**, *542*, 81–90. [CrossRef]
10. Wu, D.P.; Wang, Y.X.; Wang, F.J.; Wang, H.J.; An, Y.P.; Gao, Z.Y.; Xu, F.; Jiang, K. Oxygen-incorporated few-layer MoS$_2$ vertically aligned on three-dimensional graphene matrix for enhanced catalytic performances in quantum dot sensitized solar cells. *Carbon* **2017**, *123*, 756–766. [CrossRef]
11. Wang, L.; Jia, F.; Wu, D.P.; Wei, Q.X.; Liang, Y.; Hu, Y.S.; Li, R.F.; Yu, G.H.; Yuan, Q.P.; Wang, J.S. In-situ growth of graphene on carbon fibers for enhanced cell immobilization and xylitol fermentation. *Appl. Surf. Sci.* **2020**, *527*, 146793. [CrossRef]
12. Chen, J.L.; Wu, D.P.; Wang, H.J.; Wang, F.J.; Wang, Y.X.; Gao, Z.Y.; Xu, F.; Jiang, K. In-situ synthesis of molybdenum sulfide/reduced graphene oxide porous film as robust counter electrode for dye-sensitized solar cells. *J. Colloid Interface Sci.* **2018**, *524*, 475–482. [CrossRef] [PubMed]
13. Zou, X.X.; Wu, D.P.; Mu, Y.F.; Xing, L.Y.; Zhang, W.C.; Gao, Z.Y.; Xu, F.; Jiang, K. Boron and nitrogen Co-doped holey graphene aerogels with rich B-N motifs for flexible supercapacitors. *Carbon* **2020**, *159*, 94–101. [CrossRef]
14. Li, R.G.; Wu, D.P.; Song, J.K.; He, Y.C.; Zhu, W.Y.; Wang, X.P.; Wang, L.X.; Dube, N.M.; Jiang, K. In situ generation of reduced graphene oxide on 3D Cu-Ni foam as high-performance electrodes for capacitive deionization. *Desalination* **2022**, *540*, 115990. [CrossRef]
15. Tabani, H.; Khodaei, K.; Moghaddam, A.Z. Introduction of graphene-periodic mesoporous silica as a new sorbent for removal: Experiment and simulation. *Chem. Intermed.* **2019**, *45*, 1795–1813. [CrossRef]
16. Wang, L.; Yin, Y.L.; Zhang, S.B.; Wu, D.P.; Lv, Y.Y.; Hu, Y.S.; Wei, Q.X.; Yuan, Q.P.; Wang, J.S. A rapid microwave-assisted phosphoric-acid treatment on carbon fiber surface for enhanced cell immobilization in xylitol fermentation. *Colloids Surf. B* **2019**, *175*, 697–702. [CrossRef]
17. Wang, L.; Shen, Y.; Zhang, Y.X.; Wei, Q.X.; Liang, Y.; Tian, H.L.; Wu, D.P.; Yuan, X.Q.; Yuan, Q.P.; Wang, J.S. A novel surface treatment of carbon fiber with Fenton reagent oxidization for improved cells immobilization and xylitol fermentation. *Microporous Mesoporous Mater.* **2021**, *325*, 111318. [CrossRef]
18. Wang, L.; Liu, N.; Guo, Z.; Wu, D.P.; Chen, W.W.; Chang, Z.; Yuan, Q.P.; Hui, M.; Wang, J.S. Nitric Acid-Treated Carbon Fibers with Enhanced Hydrophilicity for Candida tropicalis Immobilization in Xylitol Fermentation. *Materials* **2016**, *9*, 206. [CrossRef]
19. Wang, R.; Wang, H.J.; Zhou, Y.; Gao, Z.Y.; Han, Y.; Jiang, K.; Zhang, W.C.; Wu, D.P. Green synthesis of N-doped porous carbon/carbon dot composites as metal-free catalytic electrode materials for iodide-mediated quasi-solid flexible supercapacitors. *Inorg. Chem. Front.* **2022**, *9*, 2530–2543. [CrossRef]
20. Ma, Z.J.; Wu, D.P.; Han, X.Y.; Wang, H.J.; Zhang, L.M.; Gao, Z.Y.; Xu, F.; Jiang, K. Ultrasonic assisted synthesis of Zn-Ni bi-metal MOFs for interconnected Ni-N-C materials with enhanced electrochemical reduction of CO$_2$. *J. CO2 Util.* **2019**, *32*, 251–258. [CrossRef]
21. Ma, Z.J.; Zhang, X.L.; Wu, D.P.; Han, X.X.; Zhang, L.M.; Wang, H.J.; Xu, F.; Gao, Z.Y.; Jiang, K. Ni and nitrogen-codoped ultrathin carbon nanosheets with strong bonding sites for efficient CO$_2$ electrochemical reduction. *J. Colloid Interface Sci.* **2020**, *570*, 31–40. [CrossRef] [PubMed]
22. Li, H.H.; Wang, Y.X.; Chen, H.Q.; Niu, B.X.; Zhang, W.C.; Wu, D.P. Synergistic mediation of polysulfide immobilization and conversion by a catalytic and dual-adsorptive system for high performance lithium-sulfur batteries. *Chem. Eng. J.* **2021**, *406*, 126802. [CrossRef]
23. Zhang, M.M.; Huang, K.X.; Ding, Y.; Wang, X.Y.; Gao, Y.L.; Li, P.F.; Zhou, Y.; Gao, Z.; Zhang, Y.; Wu, D.P. N, S Co-Doped Carbons Derived from Enteromorpha prolifera by a Molten Salt Approach: Antibiotics Removal Performance and Techno-Economic Analysis. *Nanomaterials* **2022**, *12*, 4289. [CrossRef] [PubMed]

24. Wang, C.J.; Wu, D.P.; Wang, H.J.; Gao, Z.Y.; Xu, F.; Jiang, K. Nitrogen-doped two-dimensional porous carbon sheets derived from clover biomass for high performance supercapacitors. *J. Power Sources* **2017**, *363*, 375–383. [CrossRef]
25. Chen, C.; Han, H.J.; Liu, X.P.; Chen, Y.; Wu, D.P.; Gao, Z.Y.; Gao, S.Y.; Jiang, K. Nitrogen, phosphorus, sulfur tri-doped porous carbon derived from covalent polymer with versatile performances in supercapacitor, oxygen reduction reaction and electro-fenton degradation. *Microporous Mesoporous Mater.* **2021**, *325*, 111335. [CrossRef]
26. Wang, C.J.; Wu, D.P.; Wang, H.J.; Gao, Z.Y.; Xu, F.; Jiang, K. Biomass derived nitrogen-doped hierarchical porous carbon sheets for supercapacitors with high performance. *J. Colloid Interface Sci.* **2018**, *523*, 133–143. [CrossRef]
27. Wu, D.P.; Chen, J.L.; Zhang, W.C.; Liu, W.D.; Li, J.Z.; Cao, K.; Gao, Z.Y.; Xu, F.; Jiang, K. Sealed pre-carbonization to regulate the porosity and heteroatom sites of biomass derived carbons for lithium-sulfur batteries. *J. Colloid Interface Sci.* **2020**, *579*, 667–679. [CrossRef]
28. Wang, C.J.; Wu, D.P.; Wang, H.J.; Gao, Z.Y.; Xu, F.; Jiang, K. A green and scalable route to yield porous carbon sheets from biomass for supercapacitors with high capacity. *J. Mater. Chem. A* **2018**, *6*, 1244–1254. [CrossRef]
29. Chen, C.; Tian, M.; Han, H.J.; Wu, D.P.; Chen, Y.; Gao, Z.Y.; Gao, S.Y.; Jiang, K. N, P-dual doped carbonaceous catalysts derived from bifunctional-salt activation for effective electro-Fenton degradation on waterborne organic pollutions. *Electrochim. Acta* **2021**, *389*, 138732. [CrossRef]
30. Bai, X.G.; Ying, L.Y.; Liu, N.; Ma, N.N.; Huang, K.X.; Wu, D.P.; Yin, M.M.; Jiang, K. Humulus scandens-Derived Biochars for the Effective Removal of Heavy Metal Ions: Isotherm/Kinetic Study, Column Adsorption and Mechanism Investigation. *Nanomaterials* **2021**, *11*, 3255. [CrossRef]
31. Yin, M.M.; Bai, X.G.; Wu, D.P.; Li, F.B.; Jiang, K.; Ma, N.N.; Chen, Z.H.; Zhang, X.; Fang, L.P. Sulfur-functional group tunning on biochar through sodium thiosulfate modified molten salt process for efficient heavy metal adsorption. *Chem. Eng. J.* **2022**, *433*, 134441. [CrossRef]
32. Bai, X.G.; Zhang, M.M.; Niu, B.X.; Zhang, W.L.; Wang, X.P.; Wang, J.S.; Wu, D.P.; Wang, L.; Jiang, K. Rotten sugarcane bagasse derived biochars with rich mineral residues for effective Pb (II) removal in wastewater and the tech-economic analysis. *J Taiwan Inst. Chem. Eng.* **2022**, *132*, 104231. [CrossRef]
33. Ragauskas, A.J.; Williams, C.K.; Davison, B.H.; Britovsek, G.; Cairney, J.; Eckert, C.A.; Frederick, W.J., Jr.; Hallett, J.P.; Leak, D.J.; Liotta, C.L.; et al. The path forward for biofuels and biomaterials. *Science* **2006**, *311*, 484–489. [CrossRef] [PubMed]
34. Jain, A.; Balasubramanian, R.; Srinivasan, M.P. Hydrothermal conversion of biomass waste to activated carbon with high porosity: A review. *Chem. Eng. J.* **2016**, *283*, 789–805. [CrossRef]
35. Wang, J.; Kaskel, S. KOH activation of carbon-based materials for energy storage. *J. Mater. Chem.* **2012**, *22*, 23710. [CrossRef]
36. Keiluweit, M.; Nico, P.S.; Johnson, M.G.; Kleber, M. Dynamic molecular structure of plant biomass-derived black carbon (biochar). *Environ. Sci. Technol.* **2010**, *44*, 1247–1253. [CrossRef]
37. Long, C.; Chen, X.; Jiang, L.; Zhi, L.; Fan, Z. Porous layer-stacking carbon derived from in-built template in biomass for high volumetric performance supercapacitors. *Nano Energy* **2015**, *12*, 141–151. [CrossRef]
38. Gao, S.; Li, X.; Li, L.; Wei, X. A versatile biomass derived carbon material for oxygen reduction reaction, supercapacitors and oil/water separation. *Nano Energy* **2017**, *33*, 334–342. [CrossRef]
39. Hall, P.J.; Mirzaeian, M.; Fletcher, S.I.; Sillars, F.B.; Rennie, A.J.R.; Shitta-Bey, G.O.; Wilson, G.; Crudenb, A.; Carterb, R. Energy storage in electrochemical capacitors: Designing functional materials to improve performance. *Energy Environ. Sci.* **2010**, *3*, 1238. [CrossRef]
40. Hou, J.; Cao, C.; Idrees, F.; Ma, X. Hierarchical porous nitrogen-doped carbon nanosheets derived from silk for ultrahigh-capacity battery anodes and supercapacitors. *ACS Nano* **2015**, *9*, 2556–2564. [CrossRef]
41. Niu, J.; Shao, R.; Liang, J.; Dou, M.; Li, Z.; Huang, Y.; Wang, F. Biomass-derived mesopore-dominant porous carbons with large specific surface area and high defect density as high performance electrode materials for Li-ion batteries and supercapacitors. *Nano Energy* **2017**, *36*, 322–330. [CrossRef]
42. Gu, X.; Wang, Y.; Lai, C.; Qiu, J.; Li, S.; Hou, Y.; Martens, W.; Mahmood, N.; Zhang, S. Microporous bamboo biochar for lithium-sulfur batteries. *Nano Res.* **2014**, *8*, 129–139. [CrossRef]
43. Xu, G.; Han, J.; Ding, B.; Nie, P.; Pan, J.; Dou, H.; Lia, H.; Zhang, X. Biomass-derived porous carbon materials with sulfur and nitrogen dual-doping for energy storage. *Green Chem.* **2015**, *17*, 1668–1674. [CrossRef]
44. Sun, Y.; Shi, X.L.; Yang, Y.L.; Suo, G.Q.; Zhang, L.; Lu, S.Y.; Chen, Z.G. Biomass-Derived Carbon for High-Performance Batteries: From Structure to Properties. *Adv. Funct. Mater.* **2022**, *32*, 2201584. [CrossRef]
45. Park, S.; Kim, J.; Kown, K. A review on biomass-derived N-doped carbons as electrocatalysts in electrochemical energy applications. *Chem. Eng. J.* **2022**, *446*, 137116. [CrossRef]
46. Yuan, H.D.; Liu, T.F.; Liu, Y.J.; Nai, J.W.; Wang, Y.; Zhang, W.K.; Tao, X.Y. A review of biomass materials for advanced lithium–sulfur batteries. *Chem. Sci.* **2019**, *10*, 7484. [CrossRef]
47. Liu, P.T.; Wang, Y.Y.; Liu, J.H. Biomass-derived porous carbon materials for advanced lithium sulfur batteries. *J. Energy Chem.* **2019**, *34*, 171–185. [CrossRef]
48. Li, Q.; Liu, Y.P.; Wang, Y.; Chen, Y.X.; Guo, X.D.; Wu, Z.G.; Zhong, B.H. Review of the application of biomass-derived porous carbon in lithium-sulfur batteries. *Ionics* **2020**, *26*, 4765–4781. [CrossRef]
49. Feng, Y.; Jiang, J.; Xu, Y.; Wang, S.; An, W.; Chai, Q.; Prova, U.; Wang, C.; Huang, G. Biomass derived diverse carbon nanostructure for electrocatalysis, energy conversion and storage. *Carbon* **2023**, *211*, 118105. [CrossRef]

50. Zhao, Z.Q.; Su, Z.; Chen, H.L.; Yi, S.; Zhang, W.Y.; Niu, B.; Zhang, Y.Y.; Long, D.H. Renewable biomass-derived carbon-based hosts for lithium–sulfur batteries. *Sustain. Energy Fuels* **2022**, *6*, 5211. [CrossRef]
51. Tian, X.H.; Yan, C.Z.; Kang, J.B.; Yang, X.Y.; Li, Q.X.; Yan, J.; Deng, N.P.; Cheng, B.; Kang, W.M. Working Mechanisms and Structure Engineering of Renewable Biomass-Derived Materials for Advanced Lithium-Sulfur Batteries: A Review. *ChemElectroChem* **2022**, *9*, e202100995. [CrossRef]
52. Senthil, C.; Lee, C.W. Biomass-derived biochar materials as sustainable energy sources for electrochemical energy storage devices. *Renew. Sustain. Energy Rev.* **2021**, *137*, 110464. [CrossRef]
53. Wang, Y.Z.; Huang, X.X.; Zhang, S.Q.; Hou, Y.L. Sulfur Hosts against the Shuttle Effect. *Small Methods* **2018**, *2*, 1700345. [CrossRef]
54. Walle, M.D. Environmental Solid Waste-derived Carbon for Advanced Rechargeable Lithium-Sulfur Batteries: A Review. *ChemistrySelect* **2022**, *7*, e202200511. [CrossRef]
55. Benitez, A.; Amaro-Gahete, J.; Chien, Y.C.; Caballero, A.; Morales, J.; Brandell, D. Recent advances in lithium-sulfur batteries using biomass-derived carbons as sulfur host. Renew. *Renew. Sustain. Energy Rev.* **2022**, *154*, 111783. [CrossRef]
56. Zhang, B.; Qin, X.; Li, G.R.; Gao, X.P. Enhancement of long stability of sulfur cathode by encapsulating sulfur into micropores of carbon spheres. *Energy Environ. Sci.* **2010**, *3*, 1531–1537. [CrossRef]
57. Ji, X.L.; Lee, K.T.; Nazar, L.F. A highly ordered nanostructured carbon-sulphur cathode for lithium-sulphur batteries. *Nat. Mater.* **2009**, *8*, 500–506. [CrossRef]
58. Liang, C.D.; Dudney, N.J.; Howe, J.Y. Hierarchically Structured Sulfur/Carbon Nanocomposite Material for High-Energy Lithium Battery Chemistry of Materials. *Chem. Mater.* **2009**, *21*, 4724–4730. [CrossRef]
59. Jung, D.S.; Hwang, T.H.; Lee, J.H.; Koo, H.Y.; Shakoor, R.A.; Kahraman, R.; Jo, Y.N.; Park, M.S.; Choi, J.W. Hierarchical porous carbon by ultrasonic spray pyrolysis yields stable cycling in lithium-sulfur battery. *Nano Lett.* **2014**, *14*, 4418–4425. [CrossRef]
60. Li, Z.; Jiang, Y.; Yuan, L.X.; Yi, Z.Q.; Wu, C.; Liu, Y.; Strasser, P.; Huang, Y.H. A highly ordered meso@microporous carbon-supported sulfur@smaller sulfur core-shell structured cathode for Li-S batteries. *ACS Nano* **2014**, *8*, 9295–9303. [CrossRef]
61. Chen, C.G.; Li, D.J.; Gao, L.; Harks, P.R.M.L.; Eichel, R.-A.; Notten, P.H.L. Carbon-coated core–shell $Li_2S@C$ nanocomposites as high performance cathode materials for lithium–sulfur batteries. *J. Mater. Chem. A* **2017**, *5*, 1428–1433. [CrossRef]
62. Yang, T.; Wang, X.L.; Wang, D.H.; Li, S.H.; Xie, D.; Zhang, X.Q.; Xia, X.H.; Tu, J.P. Facile and scalable synthesis of nanosized core–shell $Li_2S@C$ composite for high-performance lithium–sulfur batteries. *J. Mater. Chem. A* **2016**, *4*, 16653–16660. [CrossRef]
63. Sun, Q.; He, B.; Zhang, X.Q.; Lu, A.H. Engineering of Hollow Core-Shell Interlinked Carbon Spheres for Highly Stable Lithium-Sulfur Batteries. *ACS Nano* **2015**, *9*, 8504–8513. [CrossRef]
64. Zhou, G.M.; Yin, L.C.; Wang, D.W.; Li, L.; Pei, S.F.; Gentle, I.R.; Li, F.; Cheng, H.M. Fibrous hybrid of graphene and sulfur nanocrystals for high-performance lithium-sulfur batteries. *ACS Nano* **2013**, *7*, 5367–5375. [CrossRef] [PubMed]
65. Zhou, G.M.; Li, L.; Ma, C.Q.; Wang, S.G.; Shi, Y.; Koratkar, N.; Ren, W.C.; Li, F.; Cheng, H.M. A graphene foam electrode with high sulfur loading for flexible and high energy Li-S batteries. *Nano Energy* **2015**, *11*, 356–365. [CrossRef]
66. Zhu, L.; Peng, H.J.; Liang, J.Y.; Huang, J.Q.; Chen, C.M.; Guo, X.F.; Zhu, W.C.; Li, P.; Zhang, Q. Interconnected carbon nanotube/graphene nanosphere scaffolds as free-standing paper electrode for high-rate and ultra-stable lithium–sulfur batteries. *Nano Energy* **2015**, *11*, 746–755. [CrossRef]
67. Song, J.X.; Gordin, M.L.; Xu, T.; Chen, S.R.; Yu, Z.X.; Sohn, H.; Lu, J.; Ren, Y.; Duan, Y.H.; Wang, D.H. Strong lithium polysulfide chemisorption on electroactive sites of nitrogen-doped carbon composites for high-performance lithium-sulfur battery cathodes. *Angew. Chem. Int. Ed. Engl.* **2015**, *54*, 4325–4329. [CrossRef]
68. Yang, C.P.; Yin, Y.X.; Ye, H.; Jiang, K.C.; Zhang, J.; Guo, Y.G. Insight into the effect of boron doping on sulfur/carbon cathode in lithium-sulfur batteries. *ACS Appl. Mater. Interfaces* **2014**, *6*, 8789–8795. [CrossRef]
69. Wang, N.N.; Xu, Z.F.; Xu, X.; Liao, T.; Tang, B.; Bai, Z.C.; Dou, S.X. Synergistically Enhanced Interfacial Interaction to Polysulfide via N,O Dual-Doped Highly Porous Carbon Microrods for Advanced Lithium-Sulfur Batteries. *ACS Appl. Mater. Interfaces* **2018**, *10*, 13573–13580. [CrossRef]
70. Chen, L.; Feng, J.R.; Zhou, H.H.; Fu, C.P.; Wang, G.C.; Yang, L.M.; Xu, C.X.; Chen, Z.X.; Yang, W.J.; Kuang, Y.F. Hydrothermal preparation of nitrogen, boron co-doped curved graphene nanoribbons with high dopant amounts for high-performance lithium sulfur battery cathodes. *J. Mater. Chem. A* **2017**, *5*, 7403–7415. [CrossRef]
71. Xu, J.; Su, D.W.; Zhang, W.X.; Bao, W.Z.; Wang, G.X. A nitrogen–sulfur co-doped porous graphene matrix as a sulfur immobilizer for high performance lithium-sulfur batteries. *J. Mater. Chem. A* **2016**, *4*, 17381–17393. [CrossRef]
72. Lee, J.; Oh, J.; Jeon, Y.; Piao, Y.Z. Multi-Heteroatom-Doped Hollow Carbon Attached on Graphene Using $LiFePO_4$ Nanoparticles as Hard Templates for High-Performance Lithium–Sulfur Batteries. *ACS Appl. Mater. Interfaces* **2018**, *10*, 26485–26493. [CrossRef] [PubMed]
73. Deng, J.; Li, M.; Wang, Y. Biomass-derived carbon: Synthesis and applications in energy storage and conversion. *Green Chem.* **2016**, *18*, 4824–4854. [CrossRef]
74. Wang, J.; Nie, P.; Ding, B.; Dong, S.; Hao, X.; Doua, H.; Zhang, X. Biomass derived carbon for energy storage devices. *J. Mater. Chem. A* **2017**, *5*, 2411–2428. [CrossRef]
75. Wu, D.P.; Liu, J.Y.; Chen, J.L.; Li, H.H.; Cao, R.G.; Zhang, W.C.; Gao, Z.Y.; Jiang, K. Promoting sulphur conversion chemistry with tri-modal porous N, O-codoped carbon for stable Li–S batteries. *J. Mater. Chem. A* **2021**, *9*, 5497. [CrossRef]
76. Li, Y.; Liu, W.L.; Li, S.l.; Meng, F.C.; Chen, Y.Z.; Wu, H.T.; Liu, J.H. From purple sweet potato to sustainable lithium-sulfur batteries. *Mater. Lett.* **2022**, *325*, 132893. [CrossRef]

77. Rojas, M.D.; Lobos, M.L.N.; Para, M.L.; María Quijón, E.G.; Cámara, O.; Barraco, D.; Moyano, E.L.; Luque, G.L. Activated carbon from pyrolysis of peanut shells as cathode for lithium-sulfur batteries. *Biomass Bioenergy* **2021**, *146*, 105971. [CrossRef]
78. Chen, H.W.; Xia, P.T.; Lei, W.X.; Pan, Y.; Zou, Y.L.; Ma, Z.S. Preparation of activated carbon derived from biomass and its application in lithium–sulfur batteries. *J. Porous Mater.* **2019**, *26*, 1325–1333. [CrossRef]
79. Lee, S.Y.; Choi, Y.J.; Kim, J.K.; Lee, S.J.; Bae, J.S.; Jeong, E.D. Biomass-garlic-peel-derived porous carbon framework as a sulfur host for lithium-sulfur batteries. *Ind. Eng. Chem. Res.* **2021**, *94*, 272–281. [CrossRef]
80. Chang, Y.G.; Ren, Y.M.; Zhu, L.K.; Li, Y.; Li, T.; Ren, B.Z. Preparation of macadamia nut shell porous carbon and its electrochemical performance as cathode material for lithium–sulfur batteries. *Electrochim. Acta* **2022**, *420*, 140454. [CrossRef]
81. Deng, Y.X.; Lei, T.Y.; Feng, Y.Y.; Zhang, B.; Ding, H.Y.; Lu, Q.; Tian, R.; Mushtaq, M.; Guo, W.J.; Yao, M.M.; et al. Biomass fallen leaves derived porous carbon for high performance lithium sulfur batteries. *Ionics* **2023**, *29*, 1029–1038. [CrossRef]
82. Yang, Y.; Yang, R.; Fan, C.J.; Huang, Y.; Yan, Y.L.; Zou, Y.M.; Xu, Y.H. Eucommia leaf residue-derived hierarchical porous carbon by KCl and $CaCl_2$ Co-auxiliary activation for lithium sulfur batteries. *Mater. Charact.* **2023**, *195*, 112522. [CrossRef]
83. Zhang, L.M.; Zhao, W.Q.; Yuan, S.H.; Jiang, F.; Chen, X.Q.; Yang, Y.; Ge, P.; Sun, W.; Ji, X.B. Engineering the morphology/porosity of oxygen-doped carbon for sulfur host as lithium-sulfur batteries. *J. Energy Chem.* **2021**, *60*, 531–545. [CrossRef]
84. Wen, X.Y.; Zhang, C.F.; Zhou, W.; Chen, H.; Xiang, K.X. Nitrogen/sulfur co-doping for biomass carbon foam as superior sulfur hosts for lithium-sulfur batteries. *Int. J. Energy Res.* **2022**, *46*, 10606–10619. [CrossRef]
85. Kim, J.K.; Choi, Y.J.; Jeong, E.D.; Lee, S.J.; Kim, H.G.; Chung, J.M.; Kim, J.S.; Lee, S.Y.; Bae, J.S. Synthesis and Electrochemical Performance of Microporous Hollow Carbon from Milkweed Pappus as Cathode Material of Lithium–Sulfur Batteries. *Nanomaterials* **2022**, *12*, 3605. [CrossRef] [PubMed]
86. Nurhilal, O.; Hidayat, S.; Sumiarsa, D.; Risdiana, R. Natural Biomass-Derived Porous Carbon from Water Hyacinth Used as Composite Cathode for Lithium Sulfur Batteries. *Renew. Energy* **2023**, *15*, 1039. [CrossRef]
87. Xue, M.Z.; Xu, H.; Tan, Y.; Chen, C.; Li, B.; Zhang, C.M. A novel hierarchical porous carbon derived from durian shell as enhanced sulfur carrier for high performance Li-S batteries. *J. Energy Chem.* **2021**, *893*, 115306. [CrossRef]
88. Xiao, Q.H.Q.; Li, G.R.; Li, M.J.; Liu, R.P.; Li, H.B.; Ren, P.F.; Dong, Y.; Feng, M.; Chen, Z.W. Biomass-derived nitrogen-doped hierarchical porous carbon as efficient sulfur host for lithium–sulfur batteries. *J. Energy Chem.* **2020**, *44*, 61–67. [CrossRef]
89. Ma, Z.W.; Sui, W.H.; Liu, J.; Wang, W.J.; Li, S.M.; Chen, T.T.; Yang, G.L.; Zhu, K.X.; Li, Z.J. Pomelo peel-derived porous carbon as excellent LiPS anchor in lithium-sulfur batteries. *J. Solid State Electrochem.* **2022**, *26*, 973–984. [CrossRef]
90. Wen, Y.T.; Wang, X.B.; Huang, J.Y.; Li, Y.; Li, T.; Ren, B.Z. Coffee grounds derived sulfur and nitrogen dual-doped porous carbon for the cathode material of lithium-sulfur batteries. *Carbon Lett.* **2023**. [CrossRef]
91. Liu, Y.; Lee, D.J.; Cho, K.K.; Zou, Y.M.; Ahn, H.J.; Ahn, J.H. Promoting long cycle life with honeycomb-like tri-modal porous carbon for stable lithium-sulfur polymer batteries. *J. Alloys Compd.* **2023**, *932*, 167704. [CrossRef]
92. Salimi, P.; Venezia, E.; Taghavi, S.; Tieuli, S.; Carbone, L.; Prato, M.; Signoretto, M.; Qiu, J.F.; Zaccaria, R.P. Lithium-Metal Free Sulfur Battery Based on Waste Biomass Anode and Nano-Sized Li_2S Cathode. *Energy Environ. Mater.* **2023**, e12567. [CrossRef]
93. Choi, J.R.; Kim, E.; Park, B.I.; Choi, I.; Park, B.H.; Lee, S.B.; Lee, J.L.; Yu, S. Meringue-derived hierarchically porous carbon as an efficient polysulfide regulator for lithium-sulfur batteries. *J. Ind. Eng. Chem.* **2022**, *115*, 355–364. [CrossRef]
94. Feng, L.J.; Lu, M.; Shen, W.N.; Qiu, X.Y. N/O dual-doped hierarchical porous carbon boosting cathode performance of lithium-sulfur batteries. *Mater. Express* **2022**, *12*, 337–346. [CrossRef]
95. Cui, J.Q.; Liu, J.; Chen, X.; Meng, J.S.; Wei, S.Y.; Wu, T.; Wang, Y.; Xie, Y.M.; Lu, C.Z.; Zhang, X.C. Ganoderma Lucidum-derived erythrocyte-like sustainable materials. *Carbon* **2022**, *196*, 70–77. [CrossRef]
96. Liang, J.F.; Xu, Y.Q.; Li, C.; Yan, C.; Wang, Z.W.; Xu, J.F.; Guo, L.L.; Li, Y.F.; Zhang, Y.G.; Liu, H.T.; et al. Traditional Chinese medicine residue-rich micropore-rich porous carbon frameworks as efficient sulfur hosts for high-performance lithium–sulfur batteries. *Dalton Trans.* **2022**, *51*, 129. [CrossRef]
97. Li, X.; Cheng, X.; Gao, M.; Ren, D.; Liu, Y.; Guo, Z.; Shang, C.; Sun, L.; Pan, H. Amylose-Derived Macrohollow Core and Microporous Shell Carbon Spheres as Sulfur Host for Superior Lithium-Sulfur Battery Cathodes. *ACS Appl. Mater. Interfaces* **2017**, *9*, 10717–10729. [CrossRef]
98. Li, J.; Qin, F.; Zhang, L.; Zhang, K.; Li, Q.; Lai, Y.; Zhang, Z.; Fang, J. Mesoporous carbon from biomass: One-pot synthesis and application for Li-S batteries. *J. Mater. Chem. A* **2014**, *2*, 13916. [CrossRef]
99. Zhong, Y.; Xia, X.; Deng, S.; Zhan, J.; Fang, R.; Xia, Y.; Wang, X.; Zhang, Q.; Tu, J. Popcorn Inspired Porous Macrocellular Carbon: Rapid Puffing Fabrication from Rice and Its Applications in Lithium-Sulfur Batteries. *Adv. Energy Mater.* **2018**, *8*, 1701110. [CrossRef]
100. Ren, G.; Li, S.; Fan, Z.-X.; Warzywodac, J.; Fan, Z. Soybean-derived hierarchical porous carbon with large sulfur loading and sulfur content for high-performance lithium-sulfur batteries. *J. Mater. Chem. A* **2016**, *4*, 16507–16515. [CrossRef]
101. Li, Q.; Liu, Y.P.; Yang, L.W.; Liu, Y.H.; Chen, Y.X.; Guo, X.D.; Wu, Z.G.; Zhong, B.H. N, O co-doped chlorella-based biomass carbon modified separator for lithium-sulfur battery with high capacity and long cycle performance. *J. Colloid Interface Sci.* **2021**, *585*, 43–50. [CrossRef] [PubMed]
102. Zhu, L.; Li, J.N.; Xie, H.B.; Shen, X.Q. Biomass-derived high value-added porous carbon as the interlayer material for advanced lithium–sulfur batteries. *Ionics* **2022**, *28*, 3207–3215. [CrossRef]
103. Wang, X.; Yang, L.W.; Li, R.; Chen, Y.X.; Wu, Z.G.; Zhong, B.H.; Guo, X.D. Heteroatom-doped Ginkgo Folium porous carbon modified separator for high-capacity and long-cycle lithium-sulfur batteries. *Appl. Surf. Sci.* **2022**, *602*, 154342. [CrossRef]

104. Guo, Y.C.; Chen, L.X.; Wu, Y.; Lian, J.L.; Tian, Y.; Zhao, Z.Y.; Shao, W.Y.; Ye, Z.Z.; Lu, J.G. Rotten albumen derived layered carbon modified separator for enhancing performance of Li-S batteries. *J. Electroanal. Chem.* **2021**, *895*, 115511. [CrossRef]
105. Shao, H.; Ai, F.; Wang, W.; Zhang, H.; Wang, A.; Feng, W.; Huang, Y. Crab shell-derived nitrogen-doped micro-/mesoporous carbon as an effective separator coating for high energy lithium–sulfur batteries. *J. Mater. Chem. A* **2017**, *5*, 19892–19900. [CrossRef]
106. Gu, X.; Lai, C.; Liu, F.; Yang, W.; Hou, Y.; Zhang, S. A conductive interwoven bamboo carbon fiber membrane for Li-S batteries. *J. Mater. Chem. A* **2015**, *3*, 9502–9509. [CrossRef]
107. Yang, J.; Chen, F.; Li, C.; Bai, T.; Longa, B.; Zhou, X. A free-standing sulfur-doped microporous carbon interlayer derived from luffa sponge for high performance lithium–sulfur batteries. *J. Mater. Chem. A* **2016**, *4*, 14324–14333. [CrossRef]
108. Wei, Y.B.; Cheng, W.H.; Huang, Y.D.; Liu, Z.J.; Sheng, R.; Wang, X.C.; Jia, D.Z.; Tang, X.C. P-Doped Cotton Stalk Carbon for High-Performance Lithium-Ion Batteries and Lithium–Sulfur Batteries. *Langmuir* **2022**, *38*, 11610–11620. [CrossRef]
109. Sabet, S.M.; Sapkota, N.; Chiluwal, S.; Zheng, T.; Clemos, C.M.; Rao, A.M.; Pilla, S. Sulfurized Polyacrylonitrile Impregnated Delignified Wood-Based 3D Carbon Framework for High-Performance Lithium–Sulfur Batteries. *ACS Sustain. Chem. Eng.* **2023**, *11*, 2314–2323. [CrossRef]
110. Liu, H.; Liu, W.L.; Meng, F.C.; Jin, L.Y.; Li, S.L.; Cheng, S.; Jiang, S.D.; Zhou, R.L.; Liu, J.H. Natural nori-based porous carbon composite for sustainable lithium-sulfur batteries. *Sci. China Technol. Sci.* **2022**, *65*, 2380–2387. [CrossRef]
111. Zhu, M.L.; Wu, J.; Li, S.Q. Flower-like Ni/NiO microspheres decorated by sericin-derived carbon for high-rate lithium-sulfur batteries. *Ionics* **2021**, *27*, 5137–5145. [CrossRef]
112. Moreno, N.; Caballero, Á.; Morales, J. Improved performance of electrodes based on carbonized olive stones/S composites by impregnating with mesoporous TiO_2 for advanced Li-S batteries. *J Power Sources* **2016**, *313*, 21–29. [CrossRef]
113. Pang, Q.; Kundu, D.; Cuisinier, M. Surface-enhanced redox chemistry of polysulphides on a metallic and polar host for lithium-sulphur batteries. *Nat. Commun.* **2014**, *5*, 4759. [CrossRef] [PubMed]
114. Luna-Lama, F.; Hernández-Rentero, C.; Caballero, A. Biomass-derived carbon/g-MnO_2 nanorods/S composites prepared by facile procedures with improved performance for Li/S batteries. *Electrochim. Acta* **2018**, *292*, 522–531. [CrossRef]
115. Wang, J.; Wu, L.; Shen, L. CoO embedded porous biomass-derived carbon as dual-functional host material for lithium-sulfur batteries. *J. Colloid Interface Sci.* **2023**, *640*, 415–422. [CrossRef] [PubMed]
116. Lama, F.L.; Caballero, Á.; Morales, J. Synergistic effect between PPy: PSS copolymers and biomass-derived activated carbons: A simple strategy for designing sustainable highperformance Li-S batteries. *Sustain. Energy Fuels* **2022**, *6*, 1568–1586. [CrossRef]
117. Chen, R.X.; Zhou, Y.C.; Li, X.D. Cotton-Derived Fe/Fe_3C-Encapsulated Carbon Nanotubes for High-Performance Lithium–Sulfur Batteries. *Nano Lett.* **2022**, *22*, 1217–1224. [CrossRef]
118. Zhong, M.; Sun, J.C.; Shu, X.Q.; Guan, J.D.; Tong, G.S.; Ding, H.; Chen, L.Y.; Zhou, N.; Shuai, Y. N, P, O-codoped biochar from phytoremediation residues: A promising cathode material for Li-S batteries. *Nanotechnology* **2022**, *33*, 215403. [CrossRef]
119. Tesio, A.Y.; Gómez-Camer, J.L.; Morales, J. Simple and sustainable preparation of non-activated porous carbon from brewing waste for high-performance lithium–sulfur batteries. *ChemSusChem* **2020**, *13*, 3439–3446. [CrossRef]
120. Benítez, A.; Márquez, P.M.; Martín, Á. Simple and Sustainable Preparation of Cathodes for Li–S Batteries: Regeneration of Granular Activated Carbon from the Odor Control System of a Wastewater Treatment Plant. *ChemSusChem* **2021**, *14*, 3915–3925. [CrossRef]
121. Páez Jerez, A.L.; Mori, M.F.; Flexer, V. Water Kefir Grains—Microbial Biomass Source for Carbonaceous Materials Used as Sulfur-Host Cathode in Li-S Batteries. *Materials* **2022**, *15*, 8856. [CrossRef] [PubMed]
122. Benítez, A.; Morales, J.; Caballero, Á. Pistachio Shell-Derived Carbon Activated with Phosphoric Acid: A More Efficient Procedure to Improve the Performance of Li-S Batteries. *Nanomaterials* **2020**, *10*, 840. [CrossRef] [PubMed]
123. Liu, L.Z.; Xia, G.H.; Wang, D. Biomass-derived self-supporting sulfur host with NiS/C composite for high-loading Li-S battery cathode. *Sci. China Technol. Sci.* **2023**, *66*, 181–192. [CrossRef]
124. Lama, F.L.; Marangon, V.; Caballero, Á. Diffusional Features of a Lithium-Sulfur Battery Exploiting Highly Microporous Activated Carbon. *ChemSusChem* **2023**, *16*, e202202095. [CrossRef] [PubMed]
125. Liu, L.Z.; Xia, G.H.; Wang, D. Self-supporting Biomass Li–S Cathodes Decorated with Metal Phosphides–Higher Sulfur Loading, Better Stability, and Longer Cycle Life. *ACS Appl. Energy Mater.* **2022**, *5*, 15401–15411. [CrossRef]

Disclaimer/Publisher's Note: The statements, opinions and data contained in all publications are solely those of the individual author(s) and contributor(s) and not of MDPI and/or the editor(s). MDPI and/or the editor(s) disclaim responsibility for any injury to people or property resulting from any ideas, methods, instructions or products referred to in the content.

Review

Recent Advances in Biomass-Derived Carbon Materials for Sodium-Ion Energy Storage Devices

Mengdan Yan, Yuchen Qin *, Lixia Wang, Meirong Song, Dandan Han, Qiu Jin, Shiju Zhao, Miaomiao Zhao, Zhou Li, Xinyang Wang, Lei Meng * and Xiaopeng Wang *

College of Science, Henan Agricultural University, Zhengzhou 450001, China; yan024711@163.com (M.Y.); wanglixia@henau.edu.cn (L.W.); smr770505@henau.edu.cn (M.S.); handd@henau.edu.cn (D.H.); jinqiukl@henau.edu.cn (Q.J.); zhaoshiju008@henau.edu.cn (S.Z.); zmm2581472022@163.com (M.Z.); lizhou_1995@163.com (Z.L.); wxy1656135@163.com (X.W.)
* Correspondence: qinyuchen@henau.edu.cn (Y.Q.); menglei@henau.edu.cn (L.M.); xpwang@henau.edu.cn (X.W.)

Abstract: Compared with currently prevailing Li-ion technologies, sodium-ion energy storage devices play a supremely important role in grid-scale storage due to the advantages of rich abundance and low cost of sodium resources. As one of the crucial components of the sodium-ion battery and sodium-ion capacitor, electrode materials based on biomass-derived carbons have attracted enormous attention in the past few years owing to their excellent performance, inherent structural advantages, cost-effectiveness, renewability, etc. Here, a systematic summary of recent progress on various biomass-derived carbons used for sodium-ion energy storage (e.g., sodium-ion storage principle, the classification of bio-microstructure) is presented. Current research on the design principles of the structure and composition of biomass-derived carbons for improving sodium-ion storage will be highlighted. The prospects and challenges related to this will also be discussed. This review attempts to present a comprehensive account of the recent progress and design principle of biomass-derived carbons as sodium-ion storage materials and provide guidance in future rational tailoring of biomass-derived carbons.

Keywords: biomass-derived carbon; energy storage; sodium-ion battery; sodium-ion capacitor

1. Introduction

In recent years, the increasing demand for renewable and cleaner energy re-sources such as wind, solar, and wave, to replace traditional fossil energy, has required the development of cost-effective, high-performance, large-scale energy-storage systems. Lithium-ion batteries (LIBs) with the advantages of high energy density have been widely used in the field of energy-storage systems. However, global lithium resources are limited and unevenly distributed, which will make the cost of LIBs dramatically increase shortly. Therefore, low-cost energy-storage systems using naturally abundant raw materials have attracted extensive attention. Na-ion energy storage devices (SESDs), including sodium-ion batteries (SIBs) and sodium ion capacitors (SICs), are recognized as alternatives to LIBs due to the high overall abundance of precursors and better cycle stability and power density [1,2].

The sodium resource is rich in terms of reserves (2.74% of the earth's crust) and ranks fourth among metal elements, has an even geographical distribution, and has a distinctly lower cost than lithium. Sodium and lithium, located in the same main group, have similar physical and chemical properties [3]. Therefore, sodium has attracted increasing attention as a potential alternative to lithium in electrochemical energy-storage systems, especially for grid storage [4–9]. Many fundamental understandings of LIBs and Li-ion capacitors (LICs) provide much experience for the research of sodium-ion energy storage devices. Although sodium and lithium are chemically similar, they still have some differences. For example, Na^+ ions are larger and heavier than Li^+ ions (ion radius 1.02 Å vs. 0.76 Å, weight

23 g mol^{-1} vs. 7 g mol^{-1}) [2,10–12], resulting in worse diffusion kinetics in most host materials and inferior gravimetric/volumetric capacity. The redox potential of Na$^+$/Na is 0.3 V above that of Li$^+$/Li, which reduced operating voltage and energy density [13,14]. Thus, promising electrode materials with suitable working voltage and sodium storage performance can be one of the potential solutions to make up for the loss. In previous studies, great progress was made in the exploration of cathode materials. A large variety of cathode materials (e.g., oxides and polyanionic compounds) can effectively store sodium ions [15,16], although there is the problem of the bottleneck at the anode. Present anode materials used for sodium-ion storage mainly include alloys, metal oxides/sulphides, organic compounds, titanium-based materials, and carbonaceous materials [1]. Among many anode materials, carbon-based materials offer multiple advantages such as higher capacity, lower average sodium storage potential, good conductivity, being non-toxic, and having a low price and hence are recognized as competitive candidates for SESDs [17,18].

Unlike lithium-ion storage, graphite is unfavorable for hosting sodium-ion in graphene layers [19,20]. Various types of amorphous carbon, including soft carbon and hard carbon, have been examined that can be used as potential anode materials for SESDs [19,21–23]. Amorphous carbons with a large interlayer spacing of 0.36–0.4 nm make it a suitable Na hosting. However, soft carbon generally exhibits low initial coulombic efficiency (ICE) and a specific capacity. Hard carbon contains random stacked graphitic layers and a tortuous structure, which allow it to effectively store Na-ion in the micropores and exhibit high reversible capacity and good kinetic performance [24–26]. Owing to the renewability, low cost, and diverse inherent structure of biomass, its use as a precursor for producing amorphous carbon to reduce costs and improve electrochemical performance has become an important research direction to satisfy the requirements for its practical application in the SESDs field.

Biomass refers to widely distributed living and growing organic materials, such as plants [27–29], microorganisms, and animals [30], that have been pyrolyzed at high temperatures to become biomass-derived carbon materials. Reasonable utilization of biomass will realize "turning waste into treasure". Most importantly, biomasses have their unique microstructures and compositions, and the resultant biomass-derived carbons usually retain the diversity of their structures and compositions after pyrolysis (Figure 1) [31–35]. Different structures (e.g., hard carbons, soft carbons, and hybrid carbons), different compositions (e.g., N-doped carbons and other atom doped carbons), and different morphologies (e.g., 1D, 2D and 3D hierarchical structures) of biomass-derived carbons greatly affect their electrochemical performance in SESDs. Table 1 shows the application potential of different types of carbon materials in the field of Na-ion storage. The various morphology, structure, and electrochemical performances of carbonaceous materials have been compared [36–56]. The influence of different types of biomass-based carbon materials on the electrochemical performance of SESDs should be further systematically summarized. The understanding of the biomass-derived carbons and their storage mechanism can be reviewed to guide a rational design for effective electrode materials for SESDs.

Herein, we attempt to provide a comprehensive summary of the latest developments of various biomass-derived carbons used in SESDs, including the principle of sodium ion storage in SIBs and SICs, and the classification of biomass carbon with different structures and compositions. The recent progress and electrochemical performance of different types of biomass-derived carbons will be introduced in detail. This review focuses on the influence of different micromorphology and compositions of biomass-derived carbon on electrochemical performance. Finally, the challenges and perspectives for SESDs have also been proposed. We hope that this literature review can provide references for the rational design of carbon materials toward high-performance SESDs.

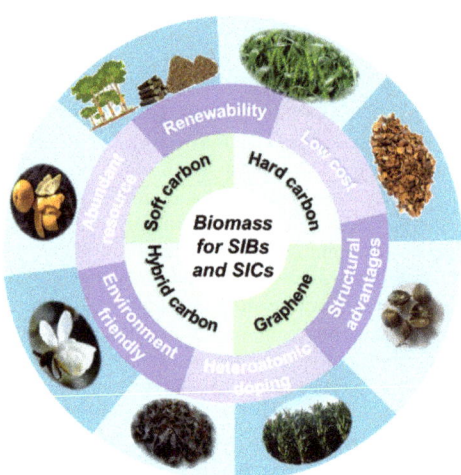

Figure 1. The schematic diagram of biomass-based electrode materials from different precursors, and their inherent advantages.

Table 1. Physical and electrochemical properties of various carbons.

Classification	Precursor	Yield [%]	Morphology	Application	Capacity [mAh g^{-1}] /Current density [mA g^{-1}]	ICE [a] [%]	Refs.
Graphite	/	/	sheets	SIB	284/20	49.53	[36]
	/	/	spherical	SIC	221 [b]/500	/	[37]
Graphene	Graphite	/	sponges	SIB	372.0/50	67.4	[38]
	Graphite	/	nanosheets	SIB	240/200	52	[39]
	Cellulose/chitosan/GO	/	Layers	SIB	395/100	/	[40]
	Graphite	/	folded texture	SIC	115.6/100	/	[41]
	Graphite	/	porous	SIC	420/100	/	[42]
Soft carbon	Coal	/	porous	SIB	267/500	34.0	[43]
	Pitch	70	porous	SIB	268.3/100	82	[44]
Hard Carbon	Kapok	<10	tube	SIB	290/30	80	[45]
	Cucumber stems	/	porous	SIB	337.9/50	64.9	[46]
	Cherry petals	/	nanosheets	SIB	310/20	67.3	[47]
	Pine pollen	/	porous	SIB	370/100	59.8	[48]
	Longan shell	/	porous	SIB	345/100	73	[49]
	Leonardite humic acid	60.73	flakes	SIB	345/100	73	[50]
	gelatin	/	nanosheets	SIB	309/200	84.1	[51]
	Mushroom stalk	/	porous	SIB	305/100	33.8	[5]
	Samara	/	porous	SIB	333.2/100	35.7	[52]
	Chlorella	/	nanoparticle	SIB	436/100	51	[53]
	Carrageenan	/	double-helix	SIB	380/100	56.3	[54]
	Enteromorpha	/	sponge	SIC	362/100	/	[55]
	Carboxymethyl cellulose	/	porous	SIC	322/50	/	[56]

[a] ICE = initial Coulombic efficiency. [b] F g^{-1}.

2. Sodium-Ion Storage Mechanism in Carbonaceous Materials for SESDs

2.1. Configuration and Mechanism of Sodium-Ion Batteries

Sodium-ion batteries have come back into the spotlight, due to their potential cost advantages, since 2010. SIBs consist of cathode materials and anode materials separated by the electrolyte. The energy storage of SIBs is realized through sodium ions shuttling between cathode and anode materials in the charge/discharge process. During the charging process, sodium ions are extracted from cathodes and accommodated into anodes transport through the electrolyte. The reverse reaction occurs in the discharge process. The widely-accepted theory suggests that SIBs operate on a similar intercalation mechanism to LIBs. However, sodium ion storage mechanism in anode materials is more complex than that of lithium insertion in graphite. Graphite is usually not suitable for sodium intercalation in traditional ester-based electrolytes, which leads to unsatisfactory capacity and cyclability of SIBs [57]. Thus, many new materials such as carbonaceous materials, alloys, metal oxides/sulfides, and titanium-based compounds have received widespread attention as anodes for SIBs [58–62]. Amorphous carbons, especially hard carbons, exhibited greatly improved performance and industrial feasibility for SIBs. They are competitive enough for practical application in large-scale grids, compared with other non-carbonaceous materials.

The current carbonaceous materials that have been widely investigated for SIBs mainly include graphite-based carbon materials, soft-carbons, and hard-carbons (Figure 2a–c). Theoretically, it is difficult to intercalate Na ions into graphite interlayer (Figure 2d). The formation energies of sodium-graphite intercalation compounds are not stable [63]. However, taking advantage of the co-intercalation effect of solvated Na ions, natural graphite in some linear ether-based electrolytes exhibited unexpected rate capability and cyclability (100 mAh g^{-1} at 10 A g^{-1}, and 95% capacity retention after 6000 cycles) [64]. The mechanism of Na-ions storage has been proposed so that solvated ions might be adsorbed in graphite lattice rather than atomically bonded to carbon. The expanded graphite with enlarged interlayer distance showed reversible capacity of 284 mA g^{-1} [36]. The experiment confirmed the efficient Na-ions insertion/extraction mechanism in lattice of expanded graphite, which is different from other carbonaceous materials. Reduced graphene oxide has been studied as the anode materials showed a reversible capacity ~450 mAh g^{-1} at a current density of 25 mA g^{-1} [65]. An adsorption mechanism of graphene has been demonstrated for Na-ions storage [56,66]. The structure engineering of graphene to control the defect and surface area is an effective strategy for Na-ions storage.

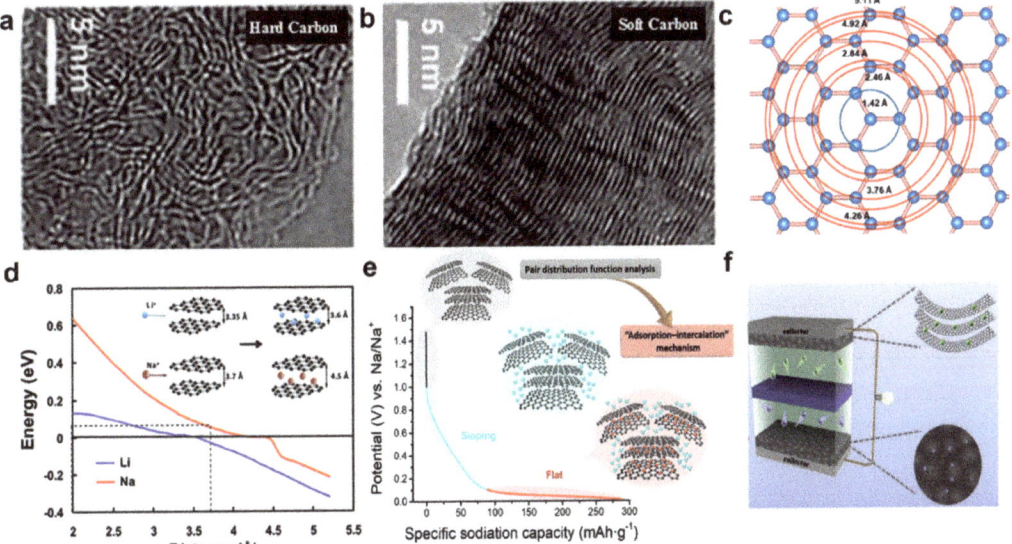

Figure 2. TEM image of hard carbon (a) and soft carbon (b). Reprinted with permission from Ref. [67]. Copyright 2017 American Chemical Society. (c) Schematic of the atomic structure of graphene and near-neighbor interatomic distances. Reprinted with permission from Ref. [68]. Copyright 2019 American Chemical Society. (d) Theoretical energy cost for Na (red curve) and Li (blue curve) ions insertion into carbon as a function of carbon interlayer distance. Reprinted with permission from Ref. [69]. Copyright 2012 American Chemical Society. (e) Schematic illustration of the mechanism for Na storage in hard carbon. Reprinted with permission from Ref. [68]. Copyright 2019 American Chemical Society. (f) Schematic of SIBs. Reprinted with permission from Ref. [70]. Copyright 2017 Wiley-VCH.

Soft carbon is non-graphitic carbon that can be graphitized at high temperatures, whose graphitizable degrees and interlayer distance can be tuned by thermal treatment. Soft carbon, similar to graphite, is always smooth and exhibits less curvature in the graphitic layer. The Na-ion storage capacity of soft carbon was first demonstrated by Doeff and colleagues to show about 90 mAh g^{-1}. The improved capacities and good rate performance have been reported by many groups. For instance, using anthracite and aromatic compounds as precursors, the prepared soft carbons showed high reversible capacities above 200 mAh g^{-1} [43]. However, the Na-storage mechanism of soft carbon revealed that Na-ions adsorption on isolated graphene sheets of soft carbon contributes to the general sloping potential profile and the lack of low-potential plateau, which could cause negative effects on energy density and potential safety of SIBs.

Compared with soft carbon, hard carbon has received more attention as anode in SIBs due to its good kinetic performance [24,46,71]. Hard carbon is non-graphitizable carbon that has curved and unaligned graphitic layers. The crosslink graphitic layers and disordered structures are generally considered to be favorable for Na-storage performance. The typical charge-discharge curves of hard carbon include high-potential sloping region and low-potential plateau region (Figure 2e). These potential regions indicate the reaction of Na-ions with different structures. In 2000, Stevens and Dahn [72] underly the mechanism of Na-ions insertion into hard carbon, suggesting that the high-potential sloping region corresponded to the insertion of Na ions inside graphitic layers and adsorption on the defect sites, whereas Na ions were inserted into nanopores along the low-potential plateau. The alternative storage mechanism is also proposed by Cao and colleagues [69]. They conclude that Na-ions are stored at defect sites of the surface in the sloping part of the potential curve and inserted into graphitic layers and pores in the low-potential plateau region. Based on the understanding of existing sodium storage mechanisms, rational, structural, and defect engineering will be effective strategies for the design of hard carbon with higher performance.

2.2. Configuration and Mechanism of Sodium-Ion Capacitors

Among the sodium-ions based energy storage devices, SIBs and SICs are currently prominent. The battery utilizes intercalation mechanism, leading to high energy density and limited power density. Supercapacitor with adsorption-desorption mechanism provides high power density but limited energy density. SICs have attracted much attention due to their combined advantages of battery and supercapacitor [73,74]. A typical SIC is generally composed of a capacitive cathode and battery-type anode and is the most commonly used type. On the contrary, other configurations of SICs include a battery-type cathode and capacitive anode. The configuration of SICs determines the charge-storage mechanism. This section will focus on the charge-storage behavior of carbonaceous materials in the first widely used configuration.

Dual-carbon SICs were first reported by Kuratanni and colleagues [75]. This SIC includes a battery-type anode of hard carbon and a capacitive cathode of activated carbon. In the case of dual-carbon SICs, the capacitive cathode of activated carbon stores charges through a non-faradaic surface ion adsorption mechanism on the interface of electrode and electrolyte. The faradaic reaction on the surface or near-surface of battery-type anodes, such as hard carbon, provided a higher capacity for SICs. In general, the two kinds of energy storage mechanisms, including electrochemical double-layer capacitance (EDLC) and pseudocapacitive behavior, are responsible for electrochemical reactions of SICs. The EDLC operates on the mechanism of electrolyte ions' adsorption/desorption on the surface of the electrodes. In the case of pseudocapacitive behavior, charge storage mainly originates from the electron-transfer rather than the adsorption of ions. Different from the storage mechanism of SIBs, when carbonaceous materials are used as anode in SICs, the faradaic redox reaction occurs only on the surface of the electrode, and pseudocapacitive intercalation does not produce phase transition. Therefore, SICs based on this hybrid mechanism provide an effective route for integrating high energy and power performance.

3. Diverse Morphology of Biomass-Derived Carbons for SESDs

An in-depth understanding of sodium-ion storage mechanisms in SESDs, especially using carbon-based materials as electrodes, can greatly improve the rational design of better electrode materials. Exciting theoretical studies suggest that morphological and composition engineering (such as defect sites, nano-porosity, and heteroatom doping) are the potential strategies to develop effective electrode materials. Biomass-derived carbons have attracted much attention for application in SESDs in recent years. Different treatment methods and precursor materials will achieve different carbonaceous materials with inherent macroscopic morphologies and structures, resulting in a variety of electrochemical behaviors [76,77]. This section mainly introduces the recent advances of diverse morphologies and structures of biomass-derived carbons used for SESDs. The relationship between the structure/morphology and electrochemical performance of different biomass-derived carbons will be discussed in detail.

Biomass materials naturally possess abundant and diverse macrostructures ranging from zero to three dimensions that can be inherited and evolved by corresponding biomass-derived carbon materials. So far, various nanostructured carbon materials with zero-dimensional (0D) spherical structures, one-dimensional (1D) nanofibers/nanotubes, two-dimensional (2D) nanosheets, and three-dimensional (3D) hierarchical structures have been synthesized. The main motivations to create electrode materials with different dimensional nanostructures are to enlarge exposed active surface areas, broaden activated ionic channels, and accelerate electron conductivity, all of which can significantly promote the electrochemical performance of SESDs. The recent progress and simple classification of biomass-derived carbons based on dimensions are shown below.

3.1. Tubular and Fiber-Shaped Biomass-Derived Carbons

As 1D nanostructures, tubular and fiber-shaped carbon precursors are widely distributed in nature, such as plant tissues [45,78] and bacterial secretions [53]. 1D carbon materials have high aspect ratio to provide fast channels for electron and ion transport. Especially, the tubular structure of carbon also forms an effective permeable inner surface structure that is conducive to Na^+ adsorption, shortening ion diffusion. The path accelerates the diffusion and migration of electrolyte ions from the electrolyte to the inside of the surface of the electrode during charge and discharge [46]. Carbon materials with tubular structures are considered as viable structures for high-performance sodium-ion energy-storage applications.

Li and co-workers [46] reported the preparation of hard carbon materials that maintain uniform microtubule shapes using renewable natural cotton biomass as a precursor (Figure 3a,b). The hollow tubular structure of the hard carbon material is beneficial to the migration of the electrolyte, reduces the diffusion distance of Na^+ ions, and improves the electrochemical performance of the hard carbon. Yu et al. [47] prepared the carbon with micro/nanotubular structure from low-cost *kapok* fibers. During the carbonization process, the *kapok* fibers underwent aromatization, polycondensation, and the formation of short graphite layers, maintaining good morphology with a highly specific surface area. The gradual reduction of carbon micro-nanotubes with the increase of carbonization temperature is expected to reduce the formation of SEI; improve the initial Coulombic efficiency; and finally yield carbon micro-nanotube samples with high reversible capacity, high initial Coulombic efficiency, and excellent rate performance at a carbonization temperature of 1400 °C. Tubular biomass carbon is considered one of the most promising anode candidates for sodium-ion batteries (SIBs) due to its abundant natural resources, low cost, and sustainability, to prepare high-performance sodium storage media with excellent microstructure and morphology. Liu and co-workers [79] proposed to prepare interconnected porous carbon frameworks that maintain a good tubular hierarchical porous structure through KOH activation and Co^{2+}-assisted graphitization (Figure 3c). The interconnected macropores improve the mass transfer between the electrolyte and the active material and shorten the diffusion distance of Na^+ to the inner surface of the carbon framework and improve the excellent electrochemical performance of carbonaceous materials. The above studies provided a reference for fully exploiting the advantages of the inherent 1D morphology of biomass precursors and highlighted the importance of tubular structures in sodium-ion storage.

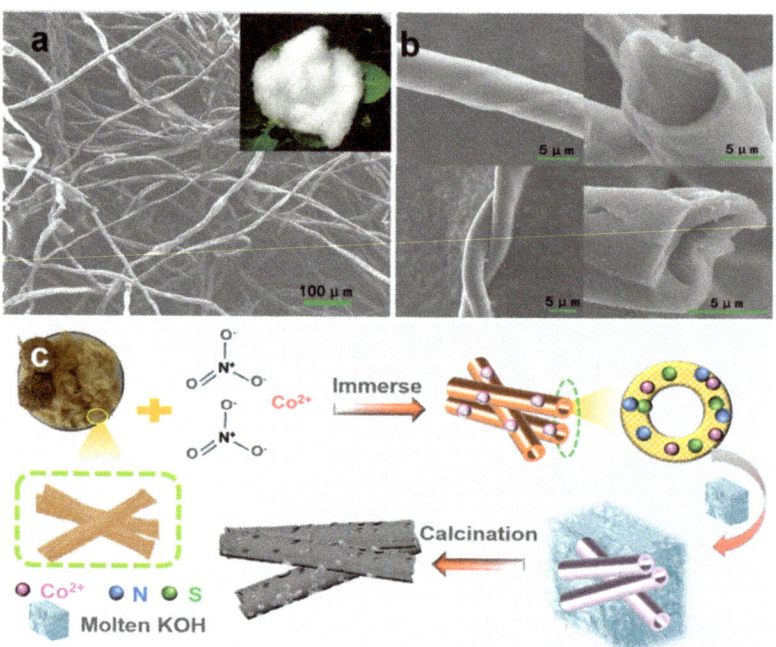

Figure 3. (a) SEM image and photograph of cotton. (b)The magnified SEM images of the carbonized cotton with the detailed structure information. Reprinted with permission from Ref. [46]. Copyright 2017 Wiley-VCH. (c) Illustration of the preparation process of the cross-linked porous sample. Reprinted with permission from Ref. [79]. Copyright 2021 American Chemical Society.

3.2. Sheet-Shaped, Biomass-Derived Carbons

Different precursors and pretreatment methods often have a serious impact on the properties of hard carbons (Figure 4). Using cucumber stems as precursor [51], the carbonization temperature was adjusted to 1000 °C to prepare sheet-like hard carbon anode materials. A reversible capacity of 337.9 mAh g^{-1} and Coulombic efficiency of 99–100% after 500 cycles were obtained. The coiled hard carbon materials were successfully prepared by a two-step method of hydrothermal treatment and pyrolysis at different temperatures using biomass templates. Flake-like hard carbon extracted from pistachio shell precursors were prepared with coiled hard carbon materials at different temperatures [80]. The anode provided a high capacity of 317 mAh g^{-1} with larger interlayer spacing when carbonized at 1000 °C. It showed that carbonization temperature and morphology control have a great influence on the electrochemical performance of hard carbon materials. A hard carbon nanosheet anode made of cherry petal [81] precursor was synthesized at 1000 °C. At a current density of 20 mA g^{-1}, its relatively stable capacity is 300.2 mAh g^{-1} and the initial Coulombic efficiency is 67.3%. The synergistic effects of the mesoporous structure and the increased interlayer distance during the pyrolysis of the precursor of *cherry petals* enhanced the storage capacity of sodium ions. In addition, sheet-like structured hard carbon anode materials were prepared using *oat flakes* [82], biomass-based gelatin [83], and *maple* [84], as precursors to study their sodium storage properties.

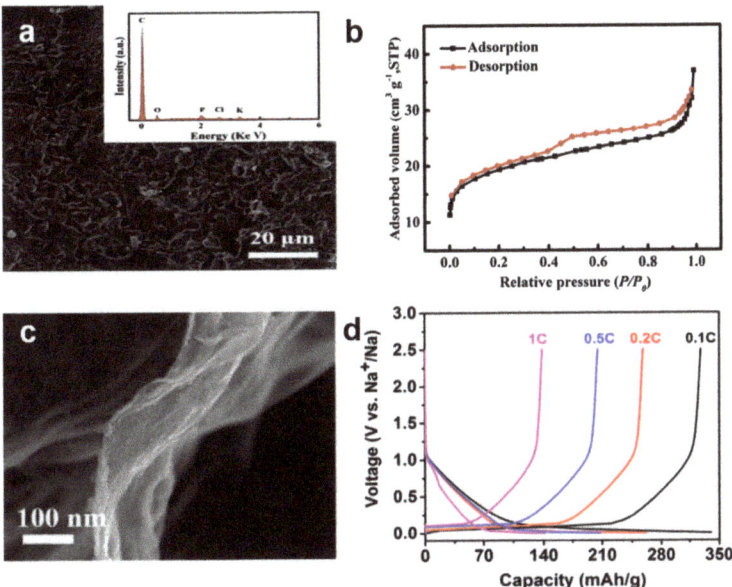

Figure 4. (**a**) Scanning electron microscopy (SEM) image and (**b**) N$_2$ adsorption–desorption isothermal curves of hard-carbon nanosheets from the pyrolysis of oat flakes. Reprinted with permission from Ref. [82]. Copyright 2019 Elsevier. (**c**) SEM image of hard-carbon nanosheets from the pyrolysis of biomass-based gelatin. Reprinted with permission from Ref. [83]. Copyright 2020 Wiley-VCH and (**d**) galvanostatic charge/discharge cycling profiles of maple-derived hard carbon. Reprinted with permission from Ref. [84]. Copyright 2019 Elsevier. 2D nanostructured carbons, with their highly specific surface areas, continuous electron conduction paths, and ability to maintain volume changes during charge and discharge, have attracted much attention for Na-ion storage.

Using oat flakes as the precursor [82], two-dimensional hard carbon was obtained by carbonization (Figure 4a). According to the BET in Figure 4b, the hard carbon contained a large number of mesopores. Compared with other carbon materials with highly specific surface area, the formation of SEI film is limited, thereby improving the Coulombic efficiency. Jin synthesized N, B co-doped carbon nanosheets [83] using biomass-based gelatin as the precursor and boronic acid as the template (Figure 4c). The synergistic effect of heteroatom doping and 2D structure with highly specific surface area enhances the capacity and rate performance of Na-ion batteries. Wang et al. [84] developed a hard carbon extracted from maple tree as the anode of the battery, as shown in Figure 4d, which achieved a capacity of 337 mAh g^{-1} at 0.1 C. The initial Coulombic efficiency was as high as 88.0%. The capacity remained at 92.3% after 100 cycles at 0.5 C. Again, it is proved that biomass-derived hard carbon has the advantages of large capacity and high Coulombic efficiency. The hard carbon materials showed interlayer spacing suitable for Na ion insertion, with highly defective sites and specific surface area, which can effectively improve the weight/volume capacity and superior cycling stability.

3.3. 3D Hierarchical Structures of Biomass-Derived Carbon

3D carbons have highly interconnected network, abundant active edges, defects, shortened ion/electron channels, accelerated dynamic ion transfer, and good electrical contacts, thus generally possessing excellent electrochemical performance [85,86]. The 3D structure can not only provide a continuous electron path but also allows the electrolyte to penetrate the whole structure and facilitate the sodium ion transport between the electrode/electrolyte interface by shortening the diffusion path to ensure good electrical contact to facilitate ion transport [87]. However, due to the natural hydrophobicity of the carbon matrix surface, its effective specific surface area is greatly limited by the highly developed single microporous structure. Therefore, improving the accessible specific surface area is an effective way to improve the electrochemical performance.

Hierarchical porous structures including macropores, mesopores, and micropores can increase the contact area of the electrolyte with the electrode material [48,88]. Among them, macropores (pore size > 50 nm) act as ion buffer reservoirs during the charging and discharging process, which is conducive to the transport of substances and the proximity of ions to adsorption sites, effectively shortening the diffusion path of electrolyte ions and ensuring the rapid transmission and diffusion of electrolyte ions. The mesopores (50 nm > pore size > 2 nm) increase the contact area between the electrode and the electrolyte, which can act as a reservoir for the electrolyte, reducing the ion diffusion resistance and speeding up the ion transfer pathway. Micropores (pore size < 2 nm) can generate a higher surface area, provide a large number of active sites, and thus increase the specific capacitance. This unique hierarchical pore structure carbon with a highly specific surface area provides an efficient route for the penetration and transport of electrolyte ions and is expected to be an excellent anode material for Na-ion energy-storage systems.

Zhang synthesized a unique "honeycomb" structure carbon (Figure 5a) that used pine pollen as a precursor [89]. Their hollow structure and robust framework reduced volumetric strain during Na-ions intercalation/deintercalation and rapidly accommodated ions/electrons for better rate performance. The initial discharge capacity can reach 370 mA h g^{-1} at a current density of 0.1 Ag^{-1}. After cycling 200 times, the reversible capacity also stabilized at 203.3 mA h g^{-1} with a retention rate of 98%. The high capacity and long lifetime of SIBs mainly benefit from the biomimetic honeycomb structure and robust carbon framework. These properties can accelerate ion transport, shorten charge diffusion paths, and effectively buffer volume expansion. In addition, the carbon film prepared by carbonization of *Osmanthus fragrans* leaves provided a conductive framework and provided better nucleation conditions for the in-situ growth of transition metal phosphides. The Fe-doped CoP has a flower-like structure composed of intersecting nanoflakes of 200–300 nm (Figure 5b). Due to the large surface area of the flower shape, which provided more active sites for the intercalation of Na ions and the strong coupling between Fe-doped CoP and the carbon film, the Na ion storage performance was significantly improved. During the electrochemical reaction, the carbon film with high conductivity was beneficial to electron transfer. Fe-doped CoP/C maintained a specific capacity of 324 mAh g^{-1} after 500 cycles, with a Coulombic efficiency of about 99%.

Lang and co-workers [52] constructed a novel sodium-ion hybrid battery (SHB) by introducing adsorption-type hierarchical porous amorphous carbon (HPAC) as the anode material. SHB has good rate performance and long-cycle cycling performance at 2C, and the capacity retention is 87% after 1000 cycles at 10C. Qin et al. [90] studied the pore size distribution of wheat straw after carbonization at different temperatures. At 900 °C, the pore size distribution was mainly about 3.8 nm, and the specific surface area was 1295.21 m^2 g^{-1}. Pore-rich biochar facilitates electrolyte diffusion and Na ion transport and can expand the interlayer spacing of graphite to de/intercalate Na ions, improving battery performance with higher stable reversible capacity. In addition, Luo and colleagues [88] prepared a novel hierarchically structured porous carbon material (Figure 5c–e) with macropores, mesopores, and micropores by activating longan shells. The highly specific surface area and excellent porous structure ensure its good sodium-ion storage and cycling performance.

Biochar has a variety of microstructures, and different microstructures largely affect the electrochemical properties of electrochemically active sites and surfaces [91]. The performance of Na-ion batteries can be tuned by increasing the specific surface area by controlling the microstructure of biomass carbon [30,68,92–94]. Hu et al. [95] prepared carbon nanosheets with pinecone shells as a precursor. Under the synergistic effect of KOH and melamine, discrete carbon nanosheets with large specific surface area and rich porosity can be prepared. This structure ensures its excellent energy storage, showing excellent rate performance and excellent cycle performance. A porous tubular carbon material was synthesized by using sycamore single villi as a precursor [96]. This electrode material had a specific capacitance of 836.4 F g^{-1} at a current density of 0.2 A g^{-1} and retained a specific capacitance of 92.96% after 10,000 cycles at a current density of 10 A g^{-1}. Wang et al. prepared a hierarchical porous carbon material containing a large number of micropores and a small number of mesopores using Paulownia husk as a raw material. The obtained biomass char has a surface area of 1914.4 m^2 g^{-1}. The porous structure exhibits excellent rate capability, with a discharge capacitance of 100 mAh g^{-1} at a current density of 1 A g^{-1} after 100 cycles. Junke Ou and co-workers [97] used human hair as raw material to prepare nitrogen-doped porous carbon, which can provide a high capacity of 308 mAhg^{-1} at a current density of 100 mAg^{-1}. In conclusion, the unique hierarchical microstructure increases the electrode-electrolyte contact area, modulates the volume expansion during cycling, and significantly improves the electrochemical performance.

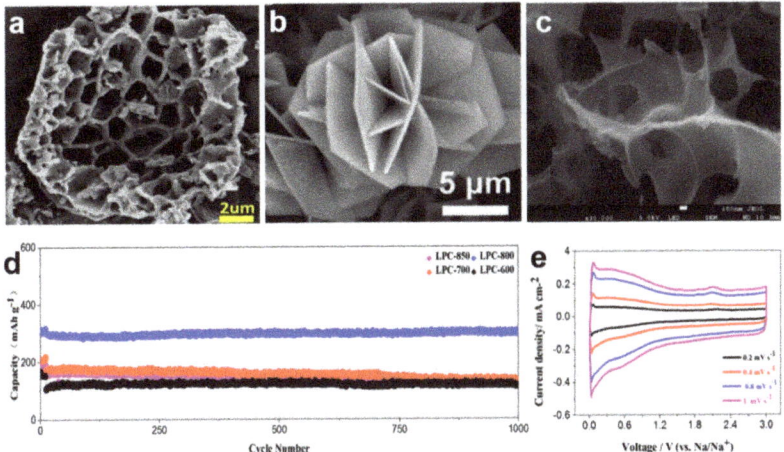

Figure 5. (a) SEM images of "honeycomb" structure carbon. (b) SEM images of CoP/C. Reprinted with permission from Ref. [89]. Copyright 2018 American Chemical Society. (c) SEM images of porous carbons from longan shells. (d) Cycling stability performance of different samples at a current density of 5 A g^{-1}. (e) CV curves of porous carbons. Reprinted with permission from Ref. [88]. Copyright 2018 Elsevier.

4. Different Structures and Components of Biomass-Derived Carbons for SESDs

4.1. Degree of Graphitization

Carbon materials are mainly divided into graphite, graphene, soft carbon, and hard carbon. Graphite is electrochemically less active for Na storage due to thermodynamic problems. To further improve electrochemical sodium-storage properties, graphene was applied as the anode material and performed better than graphite. Nitrogen-doped 3D graphene foams have been prepared to deliver a high initial reversible capacity of 852.6 mAh g^{-1} at 1 C [98]. However, the low initial Coulombic efficiency (~18.5%) of graphene owing to the irreversible Na$_2$O formation on graphene surface limits its practical application [46]. Non-graphitic carbon materials, including soft carbon and hard carbon, have been widely used as anode materials for SIBs. However, the capacity of soft carbon is lower than that of hard carbon. Among the many carbon materials, hard carbon has attracted extensive attention. Amorphous regions in hard carbon materials are often embedded in graphite layers, forming a strong cross-linked network that makes the structure more rigid. With pores between randomly arranged graphite crystallites, the structure affects storage sites, and diffusion kinetics. So, the electrochemical performance can be changed by the degree of graphitization [99–101]. Many studies have reported that with the increase of carbonization temperature [28,102–106], the degree of graphitization will increase, the defects of hard carbon will decrease, and the structure will gradually become ordered.

In 2019, Stevanus [105] carbonized fir wood under different high-temperature conditions in Figure 6. With the increase of carbonization temperature, the hard carbon gradually became ordered from high defects, and the spacing became smaller, which was not conducive to the insertion of Na$^+$ and was not thermodynamically stable. In addition, biomass-derived carbon contains randomly arranged graphite layers and disordered layered nanodomains, which cannot be fully graphitized even at temperatures above 3000 °C. Even if the carbonization temperature is increased, the material will have defects, but the number of defects will decrease. Cao [93] prepared rapeseed into layered hard carbon. When the spacing is 0.39 nm, it can ensure the insertion and extraction of Na$^+$ when it is used as a negative electrode material for SIBs, so it has excellent electrochemical performance. With the increase of temperature, the capacity increases first, because the carbon particles are gradually connected tightly, which is conducive to the transport of electrons. However, the capacity starts to drop after 700 °C, mainly because of the formation of stacked blocks, which hinder the electron transport. Although the partial carbonization of hard carbon can effectively improve the reversible capacity of Na ion intercalation, an overly high temperature will further reduce the interlayer spacing, reduce the pore volume, and cannot accommodate Na ion insertion or adsorption to the pore surface, resulting in a decrease in capacity [107,108].

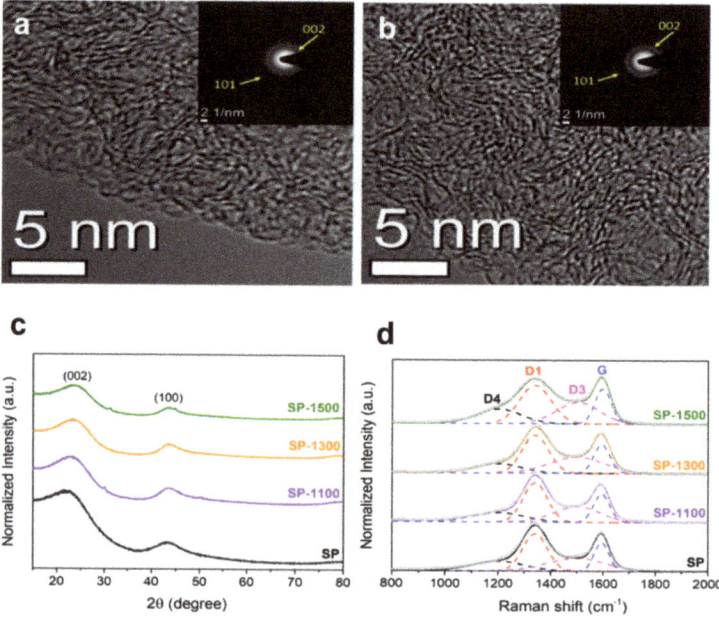

Figure 6. TEM images (**a**,**b**), XRD patterns (**c**) and Raman spectra (**d**) of hard carbons derived from fir wood. Reprinted with permission from Ref. [105]. Copyright 2019 Elsevier.

4.2. Heteroatom Doping

In order to improve the Na storage of carbon materials to meet the needs of energy storage in various aspects, using carbon materials doped with heteroatoms (N, S, P, B, O, etc.) is an effective strategy [48,88,92,109]. Heteroatom doping usually can improve conductivity, increase active sites, and expand interlayer spacing. The intercalation/deintercalation of sodium ions in the electrochemical process is promoted, and the reversible capacity of bio-based carbon is several times larger than the theoretical capacity of graphite [110].

Nitrogen-rich doped carbon spheres were synthesized using onion waste as the precursor [111]. Nitrogen doping enhances the extension of the interlayer distance, which is favorable for the insertion/extraction of large Na ions (Figure 7a,b). The considerable amorphous structure and heteroatom doping enhance the electrical conductivity and active sites of the material; the reversible capacity is also enhanced, and the structural deformation during cycling is alleviated. In addition, the synergistic effect of binary/multiple heteroatoms can not only obtain larger interlayer spacing and provide additional charge storage capacity by Na^+ binding to relevant defects or functional groups, it can also contribute to the conduction band of carbon by providing additional free electrons, resulting in higher electrical conductivity and improved electrochemical performance [92,97,112,113]. Jin synthesized N, B-doped carbon nanosheets by a one-step carbonization method using biomass gelatin as the precursor and boronic acid as the template [83]. The addition of N, B will produce more defects and disordered structures. The differential charge density and density of states are calculated by building a heteroatom doping model, indicating superior electrochemical performance. In addition, Liu et al. [5] synthesized a N, O co-doped porous carbon with uniform ultra-micropores. The presence of N atoms helps to improve the electrical conductivity, while the oxygen functional group can improve the wettability of the electrode material and promote better contact between the active material and the electrolyte ions to improve the electrochemical performance (Figure 7c–h). Overall, heteroatom-doped carbon materials are considered promising anode candidates for SESDs.

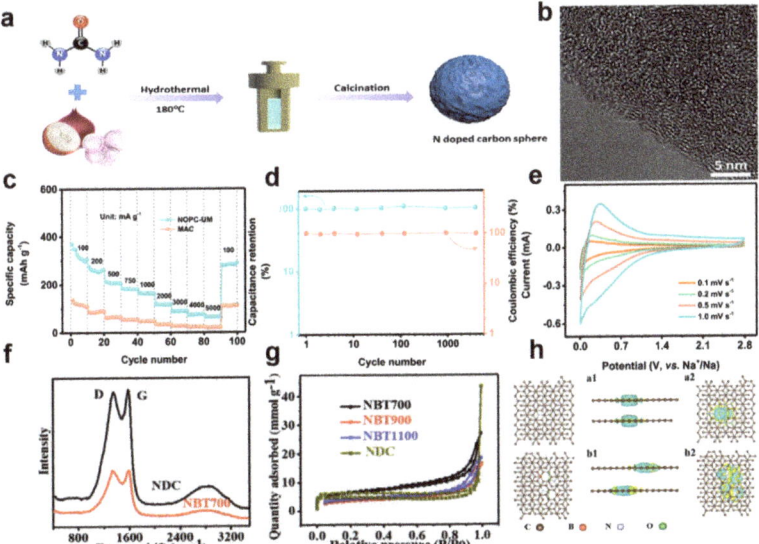

Figure 7. (a) Schematic illustration of the synthesis of N-doped carbon sphere. (b) TEM image of carbon sphere. Reprinted with permission from Ref. [111]. Copyright 2020 Elsevier. (c) The rate performance of NOPC and MAC anodes at various current densities. (d) Cycling stability of hard carbon anode measured at 1000 mA g^{-1}. (e) CV curves measured of hard carbon anode. Reprinted with permission from Ref. [5]. Copyright 2020 Elsevier. (f) The Raman spectra of NDC and NBT. (g) The N_2 adsorption–desorption isotherms at different temperature. (h) The N doping model for NDC and N, B co-doping model for NBT, respectively. Reprinted with permission from Ref. [83]. Copyright 2021 Wiley-VCH.

4.3. Hybridization of Biomass-Derived Carbon and Metal Compounds

Transition metal oxides, sulfides, and phosphides have high theoretical capacities [114]. However, as electrode materials for sodium-ion energy-storage systems, the volume changes during the charge and discharge process, and the electrodes are severely pulverized, resulting in low energy storage density and poor cycle performance. To improve these problems, the most effective method is to hybridize transition metal oxides, sulfides, or phosphides with carbon materials [53,54,115,116]. This not only provides a conductive network for electron transfer but also acts as a stable structural matrix to accommodate volume changes during cycling. Among carbon materials, biomass carbon has high thermal/chemical stability, unique morphological structure, and high electrical conductivity, especially biomass containing different functional groups, such as hydroxyl and amino. These functional groups are easily combined with metals, so biomass carbon becomes the best candidate for metal composites.

In 2016, Yang and his team first reported the double-helix three-dimensional metal sulfide/carbon aerogel nanostructures combined with carrageenan-metal hydrogel for high-performance sodium-ion storage (Figure 8a). Using it as an electrode material, it showed a high reversible specific capacity of 280 mAh g^{-1}, even after 200 cycles at a current density of 0.5 Ag^{-1} [117]. The carbon skeleton in this nanostructure not only facilitates the fast charge transfer reaction but also enhances the mechanical properties of FeS nanoparticles and buffers their volume changes, thereby extending the electrode cycle life. Moreover, Ni_3S_4 nanoparticles were embedded in porous carbon (Figure 8b,c) [118]. As a negative electrode for Na-ion batteries, it maintained a capacity of 297 mAh g^{-1} for 100 cycles at a current density of 1 A g^{-1}. Its excellent electrochemical performance benefits from porous carbon inhibit the accumulation of Ni_3S_4 nanoparticles during the synthesis. In addition, the addition of Ni_3S_4 nanocrystals accelerates the transport of sodium ions, thereby improving the capacity and reaction kinetics. In conclusion, metals are intercalated into biomass-derived carbon as active materials, providing more active sites, while biomass carbon limits the volume change during the intercalation/deintercalation of sodium ions through internal stress. The synergy between the two together improves the stability and electrochemical performance of Na-ion batteries [32,114,118–120].

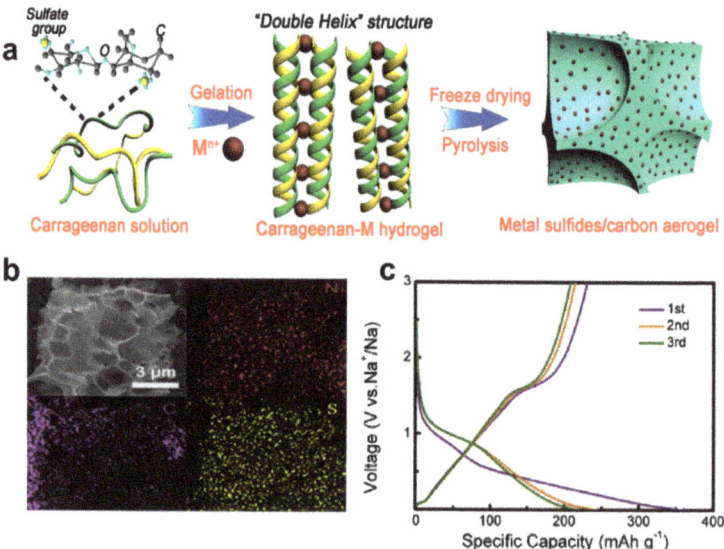

Figure 8. (a) Schematic illustration of double helix structure. Reprinted with permission from Ref. [117]. Copyright 2021 Wiley-VCH. (b) The EDS mappings of Ni, C, and S elements. (c) The Ni_3S_4/C cycled at the 1st, 2nd, and 3rd from 0.01 V to 3.0 V (vs. Na$^+$/Na) at a current density of 1 A g^{-1}. Reprinted with permission from Ref. [118]. Copyright 2019 Elcevier.

5. Conclusions, Challenges, and Outlook

Different biomass carbon materials with their inherent structure and chemical advantages have opened up a new key field for the design and preparation of electrodes for SESDs. Due to its wide range of sources, non-toxicity, and chemical stability, the application potential of biomass carbon materials in sodium-ion energy-storage systems is believed to help meet future environmental needs. This article reviews the latest developments in the application of sodium-ion batteries and sodium-ion capacitors with biochar materials of various structures, morphologies, and chemical compositions, and the factors that affect electrochemical performance. This provides references for the future tailoring of advanced carbon materials for SESDs.

Although biomass carbon electrodes have great potential in SIBs and SICs, there are still some problems that need to be solved before they can be successfully commercialized and widely used. Their further challenges are mainly as follows. 1. Biomass carbon source materials are difficult to use to achieve high-quality, uniform mass production, owing to the diversity of geography and environment. 2. The relatively low carbon yield of precursors limits its industrialization [37–39,41,50]. For this case, industrial products derived from biomass can be utilized for synthesizing uniform electrode materials for SESDs. This will be one of the feasible methods for practical production. 3. The impurities in biomass carbon are usually detrimental to the electrochemical performance of SESDs. Thus, leaching combined with rinsing is an effective strategy to decrease the impurities of biomass carbon. 4. Due to the limitations of synthesis equipment and technologies such as impurity cleaning and vacuum filtration, the continuous preparation of large-scale biomass electrodes is still a challenge worthy of attention. Therefore, the development of scale-up technology for preparation is an important issue as well. 5. An even more challenging aspect is the electrochemical shortcomings, such as the low initial Coulombic efficiency in SIBs, routine cycle performance, unsuitable voltage plateau, and poor energy density. Nanostructured strategies, such as structure/composition engineering, doping, and hybridization with active materials, are demonstrated to be the best potential choices to enhance the performance in SIBs and SICs. It is important to further elucidate the Na-ions storage mechanisms and better explore biomass-based materials with controllable microstructures.

Lastly, the existing studies of new biomaterial systems and synthesis strategies have provided a new platform for the development of SESDs, and a lot of work is still needed in the future.

Author Contributions: Writing—original draft preparation, M.Y. and X.W. (Xiaopeng Wang); formal analysis, L.W., Z.L., M.S., and D.H.; literature search, M.Y., Q.J., S.Z., and X.W. (Xinyang Wang); investigation, M.Z. and X.W. (Xiaopeng Wang); resources, Y.Q. and X.W. (Xiaopeng Wang); data curation, M.Y.; writing—review and editing, L.M. and X.W. (Xiaopeng Wang); supervision, Y.Q. and X.W. (Xiaopeng Wang); project administration, L.W. and X.W. (Xiaopeng Wang); funding acquisition, L.W. and X.W. (Xiaopeng Wang). All authors have read and agreed to the published version of the manuscript.

Funding: This work was funded by National Natural Science Foundation of China (grant numbers 21805072), the Top-Notch Talent Program of Henan Agricultural University (grant number 30500738), and the Young Talent Program of Henan Agricultural University (grant number 30500601).

Data Availability Statement: Not applicable.

Acknowledgments: We acknowledge the support from Henan Agricultural University.

Conflicts of Interest: The authors in this paper declare that we have no known competing financial interests or personal relationships that could have influenced the work reported in this paper.

References

1. Chen, S.; Chao, W.; Shen, L.; Zhu, C.; Huang, Y.; Kai, X.; Maier, J.; Yan, Y. Challenges and Perspectives for NASICON-Type Electrode Materials for Advanced Sodium-Ion Batteries. *Adv. Mater.* **2017**, *29*, 1700431. [CrossRef] [PubMed]
2. Peng, Z.; Qin, F.; Lei, Z.; Wang, M.; Kai, Z.; Lai, Y.; Jie, L. Few-layered MoS$_2$/C with expanding d-spacing as a high-performance anode for sodium-ion batteries. *Nanoscale* **2017**, *9*, 12189–12195.
3. Palomares, V.; Serras, P.; Villaluenga, I.; Hueso, K.B.; Carretero-González, J.; Rojo, T. Na-ion batteries, recent advances and present challenges to become low cost energy storage systems. *Energy Environ. Sci.* **2012**, *5*, 5884–5901. [CrossRef]
4. Ge, P.; Zhang, C.; Hou, H.; Wu, B.; Zhou, L.; Li, S.; Wu, T.; Hu, J.; Mai, L.; Ji, X. Anions induced evolution of Co$_3$X$_4$ (X=O, S, Se) as sodium-ion anodes: The influences of electronic structure, morphology, electrochemical property. *Nano Energy* **2018**, *48*, 617–629. [CrossRef]
5. Yu, P.; Zhang, W.; Yang, Y.; Zheng, M.; Hu, H.; Xiao, Y.; Liu, Y.; Liang, Y. Facile construction of uniform ultramicropores in porous carbon for advanced sodium-ion battery. *J. Colloid Interface Sci.* **2021**, *582*, 852–858. [CrossRef]
6. Wang, L.; Wei, Z.; Mao, M.; Wang, H.; Li, Y.; Ma, J. Metal Oxide/Graphene Composite Anode Materials for Sodium-Ion Batteries. *Energy Storage Mater.* **2018**, *16*, 434–454. [CrossRef]
7. Cao, X.; Pan, A.; Yin, B.; Fang, G.; Wang, Y.; Kong, X.; Zhu, T.; Zhou, J.; Cao, G.; Liang, S. Nanoflake-constructed porous Na$_3$V$_2$(PO$_4$)$_3$/C hierarchical microspheres as a bicontinuous cathode for sodium-ion batteries applications. *Nano Energy* **2019**, *60*, 312–323. [CrossRef]
8. Chen, J.; Fan, X.; Ji, X.; Gao, T.; Hou, S.; Zhou, X.; Wang, L.; Wang, F.; Yang, C.; Chen, L. Intercalation of Bi nanoparticles into graphite results in an ultra-fast and ultra-stable anode material for sodium-ion batteries. *Energy Environ. Sci.* **2018**, *11*, 1218–1225. [CrossRef]
9. Wang, Q.; Zhao, C.; Lu, Y.; Li, Y.; Zheng, Y.; Qi, Y.; Rong, X.; Jiang, L.; Qi, X.; Shao, Y.; et al. Advanced Nanostructured Anode Materials for Sodium-Ion Batteries. *Small* **2017**, *13*, 1701835. [CrossRef]
10. Wei, C.; Liang, H.; Qi, Z.; Shao, L.; Wang, Z. Enhanced electrochemical properties of lithium cobalt titanate via lithium-site substitution with sodium. *Electrochim. Acta* **2015**, *174*, 1202–1215.
11. Luo, J.Y.; Wang, Y.G.; Xiong, H.M.; Xia, Y.Y. Ordered Mesoporous Spinel LiMn$_2$O$_4$ by a Soft-Chemical Process as a Cathode Material for Lithium-Ion Batteries. *J. Cheminfom.* **2007**, *38*, 4791–4795.
12. Padhi, A.K. Phospho-olivines as Positive-Electrode Materials for Rechargeable Lithium Batteries. *J. Electrochem. Soc.* **1997**, *144*, 1188–1194. [CrossRef]
13. Xiang, X.; Zhang, K.; Chen, J. Recent Advances and Prospects of Cathode Materials for Sodium-Ion Batteries. *Adv. Mater.* **2015**, *27*, 5343–5364. [CrossRef]
14. Liu, H.; He, B.; Xiang, W.; Li, Y.C.; Guo, X. Synergistic effect of uniform lattice cation/anion doping improved structural and electrochemical performance stability for Li-rich cathode materials. *Nanotechnology* **2020**, *31*, 455704. [CrossRef]
15. Gezovi, A.; Vujkovi, M.J.; Milovi, M.; Grudi, V.; Mentus, S. Recent developments of Na$_4$M$_3$(PO$_4$)$_2$(P$_2$O$_7$) as the cathode material for alkaline-ion rechargeable batteries: Challenges and outlook. *Energy Storage Mater.* **2021**, *37*, 243–273. [CrossRef]
16. Xian, L.; Li, M.; Qiu, D.; Qiu, C.; Yue, C.; Wang, F.; Yang, R. P3-type layered Na$_{0.26}$Co$_1$-xMn$_x$O$_2$ cathode induced by Mn doping for high-performance sodium-ion batteries. *J. Alloys Compd.* **2022**, *905*, 163965. [CrossRef]
17. Peters, J.; Buchholz, D.; Passerini, S.; Weil, M. Life cycle assessment of sodium-ion batteries. *Energy Environ. Sci.* **2016**, *9*, 1744–1751. [CrossRef]
18. Zhang, J.; Chen, Z.; Wang, G.; Hou, L.; Yuan, C. Eco-friendly and scalable synthesis of micro-/mesoporous carbon sub-microspheres as competitive electrodes for supercapacitors and sodium-ion batteries. *Appl. Surf. Sci.* **2020**, *533*, 147511. [CrossRef]
19. Thomas, P.; Billaud, D. Electrochemical insertion of sodium into hard carbons. *Electrochim. Acta* **2003**, *47*, 3303–3307. [CrossRef]
20. Divincenzo, D.P.; Mele, E.J. Cohesion and structure in stage-1 graphite intercalation compounds. *Phys. Rev. B* **1985**, *32*, 2538–2553. [CrossRef]
21. Zhou, X.; Guo, Y.G. Highly Disordered Carbon as a Superior Anode Material for Room-Temperature Sodium-Ion Batteries. *ChemElectroChem* **2014**, *1*, 83–86. [CrossRef]
22. Xiao, L.; Cao, Y.; Henderson, W.A.; Sushko, M.L.; Shao, Y.; Xiao, J.; Wang, W.; Engelhard, M.H.; Nie, Z.; Liu, J. Hard carbon nanoparticles as high-capacity, high-stability anodic materials for Na-ion batteries. *Nano Energy* **2016**, *19*, 279–288. [CrossRef]
23. Bo, H.; Kan, W.; Liheng, W.; Shu-Hong, Y.; Markus, A.; Maria-Magdalena, T. Engineering carbon materials from the hydrothermal carbonization process of biomass. *Adv. Mater.* **2020**, *22*, 813–828.
24. Lu, H.; Ai, F.; Jia, Y.; Tang, C.; Zhang, X.; Huang, Y.; Yang, H.; Cao, Y. Exploring Sodium-Ion Storage Mechanism in Hard Carbons with Different Microstructure Prepared by Ball-Milling Method. *Small* **2018**, *14*, 1802694. [CrossRef]
25. Wu, Z.Y.; Ma, C.; Bai, Y.L.; Liu, Y.S.; Wang, S.; Wei, X.; Wang, K.X.; Chen, J.S. Rubber-Based Carbon Electrode Materials Derived from Dumped Tires for Efficient Sodium-ion Storage. *Dalton Trans.* **2018**, *47*, 4885–4892. [CrossRef]
26. Xiao, L.; Lu, H.; Fang, Y.; Sushko, M.L.; Cao, Y.; Ai, X.; Yang, H.; Liu, J. Low-Defect and Low-Porosity Hard Carbon with High Coulombic Efficiency and High Capacity for Practical Sodium Ion Battery Anode. *Adv. Eng. Mater.* **2018**, *8*, 1703238. [CrossRef]
27. Mohan, E.H.; Anandan, S.; Appa Rao, B.V. Neem Leaf-derived Micro and Mesoporous Carbon as an Efficient Polysulfide Inhibitor for Sulfur Cathode in a Li-S Battery. *Chem. Lett.* **2019**, *48*, 62–64. [CrossRef]
28. Izanzar, I.; Dahbi, M.; Kiso, M.; Doubaji, S.; Komaba, S.; Saadoune, I. Hard carbons issued from date palm as efficient anode materials for sodium-ion batteries. *Carbon* **2018**, *137*, 165–173. [CrossRef]

29. Wang, H.; Yu, W.; Jing, S.; Mao, N.; Chen, S.; Wei, L. Biomass derived hierarchical porous carbons as high-performance anodes for sodium-ion batteries. *Electrochim. Acta* **2016**, *188*, 103–110. [CrossRef]
30. Deng, J.; Li, M.; Wang, Y. Biomass-derived carbon: Synthesis and applications in energy storage and conversion. *Green Chem.* **2016**, *18*, 4824–4854. [CrossRef]
31. Yin, Y.; Yan, S.; Ni, Z.; Jin, C.; Zhao, L. Economical synthesized Mn_3O_4/biomass-derived carbon from vegetable sponge composites and its excellent supercapacitive behavior. *Biomass Convers. Biorefin.* **2021**, 1–10. [CrossRef]
32. Chen, K.; Li, G.; Wang, Y.; Chen, W.; Mi, L. High loading FeS_2 nanoparticles anchored on biomass-derived carbon tube as low cost and long cycle anode for sodium-ion batteries. *Green Energy Environ.* **2020**, *5*, 50–58. [CrossRef]
33. Liu, N.; Wang, Y.; Zhang, X.; He, E.; Zhang, Z.; Yu, L. Litchi-like porous carbon nanospheres prepared from crosslinked polymer precursors for supercapacitors and electromagnetic wave absorption. *Chem. Eng. J.* **2021**, *416*, 128926. [CrossRef]
34. Yang, M.; Dai, J.; He, M.; Duan, T.; Yao, W. Biomass-derived carbon from Ganoderma lucidum spore as a promising anode material for rapid potassium-ion storage. *J. Colloid Interface Sci.* **2020**, *567*, 256–263. [CrossRef]
35. Yu, Z.; Zhao, Z.; Peng, T. Coralloid carbon material based on biomass as a promising anode material for lithium and sodium storage. *New J. Chem.* **2021**, *45*, 7138–7144. [CrossRef]
36. Yang, W.; Kai, H.; Zhu, Y.; Han, F.; Wang, C. Expanded graphite as superior anode for sodium-ion batteries. *Nat. Commun.* **2014**, *5*, 4033.
37. Zahra, H.; Sawada, D.; Guizani, C.; Ma, Y.; Kumagai, S.; Yoshioka, T.; Hummel, M. Close packing of cellulose and chitosan in regenerated cellulose fibers improves carbon yield and structural properties of respective carbon fibers. *Biomacromolecules* **2020**, *21*, 4326–4335. [CrossRef]
38. Sun, Y.; Lu, P.; Liang, X.; Chen, C.; Xiang, H. High-yield microstructure-controlled amorphous carbon anode materials through a pre-oxidation strategy for sodium ion batteries. *J. Alloys Compd.* **2019**, *786*, 468–474. [CrossRef]
39. Hassan, M.M.; Schiermeister, L.; Staiger, M.P. Sustainable production of carbon fiber: Effect of cross-linking in wool fiber on carbon yields and morphologies of derived carbon fiber. *ACS Sustain. Chem. Eng.* **2015**, *3*, 2660–2668. [CrossRef]
40. Yu, F.; Liu, Z.; Zhou, R.; Tan, D.; Wang, H.; Wang, F. Pseudocapacitance contribution in boron-doped graphite sheets for anion storage enables high-performance sodium-ion capacitors. *Mater. Horiz.* **2018**, *5*, 529–535. [CrossRef]
41. Zhu, Y.; Chen, M.; Li, Q.; Yuan, C.; Wang, C. High-yield humic acid-based hard carbons as promising anode materials for sodium-ion batteries. *Carbon* **2017**, *123*, 727–734. [CrossRef]
42. Li, P.; Jeong, J.Y.; Jin, B.; Zhang, K.; Park, J.H. Vertically oriented $MoS2$ with spatially controlled geometry on nitrogenous graphene sheets for high-performance sodium-ion batteries. *Adv. Energy Mater.* **2018**, *8*, 1703300. [CrossRef]
43. Wang, J.; Cui, Y.; Gu, Y.; Xu, H.; Zhuang, Q. Coal-Based modified Carbon for High Performance Sodium-Ion Battery. *Solid State Ion.* **2021**, *368*, 115701. [CrossRef]
44. Xu, X.; Zeng, H.; Han, D.; Qiao, K.; Xing, W.; Rood, M.J.; Yan, Z. Nitrogen and sulfur co-doped graphene nanosheets to improve anode materials for sodium-ion batteries. *ACS Appl. Mater. Interfaces* **2018**, *10*, 37172–37180. [CrossRef]
45. Niu, Q.; Tang, Q.; Sun, X.; Wang, L.; Gao, K. Wood-based carbon tubes with low-tortuosity and open tubular structure for energy storage application. *J. Mater. Sci.* **2022**, *57*, 5154–5166. [CrossRef]
46. Li, Y.; Hu, Y.S.; Titirici, M.M.; Chen, L.; Huang, X. Hard Carbon Microtubes Made from Renewable Cotton as High-Performance Anode Material for Sodium-Ion Batteries. *Adv. Energy Mater.* **2016**, *6*, 1600659. [CrossRef]
47. Yu, Z.E.; Lyu, Y.; Wang, Y.; Xu, S.; Cheng, H.; Mu, X.; Guo, B. Hard carbon micro-nano tubes derived from kapok fiber as anode materials for sodium-ion batteries and the sodium-ion storage mechanism. *Chem. Commun.* **2020**, *56*, 778–781. [CrossRef]
48. Chen, C.; Huang, Y.; Meng, Z.; Xu, Z.; Liu, P.; Li, T. Multi-heteroatom doped porous carbon derived from insect feces for capacitance-enhanced sodium-ion storage. *J. Energy Chem.* **2021**, *54*, 482–492. [CrossRef]
49. Wang, B.; Wang, Y.; Peng, Y.; Wang, X.; Wang, N.; Wang, J.; Zhao, J. Nitrogen-doped biomass-based hierarchical porous carbon with large mesoporous volume for application in energy storage. *Chem. Eng. J.* **2018**, *348*, 850–859. [CrossRef]
50. Gao, X.; Yu, Y.; He, Q.; Li, H.; Liu, Y. Transition metal assisted ionothermal carbonization of cellulose towards high yield and recycling. *Cellulose* **2021**, *28*, 4025–4037. [CrossRef]
51. Li, C.; Li, J.; Zhang, Y.; Cui, X.; Lei, H.; Li, G. Heteroatom-doped hierarchically porous carbons derived from cucumber stem as high-performance anodes for sodium-ion batteries. *J. Mater. Chem.* **2018**, *54*, 5641–5657. [CrossRef]
52. Lang, J.; Li, J.; Zhang, F.; Ding, X.; Zapien, J.A.; Tang, Y. Sodium-Ion Hybrid Battery Combining an Anion-Intercalation Cathode with an Adsorption-Type Anode for Enhanced Rate and Cycling Performance. *Batter. Supercaps* **2019**, *2*, 440–447. [CrossRef]
53. Li, J.; Wang, L.; Li, L.; Lv, C.; Zatovsky, I.V.; Han, W. Metal sulfides@ carbon microfiber networks for boosting lithium ion/sodium ion storage via a general metal–aspergillus niger bioleaching strategy. *ACS Appl. Mater. Interfaces* **2019**, *11*, 8072–8080. [CrossRef]
54. Li, D.; Sun, Y.; Chen, S.; Yao, J.; Zhang, Y.; Xia, Y.; Yang, D. Highly Porous FeS/Carbon Fibers Derived from Fe-Carrageenan Biomass: High-capacity and Durable Anodes for Sodium-Ion Batteries. *ACS Appl. Energy Mater.* **2018**, *10*, 17175–17182. [CrossRef]
55. Zhang, J.; Lv, W.; Zheng, D.; Liang, Q.; Wang, D.W.; Kang, F.; Yang, Q.H. The interplay of oxygen functional groups and folded texture in densified graphene electrodes for compact sodium-ion capacitors. *Adv. Energy Mater.* **2018**, *8*, 1702395. [CrossRef]
56. Dan, R.; Chen, W.; Xiao, Z.; Li, P.; Liu, M.; Chen, Z.; Yu, F. N-Doped Biomass Carbon/Reduced Graphene Oxide as a High-Performance Anode for Sodium-Ion Batteries. *Energy Fuels* **2020**, *34*, 3923–3930. [CrossRef]
57. Ge, P.; Fouletier, M. Electrochemical intercalation of sodium in graphite—ScienceDirect. *Solid State Ionics* **1988**, *28*, 1172–1175. [CrossRef]

58. Lau, W.H.; Kim, J.B.; Zou, F.; Kang, Y.M. Elucidating the charge storage mechanism of carbonaceous and organic electrode materials for sodium ion batteries. *Chem. Commun.* **2021**, *57*, 13465–13494. [CrossRef] [PubMed]
59. Lu, Y.; Shin, K.H.; Yu, Y.; Hu, Y.; Liang, J.; Chen, K.; Yuan, H.; Park, H.S.; Wang, D. Multiple Active Sites Carbonaceous Anodes for Na+ Storage: Synthesis, Electrochemical Properties and Reaction Mechanism Analysis. *Adv. Funct. Mater.* **2021**, *31*, 2007247. [CrossRef]
60. Hu, Y.; Zhang, L.; Bai, J.; Liu, F.; Wang, Z.; Wu, W.; Bradley, R.; Li, L.; Ruan, H.; Guo, S. Boosting High-Rate Sodium Storage of CuS via a Hollow Spherical Nanostructure and Surface Pseudocapacitive Behavior. *ACS Appl. Energy Mater.* **2021**, *4*, 8901–8909. [CrossRef]
61. Song, K.; Liu, C.; Mi, L.; Chou, S.; Chen, W.; Shen, C. Recent progress on the alloy-based anode for sodium-ion batteries and potassium-ion batteries. *Small* **2021**, *17*, 1903194. [CrossRef]
62. Cao, Y.; Zhang, Q.; Wei, Y.; Guo, Y.; Cui, Y. A Water Stable, Near-Zero-Strain O3-Layered Titanium-Based Anode for Long Cycle Sodium-Ion Battery. *Adv. Funct. Mater.* **2019**, *30*, 1907023. [CrossRef]
63. Nobuhara, K.; Nakayama, H.; Nose, M.; Nakanishi, S.; Iba, H. First-principles study of alkali metal-graphite intercalation compounds. *J. Power Sources* **2013**, *243*, 585–587. [CrossRef]
64. Zhu, Z.; Cheng, F.; Zhe, H.; Niu, Z.; Chen, J. Highly stable and ultrafast electrode reaction of graphite for sodium ion batteries. *J. Power Sources* **2015**, *293*, 626–634. [CrossRef]
65. Wan, J.; Shen, F.; Luo, W.; Zhou, L.; Dai, J.; Han, X.; Bao, W.; Xu, Y.; Panagiotopoulos, J.; Fan, X.; et al. In Situ Transmission Electron Microscopy Observation of Sodiation–Desodiation in a Long Cycle, High-Capacity Reduced Graphene Oxide Sodium-Ion Battery Anode. *Chem. Mater.* **2016**, *28*, 6528. [CrossRef]
66. Luo, X.F.; Yang, C.H.; Peng, Y.Y.; Pu, N.W.; Ger, M.D.; Hsiehd, C.T.; Chang, J.K. Graphene nanosheets, carbon nanotubes, graphite, and activated carbon as anode materials for sodium-ion batteries. *J. Mater. Chem. A* **2015**, *3*, 10320. [CrossRef]
67. Jian, Z.; Bommier, C.; Luo, L.; Li, Z.; Wang, W.; Wang, C.; Greaney, P.A.; Ji, X. Insights on the Mechanism of Na-Ion Storage in Soft Carbon Anode. *Chem. Mater.* **2017**, *29*, 2314–2320. [CrossRef]
68. Gomez-Martin, A.; Martinez-Fernandez, J.; Ruttert, M.; Winter, M.; Placke, T.; Ramirez-Rico, J. Correlation of Structure and Performance of Hard Carbons as Anodes for Sodium Ion Batteries. *Chem. Mater.* **2019**, *31*, 7288–7299. [CrossRef]
69. Cao, Y.; Xiao, L.; Sushko, M.L.; Wang, W.; Schwenzer, B.; Xiao, J.; Nie, Z.; Saraf, L.V.; Yang, Z.; Liu, J. Sodium ion insertion in hollow carbon nanowires for battery applications. *Nano Lett.* **2012**, *12*, 3783–3787. [CrossRef]
70. Yang, B.; Chen, J.; Lei, S.; Guo, R.; Li, H.; Shi, S.; Yan, X. Spontaneous Growth of 3D Framework Carbon from Sodium Citrate for High Energy-and Power-Density and Long-Life Sodium-Ion Hybrid Capacitors. *Adv. Energy Mater.* **2018**, *8*, 1702409. [CrossRef]
71. Chen, D.; Zhang, W.; Luo, K.; Song, Y.; Guo, X. Hard carbon for sodium storage: Mechanism and optimization strategies toward commercialization. *Energy Environ. Sci.* **2021**, *14*, 2244–2262. [CrossRef]
72. Stevens, D.A.; Dahn, J.R. High capacity anode materials for rechargeable sodium-ion batteries. *J. Electrochem. Soc.* **2000**, *147*, 1271–1273. [CrossRef]
73. Wang, X.; He, S.; Chen, F.; Hou, X. Nitrogen-Doped Hard Carbon as Symmetric Electrodes for Sodium-Ion Capacitor. *Energy Fuels* **2020**, *34*, 13144–13148. [CrossRef]
74. Subburam, G.; Ramachandran, K.; El-Khodary, S.A.; Zou, B.; Lian, J. Development of Porous Carbon Nanosheets from Polyvinyl Alcohol for Sodium-Ion Capacitors. *Chem. Eng. J.* **2021**, *415*, 129012. [CrossRef]
75. Kuratani, K.; Yao, M.; Senoh, H.; Takeichi, N.; Sakai, T.; Kiyobayashi, T. Kiyobayashi Na-ion capacitor using sodium pre-doped hard carbon and activated carbon. *Electrochim. Acta* **2012**, *76*, 32. [CrossRef]
76. Li, W.; Huang, J.; Feng, L. Controlled synthesis of macroscopic three-dimensional hollow reticulate hard carbon as long-life anode materials for Na-ion batteries. *J. Alloys Compd.* **2017**, *716*, 210–219. [CrossRef]
77. Zhang, L.; Zhao, W.; Jiang, F. Carbon nanosheets from biomass waste: Insights into the role of a controlled pore structure for energy storage. *Sustain. Energy Fuels* **2020**, *4*, 3552–3565. [CrossRef]
78. Qing, L.; Li, R.; Su, W.; Zhao, W.; Li, Y.; Chen, G.; Chen, J. Nanostructures of Carbon Nanofiber-Constrained Stannous Sulfide with High Flexibility and Enhanced Performance for Sodium-Ion Batteries. *Energy Fuels* **2022**, *36*, 2179–2188. [CrossRef]
79. Liu, H.; Liu, H.; Di, S.; Zhai, B.; Li, L.; Wang, S. Advantageous Tubular Structure of Biomass-Derived Carbon for High-Performance Sodium Storage. *ACS Appl. Energy Mater.* **2021**, *4*, 4955–4965. [CrossRef]
80. Xu, S.-D.; Zhao, Y.; Liu, S.; Ren, X.; Chen, L.; Shi, W.; Wang, X.; Zhang, D. Curly hard carbon derived from pistachio shells as high-performance anode materials for sodium-ion batteries. *J. Mater. Chem.* **2018**, *53*, 12334–12351. [CrossRef]
81. Zhu, Z.; Liang, F.; Zhou, Z.; Zeng, X.; Wang, D.; Dong, P.; Zhao, J.; Sun, S.; Zhang, Y.; Li, X. Expanded biomass-derived hard carbon with ultra-stable performance in sodium-ion batteries. *J. Mater. Chem.* **2018**, *6*, 1513–1522. [CrossRef]
82. Li, X.; Zeng, X.; Ren, T.; Zhao, J.; Zhu, Z.; Sun, S.; Zhang, Y. The transport properties of sodium-ion in the low potential platform region of oatmeal-derived hard carbon for sodium-ion batteries. *J. Alloys Compd.* **2019**, *787*, 229–238. [CrossRef]
83. Jin, Q.; Li, W.; Wang, K.; Li, H.; Feng, P.; Zhang, Z.; Jiang, K. Tailoring 2D heteroatom-doped carbon nanosheets with dominated pseudocapacitive behaviors enabling fast and high-performance sodium storage. *Adv. Funct. Mater.* **2020**, *30*, 1909907. [CrossRef]
84. Wang, Y.; Feng, Z.; Zhu, W.; Gariepy, V.; Gagnon, C.; Provencher, M.; Laul, D.; Veillette, R.; Trudeau, M.L.; Guerfi, A.; et al. High Capacity and High Efficiency Maple Tree-Biomass-Derived Hard Carbon as an Anode Material for Sodium-Ion Batteries. *Materials* **2018**, *11*, 1294. [CrossRef]

85. Zhang, Y.; Wang, P.; Yin, Y. Carbon coated amorphous bimetallic sulfide hollow nanocubes towards advanced sodium ion battery anode. *Carbon* **2019**, *150*, 378–387. [CrossRef]
86. Yan, Z.; Yang, Q.W.; Wang, Q. Nitrogen doped porous carbon as excellent dual anodes for Li-and Na-ion batteries. *Chin. Chem. Lett.* **2020**, *31*, 583–588. [CrossRef]
87. Xue, X.; Weng, Y.; Jiang, Z.; Yang, S.; Wu, Y.; Meng, S.; Zhang, Y. Naturally nitrogen-doped porous carbon derived from waste crab shell as anode material for high performance sodium-ion battery. *J. Anal. Appl. Pyrolysis* **2021**, *157*, 105215. [CrossRef]
88. Luo, D.; Han, P.; Shi, L.; Huang, J.; Yu, J.; Lin, Y.; Du, J.; Yang, B.; Li, C.; Zhu, C.; et al. Biomass-derived nitrogen/oxygen co-doped hierarchical porous carbon with a large specific surface area for ultrafast and long-life sodium-ion batteries. *Appl. Surf. Sci.* **2018**, *462*, 713–719. [CrossRef]
89. Zhang, Y.; Li, X.; Dong, P.; Wu, G.; Xiao, J.; Zeng, X.; Zhang, Y.; Sun, X. Honeycomb-like Hard Carbon Derived from Pine Pollen as High-Performance Anode Material for Sodium-Ion Batteries. *ACS Appl. Mater. Interfaces* **2018**, *10*, 42796–42803. [CrossRef]
90. Qin, D.; Chen, S. A sustainable synthesis of biomass carbon sheets as excellent performance sodium ion batteries anode. *J. Solid State Electrochem.* **2017**, *21*, 1305–1312. [CrossRef]
91. Zhu, Y.; Huang, Y.; Chen, C.; Wang, M.; Liu, P. Phosphorus-doped porous biomass carbon with ultra-stable performance in sodium storage and lithium storage. *Electrochim. Acta* **2019**, *321*, 134698. [CrossRef]
92. Zhang, J.; Zhang, D.; Li, K.; Tian, Y.; Wang, Y.; Sun, T. N, O and S co-doped hierarchical porous carbon derived from a series of samara for lithium and sodium storage: Insights into surface capacitance and inner diffusion. *J. Colloid Interface Sci.* **2021**, *598*, 250–259. [CrossRef]
93. Cao, L.; Hui, W.; Xu, Z.; Huang, J.; Zheng, P.; Li, J.; Sun, Q. Rape seed shuck derived-lamellar hard carbon as anodes for sodium-ion batteries. *J. Alloys Compd.* **2017**, *695*, 632–637. [CrossRef]
94. Beda, A.; Villevieille, C.; Taberna, P.L.; Simon, P.; Ghimbeu, C.M. Self-Supported Binder-Free Hard Carbon Electrodes for Sodium-Ion Batteries: Insights into the Sodium Storage Mechanisms. *J. Mater. Chem. A* **2020**, *8*, 5558–5571. [CrossRef]
95. Guan, L.; Pan, L.; Peng, T.; Gao, C.; Zhao, W.; Yang, Z.; Hu, H.; Wu, M. Synthesis of Biomass-Derived Nitrogen-Doped Porous Carbon Nanosheests for High-Performance Supercapacitors. *ACS Sustain. Chem. Eng.* **2019**, *7*, 8405–8412. [CrossRef]
96. Wang, P.; Li, X.; Li, X.; Shan, H.; Li, D.; Sun, X. Paulownia tomentosa derived porous carbon with enhanced sodium storage. *J. Mater. Res.* **2018**, *33*, 1236–1246. [CrossRef]
97. Ou, J.; Yang, L.; Xi, X. Nitrogen-rich porous carbon anode with high performance for sodium ion batteries. *J. Porous Mater.* **2016**, *24*, 189–192. [CrossRef]
98. Xu, J.; Wang, M.; Wickramaratne, N.P.; Jaroniec, M.; Dou, S.; Dai, L. High-performance sodium ion batteries based on a 3D anode from nitrogen-doped graphene foams. *Adv. Mater.* **2015**, *27*, 2042–2048. [CrossRef]
99. Wang, C.; Huang, J.; Li, Q. Catalyzing carbon surface by Ni to improve initial coulombic efficiency of sodium-ion batteries. *J. Energy Storage* **2020**, *32*, 101868. [CrossRef]
100. Yu, Y.; Ren, Z.; Li, L. Ionic liquid-induced graphitization of biochar: N/P dual-doped carbon nanosheets for high-performance lithium/sodium storage. *J. Mater. Sci.* **2021**, *56*, 8186–8201. [CrossRef]
101. Wu, F.; Zhang, M.; Bai, Y.; Wang, X.; Dong, R.; Wu, C. Lotus seedpod-derived hard carbon with hierarchical porous structure as stable anode for sodium-ion batteries. *ACS Appl Mater. Interfaces* **2019**, *11*, 12554–12561. [CrossRef] [PubMed]
102. Susanti, R.F.; Alvin, S.; Kim, J. Toward high-performance hard carbon as an anode for sodium-ion batteries: Demineralization of biomass as a critical step. *J. Ind. Eng. Chem.* **2020**, *91*, 317–329. [CrossRef]
103. Hou, H.; Yu, C.; Liu, X.; Yao, Y.; Dai, Z.; Li, D. The effect of carbonization temperature of waste cigarette butts on Na-storage capacity of N-doped hard carbon anode. *Chem. Pap.* **2019**, *73*, 1237–1246. [CrossRef]
104. Li, T.; Liu, Z.; Gu, Y. Hierarchically porous hard carbon with graphite nanocrystals for high-rate sodium ion batteries with improved initial Coulombic efficiency. *J. Alloys Compd.* **2020**, *817*, 152703. [CrossRef]
105. Alvin, S.; Yoon, D.; Chandra, C.; Susanti, R.F.; Chang, W.; Ryu, C.; Kim, J. Extended flat voltage profile of hard carbon synthesized using a two-step carbonization approach as an anode in sodium ion batteries. *J. Power Sources* **2019**, *430*, 157–168. [CrossRef]
106. Ou, J.; Yang, L.; Xi, X. Hierarchical porous nitrogen doped carbon derived from horn comb as anode for sodium-ion storage with high performance. *Electron. Mater. Lett.* **2017**, *13*, 66–71. [CrossRef]
107. Zhao, X.; Ding, Y.; Xu, Q.; Yu, X.; Liu, Y.; Shen, H. Low-Temperature Growth of Hard Carbon with Graphite Crystal for Sodium-Ion Storage with High Initial Coulombic Efficiency: A General Method. *Adv. Energy Mater.* **2019**, *9*, 1803648. [CrossRef]
108. Tang, Z.; Wang, Y.; Zheng, Z.; Luo, X. In situ dual growth of graphitic structures in biomass carbon to yield a potassium-ion battery anode with high initial coulombic efficiency. *J. Mater. Chem. A* **2021**, *9*, 9191–9202. [CrossRef]
109. Chen, C.; Huang, Y.; Zhu, Y.; Zhang, Z.; Guang, Z.; Meng, Z.; Liu, P. Nonignorable Influence of Oxygen in Hard Carbon for Sodium Ion Storage. *ACS Sustain. Chem. Eng.* **2020**, *8*, 1497–1506. [CrossRef]
110. Liu, X.; Wang, H.; Cui, Y.; Xu, X.; Zhang, H.; Lv, G.; Shi, J.; Liu, W.; Chen, S.; Wang, X. High-energy sodium-ion capacitor assembled by hierarchical porous carbon electrodes derived from Enteromorpha. *J. Mater. Sci.* **2018**, *53*, 6763–6773. [CrossRef]
111. Khan, M.; Ahmad, N.; Lu, K. Nitrogen-doped carbon derived from onion waste as anode material for high performance sodium-ion battery. *Solid State Ionics* **2020**, *346*, 115223. [CrossRef]
112. Hu, Y.; Ma, R.; Ju, Q.; Guo, B.; Yang, M.; Liu, Q.; Wang, J. S, N dual-doped porous carbon materials derived from biomass for Na ion storage and O2 electroreduction. *Microporous Mesoporous Mater.* **2020**, *294*, 109930. [CrossRef]

113. Wan, H.; Shen, X.; Jiang, H.; Zhang, C.; Jiang, K.; Chen, T.; Chen, Y. Biomass-derived N/S dual-doped porous hard-carbon as high-capacity anodes for lithium/sodium ions batteries. *Energy* **2021**, *231*, 121102. [CrossRef]
114. Jla, B.; Jian, Z.B.; Qz, B.; Xin, W.A.; Cong, G.A.; Min, L.B. Promoting the Na$^+$-storage of NiCo$_2$S$_4$ hollow nanospheres by surfacing Ni–B nanoflakes. *J. Mater. Sci. Technol.* **2021**, *82*, 114–121.
115. Zou, Y.; Gu, Y.; Hui, B. Nitrogen and sulfur vacancies in carbon shell to tune charge distribution of Co$_6$Ni$_3$S$_8$ core and boost sodium storage. *Adv. Energy Mater.* **2020**, *10*, 1904147. [CrossRef]
116. Yin, Y.; Zhang, Y.; Liu, N.; Sun, B.; Zhang, N. Biomass-Derived P/N-Co-Doped Carbon Nanosheets Encapsulate Cu$_3$P Nanoparticles as High-Performance Anode Materials for Sodium-Ion Batteries. *Front. Chem.* **2020**, *8*, 316. [CrossRef]
117. Li, D.; Yang, D.; Yang, X. Double-Helix Structure in Carrageenan–Metal Hydrogels: A General Approach to Porous Metal Sulfides/Carbon Aerogels with Excellent Sodium-Ion Storage. *Angew. Chem.* **2016**, *128*, 16157–16160. [CrossRef]
118. Guo, R.; Li, D.; Lv, C.; Wang, Y.; Zhang, H.; Xia, Y.; Yang, D.; Zhao, X. Porous Ni$_3$S$_4$/C aerogels derived from carrageenan-Ni hydrogels for high-performance sodium-ion batteries anode. *Electrochim. Acta* **2019**, *299*, 72–79. [CrossRef]
119. Wu, H.; Li, X.; Chen, L.; Dan, Y. Facile Synthesis of Hierarchical Iron Phosphide/Biomass Carbon Composites for Binder-Free Sodium-Ion Batteries. *Batter. Supercaps* **2019**, *2*, 144–152. [CrossRef]
120. Kang, B.; Wang, Y.; He, X.; Wu, Y.; Li, X.; Lin, C.; Chen, Q.; Zeng, L.; Wei, M.; Qian, Q. Facile fabrication of WS$_2$ nanocrystals confined in chlorella-derived N, P co-doped bio-carbon for sodium-ion batteries with ultra-long lifespan. *Dalton Trans.* **2021**, *50*, 14745–14752. [CrossRef]